Graduate Texts in Mathematics

Graduate Texts in Mathematics bridge the gap between passive study and creative understanding, offering graduate-level introductions to advanced topics in mathematics. The volumes are carefully written as teaching aids and highlight characteristic features of the theory. Although these books are frequently used as textbooks in graduate courses, they are also suitable for individual study.

More information about this series at http://www.springer.com/series/136

Ömer Eğecioğlu • Adriano M. Garsia

Lessons in Enumerative Combinatorics

 Springer

Ömer Eğecioğlu
Department of Computer Science
University of California, Santa Barbara
Santa Barbara, CA, USA

Adriano M. Garsia
Department of Mathematics
University of California, San Diego
La Jolla, CA, USA

ISSN 0072-5285   ISSN 2197-5612   (electronic)
Graduate Texts in Mathematics
ISBN 978-3-030-71252-5   ISBN 978-3-030-71250-1   (eBook)
https://doi.org/10.1007/978-3-030-71250-1

Mathematics Subject Classification: 05A05, 05A10, 05A17, 05A18, 05A19

This Springer imprint is published by the registered company Springer Nature Switzerland AG
The registered company address is: Gewerbestrasse 11, 6330 Cham, Switzerland

Dedicated to the generations of students who have learned these beautiful and basic methods of counting from us over the years.

Foreword

I became aware in the 1980s of a plan by Adriano Garsia and Ömer Eğecioğlu to write a book on the combinatorial foundations of computer science at the undergraduate level based on formal languages and bijective proofs. After hearing nothing further for over 30 years, I assumed that this project was moribund. Therefore, I was greatly surprised and pleased when I was asked to write this Foreword to the completed book.

Bijective proofs are ubiquitous in enumerative combinatorics because it is often very natural and important to prove or conjecture that two finite sets S and T have the same number of elements. In many cases, bijections $S \to T$, though interesting and elucidating, are rather *ad hoc* and do not fit into a larger picture. A notable exception concerns bijections based on formal languages. Many combinatorial objects can be encoded in a natural way by words in some (formal) language. Standard operations on words and languages then lead to elegant and conceptual bijections between combinatorial objects that at first sight are very different.

Garsia and Eğecioğlu have written a very friendly and accessible development of the formal language approach to bijective proofs. In additional to rather obvious topics such as lattice paths and trees, this approach is also applied to some unexpected situations, including the Cayley–Hamilton theorem and Lagrange inversion. Altogether, the authors give a clear and unified picture of an important area of combinatorics, which should become a definitive treatment of the subject matter therein.

Miami, FL, USA   Richard P. Stanley
June 2020

Preface

A combinatorial structure is a visual representation of a mathematical construct and often reveals aspects of the construct that one cannot see in the original mathematical formulation. Methods for counting these combinatorial structures is an introduction to what is known as *enumerative combinatorics*. The history of this subject is very rich, with many contributors. The original giants of the field were Euler and Gauss. After Euler, enormous developments took place and continue to take place today.

What we have put together here can be used as an introduction to enumerative combinatorics, covering selected topics for a graduate mathematics course, or an undergraduate honors course for mathematics and computer sciences majors. We aim for a simple presentation and make minimal assumptions on the prior knowledge of the students. A familiarity with single-variable calculus and basic elements of discrete mathematics is sufficient preparation for the material. As usual, the most valuable qualification one could hope for as an instructor is a degree of enthusiasm on the part of the students.

The material here is presented in nine chapters:

1. *Basic Combinatorial Structures*
2. *Partitions and Generating Functions*
3. *Planar Trees and the Lagrange Inversion Formula*
4. *Cayley Trees*
5. *The Cayley–Hamilton Theorem*
6. *Exponential Structures and Polynomial Operators*
7. *The Inclusion-Exclusion Principle*
8. *Graphs, Chromatic Polynomials, and Acyclic Orientations*
9. *Matching and Distinct Representatives*

This textbook is suitable for 10-week quarter or a 15-week semester course, as the material gives sufficient latitude to the instructor to enhance or limit the number of topics covered, and plenty of places at which to cover side material of interest before returning to the main topic.

We make use of formal languages as the common setting for modeling and counting classical combinatorial objects in Chapter 1. With this, we are only scratching the surface of formal languages and do not go deep into this subject itself. Chapter 1 can probably be covered in two weeks in an undergraduate course and perhaps in a single week in a graduate course. After that, the remaining chapters are more or less self-contained and can be covered independently of one another. Still, a rough schema of possible orderings is given in the following figure.

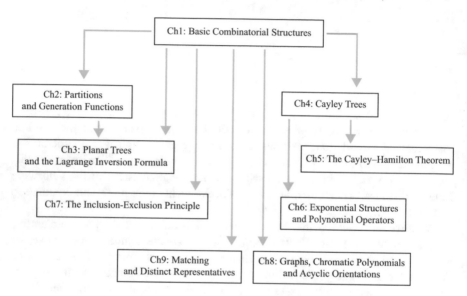

Dependencies implied in the above schema after Chapter 1 are minimal. Each chapter in this book can be covered independently of the others with a modest amount of notational preparation.

If the instructor chooses to do so, a good place to invest additional time to the study the elements of generating functions is near the end of Chapter 1, preferably before the beginning of Chapter 2. This may also be a convenient place to take additional time to study binomial identities in more detail, say after Section 1.3, covering the uses of the Chu–Vandermonde summation formula, for instance. For a graduate-oriented course, an introduction to hypergeometric series and perhaps the Pfaff–Saalschütz summation formula may be considered. As another direction for a follow-up to Chapter 1, the bijection between Dyck paths and 312-avoiding permutations easily leads into a discussion of stacks and queues for a computer science emphasis. The material can also be followed by topics on pattern avoidance in permutations and other combinatorial objects.

Generating functions are introduced in Section 2.2, only to the extent that is needed in the treatment of the topic of the chapter and a lot more can be covered at the discretion of the instructor. There are many convenient places in the text where a separate treatment of recursions and generating functions can be made with as much detail as the instructor sees fit. For example, recursions can be presented in more detail before Section 1.8.

After Chapter 2, more material related to the combinatorics behind the Jacobi triple-product identity and an introduction to the q-series can be covered for a graduate version of the course.

Chapter 3 uses generating functions, which may be covered in more detail if this chapter is presented independently of the others. As an example, Section 3.2 makes use of the Newton expansion formula. After Chapter 3, computer science applications of trees can be covered to the extent desired by the instructor.

Chapter 4 uses determinants, and they may be covered in more detail before Section 4.6, even though the basics are introduced there.

The beginning of Chapter 5, namely the combinatorial proof of the Cayley–Hamilton theorem, is a continuation of the ideas in Chapter 4 on Cayley trees. In fact, this used to be a part of Chapter 4, creating quite a long chapter. We have made it a chapter by itself by adding the most interesting applications of the theorem, which is usually seen in treatments of Linear Algebra texts. The proof in Section 5.1 can be covered independently of the material in Chapter 4. The applications given in Section 5.2 do not depend on the nature of the combinatorial proof of the theorem presented in Section 5.1 and can be covered independently if desired.

Chapter 6 starts with set partitions and permutations and uses Cayley trees that are covered in Chapter 4 as an application. It may be a good idea to cover exponential generating functions right before starting Chapter 6 if the instructor deems this necessary. In Chapter 6, Sections 6.4 and 6.5 are more technical in nature than the preceding ones and can be skipped in an undergraduate version of the course.

Chapter 7 uses generating functions and recursions. After Chapter 7, both in undergraduate and graduate versions of the course, a detailed treatment of partially ordered sets and the Möbius inversion with the standard applications can be presented.

Chapter 8 leads into the world of graph theory and can be expanded along various paths again at the discretion of the instructor. Of additional interest is Whitney's broken circuit theorem. For a computer science emphasis, the computational difficulty of coloring problems, for example, the reduction from 3-SAT, register allocation by graph coloring, planarity algorithms, and other algorithmic aspects of graphs, can be included.

Chapter 9 is completely independent and can be skipped in an undergraduate version of the course, except, of course, Philip Hall's first available matching algorithm is always amusing for the students. This can be covered by itself. This chapter can be followed by topics in matchings in graphs and network flow algorithms for a graduate and computer science-oriented approach to the course.

As a historical note, the material presented here started as lecture notes in the 1980s at the University of California at San Diego (UCSD), continued to be used there for a number of years, and were later also used at the University of California at Santa Barbara, both at the undergraduate and graduate levels.

At UCSD, the notes were used for the EECS 160A-B sequence course titled "Foundations of Computer Science." This junior year undergraduate course was a part of the upper division requirements that covered the principal elements of discrete mathematics and combinatorics, with the standard first year Calculus and Analytic Geometry sequence as prerequisites.

They have also been used in a course titled "CS-2: Foundations of Computer Science" at the College of Creative Studies (CCS) at UCSB for almost two decades. This required course for all lower-division Computing majors and interested Mathematics majors is a part of the accelerated curriculum covering mathematical foundations for aspiring computer scientists. It is worth mentioning that CCS is a unique college in all of the UC campuses and is known as a "Graduate School for Undergraduates." The material from these notes was also used as the core of a graduate topics course at UCSB.

When warranted, we cannot resist the temptation to explore—even though to a very small extent—some of the applications in a separate miscellaneous section at the end of each chapter. All nine chapters in this book have problems at the end, followed by a sample quiz based on the material covered. The sample quiz serves to gauge the minimal amount of material that undergraduate students are expected to retain from the topics presented.

The chapters themselves are followed by a short bibliography, and a necessarily incomplete list of books on related topics, which are at various levels of sophistication, focus, and pedagogical approach.

We should especially remark that further advances in the field as well as a wealth of additional references can be found in Richard P. Stanley's encyclopedic two-volume work *Enumerative Combinatorics*.

We believe that this treatment has been useful to generations of students in mathematics and computer science. We will achieve our purpose with this book if we are able to convey to the reader the sense of excitement and wonder that we felt upon being exposed to these beautiful subjects ourselves. We can also hope that perhaps the subset of combinatorial objects we introduce manages to inspire the students for further study.

Acknowledgments We would like to thank Arthur Gatin for his meticulous reworking of the figures in this book. His dedication to work and attention to detail was second to none.

We would also like to acknowledge Loretta Bartolini and the editorial staff at Springer for their valuable assistance in the publication of this book.

Santa Barbara, CA, USA   Ömer Eğecioğlu
La Jolla, CA, USA   Adriano M. Garsia
August 2020

Contents

Chapter 1
Basic Combinatorial Structures

1.1 Introduction

In this chapter, we shall familiarize ourselves with some of the most basic combinatorial structures and develop skills for constructing them and counting them.

Tradition demands that the combinatorial structures we are to study include objects of the most varied nature. To name a few:

Configurations of assorted fruits in baskets,

Latin squares,

Shuffled decks of cards,

Chessboard configurations,

Collections of pairs of identical socks,

Bridges of Königsberg,

....

the list is endless.

Perhaps these devices individually serve as colorful mnemonics of the sort that are useful for the students. However working directly with such a varied collection of objects becomes at times rather difficult and confusing. Indeed, especially for the beginner, most of the difficulty encountered in counting problems derives from a lack of a precise description of what is being talked about.

At this level, enumeration is more an art than a science, more akin to chess than to calculus.

This traditional aspect gives combinatorics, the appearance of a frivolous subject. However, a deeper familiarity with the subject reveals that the basic combinatorial mechanisms underlie all of mathematics, in fact all of our thinking processes. Moreover, we see that combinatorics contains some of the most powerful and beautiful mathematical constructs.

What has really been lacking is a vehicle that could be used to communicate and develop the subject in a precise and unifying manner.

© Springer Nature Switzerland AG 2021
Ö. Eğecioğlu, A. M. Garsia, *Lessons in Enumerative Combinatorics*, Graduate
Texts in Mathematics 290, https://doi.org/10.1007/978-3-030-71250-1_1

Fortunately, the advent of computer science and the need to communicate to the computer have brought to the fore, a most appropriate tool for the precise description of combinatorial objects: *the theory of formal languages.*

It develops that formal languages have also the desired unifying capability.

We shall thus make systematic use of languages throughout our presentation. We shall see that after we learn how to construct and count the words, we shall have no difficulty in the enumeration of more traditional combinatorial objects.

Of course, we will not develop here the complete theory of formal languages, rather we will only introduce and use some basic constructs of that theory as fit our needs. We will also try not to be overly formal and rather introduce a concept through an example than through a set of axioms. As a result, our treatment may not be as general as necessary for all needs of computer science. The additional generality may however be within easier reach after an exposure to the salient examples.

1.2 Languages

In our presentation, we shall use terms such as *words, languages,* etc., which have already a meaning in ordinary conversation. However, the meaning assigned to them here will be slightly different. Thus it is good to make this distinction precise before proceeding any further.

The basic terms are *alphabet, letter, word, string,* and *language.* They are defined as follows. First of all,

$$alphabet = any\ finite\ collection\ of\ symbols.$$

Even though reordering the elements of a set does not change it, i.e.,

$$\{a, 5, b, x, c\} = \{b, a, x, 5, c\},$$

for instance, when we view these sets as alphabets, we assume that they are ordered from left to right in the order we put them down. Thus when we write the sentence

$$Let\ \mathcal{A} = \{a, 5, b, x, c\}\ be\ an\ alphabet, \tag{1.1}$$

we shall mean not only that $\mathcal{A}$ consists of $a, 5, b, x$, and c but also that as long as we work with $\mathcal{A}$,

$$
\begin{aligned}
&a\ comes\ before\ 5\\
&5\ comes\ before\ b\\
&b\ comes\ before\ x\\
&x\ comes\ before\ c.
\end{aligned}
\tag{1.2}
$$

If we write down

Let $\mathcal{A} = \{b, a, x, 5, c\}$ be an alphabet,

we shall mean that $\mathcal{A}$ consists of $a, 5, b, x$ and as long as we work with $\mathcal{A}$

> *b comes before* a
>
> *a comes before x*
>
> *x comes before* 5
>
> *5 comes before c.*

With this convention in mind, if $\mathcal{A}$ is an alphabet, then we define

$\mathcal{A}$-letter = *any one of the symbols in $\mathcal{A}$*

$\mathcal{A}$-string = *a juxtaposition of any finite number of $\mathcal{A}$-letters*

$\mathcal{A}$-word = *a juxtaposition of any finite number of $\mathcal{A}$-letters*

$\mathcal{A}$-language = *any finite or infinite collection of $\mathcal{A}$-strings.*

Of course, when there is no danger of confusion, we shall simply drop the prefix "$\mathcal{A}$-" and simply talk about letters, strings, languages, etc.

For instance, if $\mathcal{A} = \{a, x, b\}$, then

$$xaa, bxaaa, xbbx$$

are all $\mathcal{A}$-strings as well as $\mathcal{A}$-words. We see that the terms *string* and *word* may be used interchangeably; however, it may be good to use the term *word* to refer to strings that satisfy certain restrictions. In particular, if $\mathcal{L}$ is an $\mathcal{A}$-language, we prefer to talk about *words of $\mathcal{L}$* rather than *strings of $\mathcal{L}$*.

Clearly, we are not going to study *arbitrary* languages since the latter are totally meaningless objects. We are going to put together languages by selecting certain very special strings, and it is the latter that shall be dignified with the appellative of *words*. However, this distinction is mostly philosophical and cannot be made out of context.

Here and after we shall adopt the following notations and conventions.

1.2.1 Length

If w is a string or a word, the number of letters of w will be referred to as the *length* of w and will be denoted by the symbols $l(w)$ or $|w|$.

Words or strings of length k will also be referred to as k-letter words or k-letter strings, respectively.

1.2.2   Concatenation

If w_1 and w_2 are $\mathcal{A}$-words by $w_1 \cdot w_2$ or by simply $w_1 w_2$, we mean the word of length $l(w_1) + l(w_2)$ obtained by placing w_1 and w_2 one after the other. Thus if $w_1 = accd$ and $w_2 = xxy$, then $w_1 w_2$ is the 7-letter word

$$w_1 w_2 = accdxxy \,.$$

It is customary to refer to this operation as *concatenation*. In fact, the same term is used when we juxtapose any number of words.

We should also mention that our dot notation $w_1 \cdot w_2$ refers to the fact that concatenation has the mathematical character of multiplication.

Indeed, we can trivially see that for any triplet of words w_1, w_2, w_3, we have

$$(w_1 \cdot w_2) \cdot w_3 = w_1 \cdot (w_2 \cdot w_3) \,. \tag{1.3}$$

Clearly, it is immaterial whether we concatenate w_1 and w_2 and then concatenate the resulting word with w_3 or concatenate w_1 with the result of concatenating w_2 and w_3.

We might refer to (1.3) as *associativity* of word concatenation.

1.2.3   The Empty Word

Frequently, for technical reasons, we have to make use of the special word consisting of no letters at all. This is usually referred to as the *empty word* or the *null word* and it is assigned zero length.

This word is often denoted by the symbol ϵ. Sometimes we shall do so here as well. However we can easily see that this choice may lead to confusion if ϵ happens to be also one of the letters of the alphabet with which we work. To avoid this problem, we shall have to keep at hand an alternate symbol.

Clearly, concatenating the empty word with any other word w should give w back. More precisely, we want

$$w \cdot \epsilon = \epsilon \cdot w = w \,. \tag{1.4}$$

Thus ϵ behaves with respect to concatenation very much in the same manner the familiar number 1 behaves with respect to multiplication. Thus an alternate and very appropriate notation for the empty word could be the symbol "1" itself, assuming of course that it is not one of the letters of the alphabet we are working with. Using "1" for the null word, (1.4) could be written as

$$w \cdot 1 = 1 \cdot w = w \,,$$

and this is consistent with common use of 1. We shall later see that this notation has further advantages.

1.2.4 Initial Segments

If w_1 and w are words, we shall say that w_1 is an *initial segment of w* or a *prefix of w* if for some word w_2, we have

$$w = w_1 \cdot w_2 \,.$$

Crudely speaking, w_1 is an initial segment of w if w_1 is at the *beginning* of w. For instance, ax is an initial segment of

$$axcde \quad \text{and also of} \quad axbbf \,.$$

Note that the null word ϵ and w itself are always prefixes of w.

1.2.5 $\mathcal{A}^*$ and $\mathcal{A}^+$

It is customary to denote the collection consisting of all $\mathcal{A}$-words (including the null word) by the symbol $\mathcal{A}^*$. The collection of nonnull $\mathcal{A}$-words is denoted by $\mathcal{A}^+$.

1.2.6 Lexicographic Order

If x and y are $\mathcal{A}$-letters, then we shall write

$$x <_{\mathcal{A}} y$$

to indicate that when the letters of $\mathcal{A}$ are listed in the given order, x comes before y. Of course, when there is no danger of confusion, we shall drop the subscript $\mathcal{A}$ and write simply $x < y$. Thus for the alphabet $\mathcal{A}$ given in (1.1), the statements in (1.2) may simply be written as

$$a < 5$$
$$5 < b$$
$$b < x$$
$$x < c \,.$$

This given, if w_1 and w_2 are $\mathcal{A}$-words, we shall say that w_1 *lexicographically precedes* w_2 and write

$$w_1 <_{\mathcal{A}} w_2$$

if either

1. w_1 *is an initial segment of* w_2; or we have
2. $w_1 = x_1 x_2 \cdots x_h$, $w_2 = y_1 y_2 \cdots y_k$ *with*

$$x_1 = y_1,$$
$$x_2 = y_2,$$
$$\vdots$$
$$x_{i-1} = y_{i-1}$$

and

$$x_i <_{\mathcal{A}} y_i.$$

Crudely speaking, $w_1 <_{\mathcal{A}} w_2$ means that in the first position where w_1 and w_2 disagree, w_1 either has no letter at all or has a smaller letter than w_2.

The term lexicographic order is derived from the word *lexicon*, which denotes a dictionary. Indeed, we can easily see that in the case of our customary alphabet

$$\mathcal{A} = \{a, b, c, d, \ldots, y, z\},$$

this is precisely our familiar *dictionary order.*

A collection of words listed in lexicographic order will be briefly referred to as a *lexicographic list.*

It is often convenient to think that $\mathcal{A}^*$ and any other $\mathcal{A}$-languages are given the lexicographic order. However, we shall not require this in general, and other orders will be considered.

1.2.7   Set theoretical Notation

We shall assume throughout that the reader is familiar with set theoretical notation, most particularly, the symbols

$$\in, \subseteq, \supseteq, \cup, \cap$$

Denoting, respectively,
  a member of,
  contained in or subset of,
  contains or superset of,
  union,
  intersection.

Thus, for instance, the expression

$$w \in \mathcal{A}^*$$

meaning that w *is a member of* $\mathcal{A}^*$ may be used as a shorthand for the sentence

$$w \text{ is an } \mathcal{A}\text{-word}.$$

Similarly the expression

$$\mathcal{L} \subseteq \mathcal{A}^*$$

meaning that $\mathcal{L}$ is a subset of $\mathcal{A}^*$ may be used as a shorthand for

$$\mathcal{L} \text{ is an } \mathcal{A}\text{-language}.$$

Note also that if $\mathcal{L}_1$ and $\mathcal{L}_2$ are $\mathcal{A}$-languages, then the expression

$$\mathcal{L}_1 \cup \mathcal{L}_2$$

denotes the $\mathcal{A}$-language consisting of all the words that are in $\mathcal{L}_1$ or $\mathcal{L}_2$ or both.

For a finite set S, we will denote its cardinality (the number of elements it has) by

$$|S| , \tag{1.5}$$

or equivalently by

$$\#S . \tag{1.6}$$

Throughout our treatment, there will be plenty of occasions where one of these notations will be preferred over the other. In particular we may choose to use (1.6) in the contexts where there may be confusion whether (1.5) is used to mean the length of a word or the cardinality of a set.

A special set that we will have frequent occasion to use is the first n positive integers $\{1, 2, \ldots, n\}$. We will use the shorthand $[n]$ for this set.

Another convenient notation is given by the function χ, which is defined as follows. For any statement S, we let

$$\chi(S) = \begin{cases} 1 \text{ if } S \text{ is true,} \\ 0 \text{ if } S \text{ is false.} \end{cases}$$

1.2.8   Listing Series and Algebraic Operations

One of the fundamental facts that shall guide us throughout this book is that certain basic constructions involving combinatorial objects may be translated into algebraic operations.

This is very fortunate since algebraic manipulations are very precise and almost mechanical, while *object manipulations* can be rather clumsy, difficult to describe, and ambiguous.

Languages are most suited for the algebraic treatment. To help translating operations on languages into algebraic operations, it is convenient to represent languages as *sum of words*. To this end, if $\mathcal{L}$ is a language, we shall set

$$s\mathcal{L} = \sum_{w \in \mathcal{L}} w \tag{1.7}$$

and refer to it as the *listing series for $\mathcal{L}$*.

We should emphasize that the expression on the right hand side of (1.7) is to be interpreted only as a *formal sum*, which is merely a convenient way to deal with all the words of $\mathcal{L}$ at the same time. For instance, if $\mathcal{L}$ consists of all the 2-letter words in the alphabet

$$\mathcal{A} = \{a, b, c\}, \tag{1.8}$$

then in the lexicographic order, we have

$$s\mathcal{L} = aa + ab + ac + ba + bb + bc + ca + cb + cc . \tag{1.9}$$

This may appear somewhat strange at first, especially since in general the sum in (1.7) may contain an infinite number of terms. However, we shall have plenty of occasion to see the convenience of this notation.

One of the most basic operations that can be used to produce further languages from given ones is *concatenation*. More precisely, given two languages $\mathcal{L}_1$ and $\mathcal{L}_2$, let

$$\mathcal{L}_1 \cdot \mathcal{L}_2$$

denote the language consisting of all the words obtained by concatenating a word of $\mathcal{L}_1$ with a word of $\mathcal{L}_2$. Now, simple high school algebra should suggest that we should have

$$s\mathcal{L}_1 \cdot \mathcal{L}_2 = s\mathcal{L}_1 \cdot s\mathcal{L}_2 . \tag{1.10}$$

This may be seen more clearly in a specific example. For instance, let $\mathcal{L}_1$ and $\mathcal{L}_2$ both consist of the 1-letter words from the alphabet given in (1.8) , that is,

$$s\mathcal{L}_1 = a + b + c, \quad s\mathcal{L}_2 = a + b + c.$$

In this case, (1.10) gives

$$s\mathcal{L}_1 \cdot \mathcal{L}_2 = (a + b + c)(a + b + c). \tag{1.11}$$

Now clearly, $\mathcal{L}_1 \cdot \mathcal{L}_2$ consists of all 2-letter words, and indeed, if we carry out the multiplication in (1.11), we do get precisely the right hand side of (1.9).

It should be noted that if the two languages $\mathcal{L}_1$ and $\mathcal{L}_2$ have no words in common, then

$$s\mathcal{L}_1 \cup \mathcal{L}_2 = s\mathcal{L}_1 + s\mathcal{L}_2. \tag{1.12}$$

For this reason, throughout our treatment, we shall use the symbol "+" to denote the union of two disjoint sets.

Formulas (1.10) and (1.12) are two good examples of the convenience of representing languages by their listing series. However, in some situations, the notation $s\mathcal{L}$ does turn out to be a bit bulky. Thus when notational clarity is at stake, and to keep our formulas as simple as possible, we shall usually identify a language $\mathcal{L}$ with its listing series $s\mathcal{L}$ and use the symbol $\mathcal{L}$ interchangeably for the language itself and also for its listing series $s\mathcal{L}$.

1.3 The 2-Letter Alphabet, Sets, and Lattice Paths

The simplest alphabet is of course the one consisting of a single letter. However, the words from such an alphabet are not very interesting. For instance, if $\mathcal{A} = \{a\}$, then

$$\mathcal{A}^* = \{\epsilon, a, aa, aaa, \ldots\}.$$

Thus we may write

$$s\mathcal{A}^* = 1 + a + a^2 + a^3 + \cdots \tag{1.13}$$

Note that the notation 1 for the null word is consistent with the mathematical convention that $a^0 = 1$.

From high school algebra, we get that for any x, we have

$$(1 - x)(1 + x + x^2 + x^3) = 1 + x + x^2 + x^3 - x - x^2 - x^3 - x^4 = 1 - x^4.$$

More generally, for any integer n,

$$(1 - x)(1 + x + x^2 + \cdots + x^n) = 1 - x^{n+1}. \tag{1.14}$$

We may argue that we have as well

$$(1 - x)(1 + x + x^2 + x^3 + \cdots) = 1 .$$

For this reason, here and in the following, we shall adopt the convention that if x is any expression whatever the symbol

$$\frac{1}{1 - x}$$

shall be interpreted as a shorthand for the expression

$$1 + x + x^2 + x^3 + x^4 + \cdots$$

Thus in particular, in view of (1.13), when $\mathcal{A} = \{a\}$, we may write

$$s\mathcal{A}^* = \frac{1}{1 - a} . \tag{1.15}$$

Let us now consider the 2-letter alphabet

$$\mathcal{A} = \{a, b\} . \tag{1.16}$$

For convenience, let $\mathcal{L}_n$ denote the language consisting of all $\mathcal{A}$-words of length n. In particular, we have

$$\mathcal{L}_1 = a + b, \tag{1.17}$$
$$\mathcal{L}_2 = aa + ab + ba + bb = (a + b)^2 .$$

When carrying out calculations with sums of words, we should keep in mind that the familiar *commutativity* law of ordinary multiplication does not hold for concatenation. For instance, ab and ba are two different words. Thus it is incorrect in the present context to write

$$(a + b)^2 = a^2 + 2ab + b^2 .$$

However, except for this proviso, we may freely use all other rules of high school algebra. In particular, we may use the shorthand

$$a^2 b^2 a^2$$

for the word

$$aabbaa$$

and

$$a^4 b^2$$

for the word *aaaabb*; of course, as words, $a^2 b^2 a^2 \neq a^4 b^2$.

Note now that any word of length 5 may be expressed as the concatenation of a word of length 2 by a word of length 3. Thus we may write

$$\mathcal{L}_5 = \mathcal{L}_2 \cdot \mathcal{L}_3 \, .$$

Similarly, for any 3 integers, h, k, and n such that

$$n = h + k,$$

we obtain the decomposition

$$\mathcal{L}_n = \mathcal{L}_h \cdot \mathcal{L}_k \, . \tag{1.18}$$

In particular, for any n, we have

$$\mathcal{L}_n = \mathcal{L}_1 \cdot \mathcal{L}_{n-1} \, , \tag{1.19}$$

and from (1.17), we progressively obtain

$$\mathcal{L}_3 = (a + b) \cdot \mathcal{L}_2 = (a + b)^3$$
$$\mathcal{L}_4 = (a + b) \cdot \mathcal{L}_3 = (a + b)^4$$
$$\vdots$$
$$\mathcal{L}_n = (a + b)^n \tag{1.20}$$
$$\vdots$$

Thus since

$$s\mathcal{L}^* = 1 + \mathcal{L}_1 + \mathcal{L}_2 + \mathcal{L}_3 + \cdots ,$$

we may again write

$$s\mathcal{A}^* = \frac{1}{1 - (a + b)} \, . \tag{1.21}$$

The shorthand used in (1.21) for the language $\mathcal{A}^*$ which describes its listing series is also referred to the *generating series* of this language.

Note further that (1.19) can also be written in the form

$$\mathcal{L}_n = a \cdot \mathcal{L}_{n-1} + b \cdot \mathcal{L}_{n-1} . \tag{1.22}$$

This relation has an interesting interpretation. Observe that if the words of $\mathcal{L}_{n-1}$ are given in lexicographic order and we prefix each of them by a, then we obtain the words of $a\mathcal{L}_{n-1}$ in lexicographic order. The same, of course, is true for $b\mathcal{L}_{n-1}$. Clearly, the words of $a\mathcal{L}_{n-1}$ precede lexicographically the words of $b\mathcal{L}_{n-1}$. Thus (1.22) may be interpreted as an algorithm for listing n-letter words lexicographically, given the lexicographic list of $(n-1)$-letter words.

For example, this algorithm for $n = 3$ yields

$$\mathcal{L}_3 = a(aa + ab + ba + bb) + b(aa + ab + ba + bb) \tag{1.23}$$
$$= aaa + aab + aba + abb + baa + bab + bba + bbb .$$

Our next task is to study some special classes of 2-letter words and learn how to count them. The first interesting problem we encounter is counting words that have a given number of a's and b's.

To better appreciate the following developments, the beginner should try counting the number of 8-letter words with 5 a's and 3 b's.

Now, let $\mathcal{L}_{n,k}$ denote the language consisting of all $\{a, b\}$-words of length n with k occurrences of the letter b.

For instance, inspecting (1.23), we see that

$$\mathcal{L}_{3,0} = aaa$$
$$\mathcal{L}_{3,1} = aab + aba + baa$$
$$\mathcal{L}_{3,2} = abb + bab + bba \tag{1.24}$$
$$\mathcal{L}_{3,3} = bbb .$$

Note that words of $\mathcal{L}_{4,2}$ may be obtained in two ways:

1. *by prefixing the letter a to a 3-letter word that already has 2 b's or*
2. *by prefixing the letter b to a 3-letter word that has only one b.*

This observation leads to the equation

$$\mathcal{L}_{4,2} = a\mathcal{L}_{3,2} + b\mathcal{L}_{3,1} . \tag{1.25}$$

Thus from (1.24), we derive that

$$\mathcal{L}_{4,2} = a \cdot (abb + bab + bba) + b \cdot (aab + aba + baa)$$
$$= aabb + abab + abba + baab + baba + bbaa .$$

It is interesting to observe that this procedure yielded us again a lexicographic list.

More generally, we can easily see that for any pair n, k $(0 \leq k \leq n)$, we have

$$\mathcal{L}_{n+1,k} = a\mathcal{L}_{n,k} + b\mathcal{L}_{n,k-1} . \tag{1.26}$$

Of course, in this expression, $\mathcal{L}_{n,k-1}$ should be interpreted as the empty language when $k = 0$.

For convenience, let us denote the number of words in $\mathcal{L}_{n,k}$ by $C_{n,k}$. Clearly, (1.25) gives

$$C_{4,2} = C_{3,2} + C_{3,1} = 6 , \tag{1.27}$$

and in general, from (1.26), we deduce that

$$C_{n+1,k} = C_{n,k} + C_{n,k-1} . \tag{1.28}$$

This equation essentially solves our first counting problem. Indeed, from it we can easily calculate the number of 8-letter words with 5 a's and 3 b's, that is, $C_{8,3}$ with the present notation.

To see how the calculation goes, suppose we arrange our numbers $C_{n,k}$ in a triangular table in the following manner:

$$
\begin{array}{ccccccccc}
 & & & & C_{1,0} & & C_{1,1} & & \\
 & & & C_{2,0} & & C_{2,1} & & C_{2,2} & \\
 & & C_{3,0} & & C_{3,1} & & C_{3,2} & & C_{3,3} \\
 & C_{4,0} & & C_{4,1} & & C_{4,2} & & C_{4,3} & & C_{4,4}
\end{array}
$$

Then (1.27) says that the third number in the last row is the sum of the numbers that are immediately above it. More generally, equation (1.28) yields that any entry in the table may be obtained by summing the two entries immediately above.

Using this observation, starting with $C_{1,0} = C_{1,1} = 1$, we obtain the table

$$
\begin{array}{ccccccccccc}
 & & & & & 1 & & 1 & & & \\
 & & & & 1 & & 2 & & 1 & & \\
 & & & 1 & & 3 & & 3 & & 1 & \\
 & & 1 & & 4 & & 6 & & 4 & & 1 \\
 & 1 & & 5 & & 10 & & 10 & & 5 & & 1 \\
 1 & & 6 & & 15 & & 20 & & 15 & & 6 & & 1 \\
\end{array}
\tag{1.29}
$$

$$
\begin{array}{ccccccccc}
1 & 7 & 21 & 35 & 35 & 21 & 7 & 1 \\
1 & 8 & 28 & 56 & 70 & 56 & 28 & 8 & 1
\end{array}
$$

from which we deduce that

$$C_{8,3} = 56.$$

The table in (1.29) is usually referred to as the *Pascal triangle*, since Pascal was led to it in his study of games of chance. However, there are indications that it was known much earlier; it is referred to as being an "ancient" table in a Chinese manuscript which itself was written at the beginning of the 14th century.

In Section 1.5, we shall derive a formula for $C_{n,k}$. However, such a formula has only theoretical significance since the recursion in (1.28) yields the best method available for computing $C_{n,k}$.

The developments in this section have a number of interesting implications. To begin with, note that (1.20) and (1.23) combined give

$$aaa + aab + aba + abb + baa + bab + bba + bbb = (a + b)^3.$$

Replacing each a by 1 and each b by some variable t on both sides of this identity yields the polynomial identity

$$1 + 3t + 3t^2 + t^3 = (1 + t)^3.$$

In particular, we see that the coefficient of t^2 is equal to 3 here because there are 3 words with 2 b's in $\mathcal{L}_3$, and each one of these words contributes a t^2.

More generally, we see that if we make the same replacements in the listing series of $\mathcal{L}_n$, we necessarily get the polynomial

$$C_{n,0} + C_{n,1}t + C_{n,2}t^2 + \cdots + C_{n,k}t^k + \cdots + C_{n,n}t^n.$$

On the other hand, since this listing series is also equal to $(a+b)^n$, we must conclude that

$$C_{n,0} + C_{n,1}t + C_{n,2}t^2 + \cdots + C_{n,k}t^k + \cdots + C_{n,n}t^n = (1 + t)^n. \tag{1.30}$$

This result may be stated in the following manner.

Theorem 1.3.1 *The coefficients of successive powers of t in the expansion of $(1 + t)^n$ are given by the nth row of the Pascal triangle.*

Thus the table in (1.29) yields in particular that

$$(1 + t)^4 = 1 + 4t + 6t^2 + 4t^3 + t^4,$$
$$(1 + t)^5 = 1 + 5t + 10t^2 + 10t^3 + 5t^4 + t^5, \tag{1.31}$$
$$(1 + t)^6 = 1 + 6t + 15t^2 + 20t^3 + 15t^4 + 6t^5 + t^6.$$

For this reason, the $C_{n,k}$'s are called *binomial coefficients*, and Theorem 1.3.1 is commonly referred to as the *binomial theorem*.

Many combinatorial structures can be represented by 2-letter languages. The best known ones are *sets* and *lattice paths*. We shall show how these come about in full detail since they will provide us with our first examples of representations of traditional combinatorial objects by formal languages.

1.3.1   Sets

Let S be a subset of $\{1, 2, \ldots, n\}$. Let $w(S)$ denote the n-letter word whose ith letter is b if i is in S and a if i is not in S. For instance, for $n = 8$ and $S = \{1, 3, 6\}$, our definition gives

$$w(S) = babaabaa \ .$$

Conversely, if w is an n-letter $\{a, b\}$-word, let $S(w)$ be the subset of $\{1, 2, \ldots, n\}$ obtained by placing i in S if and only if the ith letter of w is a b. For instance, if

$$w = abbababb \ ,$$

then

$$S(w) = \{2, 3, 5, 7, 8\}.$$

Clearly, for any given n, the maps $S \to w(S)$ and $w \to S(w)$ are inverses of each other. We see that these maps are bijections (one-to-one correspondences) between the k-element subsets of $\{1, 2, \ldots, n\}$ and n-letter $\{a, b\}$-words with k occurrences of the letter b. In particular, we may conclude that these two classes of objects are equinumerous. Therefore we have the following theorem.

Theorem 1.3.2 *The number of k-subsets of the n-set $\{1, 2, \ldots, n\}$ is given by the binomial coefficient $C_{n,k}$.*

In particular, from our previous calculations, we derive that there are 56 3-element subsets of $\{1, 2, \ldots, 8\}$. In the traditional literature, this fact may be found phrased in the following manner:
  there are 56 ways of selecting 3 distinct objects out of a set of 8 distinct objects.
The number of ways of selecting k distinct objects out of a set consisting of n distinct objects is traditionally denoted by the symbol

$$\binom{n}{k},$$

and it is usually referred to as

"n *choose* k."

From our previous observations, we deduce that $\binom{n}{k}$ and $C_{n,k}$ are one and the same number. Thus to conform with common usage, we shall here and after use the symbol $\binom{n}{k}$ rather than $C_{n,k}$.

Remark 1.3.1 Note that if we replace t by x/y in (1.30) and then multiply both sides by y^n, we obtain the polynomial identity

$$(x + y)^n = C_{n,0}y^n + C_{n,1}xy^{n-1} + C_{n,2}x^2y^{n-2} + \cdots + C_{n,n}x^n,$$

which with our new notation becomes

$$(x + y)^n = \binom{n}{0}y^n + \binom{n}{1}xy^{n-1} + \binom{n}{2}x^2y^{n-2} + \cdots + \binom{n}{n}x^n,$$

and this, using the summation symbol, may be rewritten as

$$(x + y)^n = \sum_{k=0}^{n} \binom{n}{k}x^k y^{n-k}. \tag{1.32}$$

This is the formula that is most commonly referred to as the *binomial identity*.

1.3.2   Lattice Paths

We recall that a polygonal path in the (x, y)-plane whose vertices are points with integer coordinates and whose edges are horizontal or vertical segments of unit length is called a *lattice path*. We imagine the horizontal edges as going from left to right and the vertical ones as going from bottom to top. If P is a lattice path, let us denote by $w(P)$ the word whose ith letter is a or b according as the ith edge of P is horizontal or vertical. For instance, if P is the path in Figure 1.1, then

$$w(P) = abaababa.$$

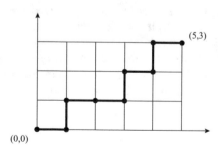

Fig. 1.1   A lattice path from the origin to the point $(5, 3)$.

Conversely, given an $\{a, b\}$-word w, let $P(w)$ be the path obtained by reversing the previous construction. That is, to get $P(w)$ start at $(0, 0)$ and, reading the letters of w successively from left to right, replace each a by a horizontal step and each b by a vertical step.

For instance, if

$$w = aabbabaa,$$

then the lattice path corresponding to w is as shown in Figure 1.2.

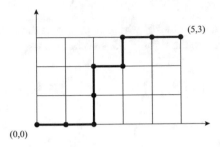

Fig. 1.2 The lattice path from the origin to $(5, 3)$ corresponding to the word $w = aabbabaa$.

Note from Figures 1.1 and 1.2 that the maps $P \rightarrow w(P)$ and $w \rightarrow P(w)$ define a bijection between the family of lattice paths joining $(0, 0)$ to $(5, 3)$ and the 8-letter words with 5 a's and 3 b's. Thus, from the table in (1.29), we deduce that there are 56 such lattice paths. The general case is given by the following theorem.

Theorem 1.3.3 *The number of lattice paths joining $(0, 0)$ to (h, k) is given by the binomial coefficient $\binom{h+k}{k}$.*

This interpretation of the binomial coefficients leads to a number of interesting identities. We shall present some of the most remarkable ones.

First of all, we notice that the table in (1.29) is symmetric. That is, for any n and k, we have

$$\binom{n}{k} = \binom{n}{n-k}. \tag{1.33}$$

This fact is easy to show from just about any interpretation of the binomial coefficients. However, the path interpretation makes it entirely obvious. For instance, the case $n = 8, k = 3$ of (1.33) is

$$\binom{8}{3} = \binom{8}{5}.$$

This corresponds to saying that there are as many lattice paths from $(0, 0)$ to $(3, 5)$ as there are from $(0, 0)$ to $(5, 3)$. However, this is clear since the former are transformed into the latter by the reflection which interchanges the x and y axes (that is, the horizontal and vertical directions).

Note further that if we start with the 1 at the end of the fourth row of the table in (1.29) and proceed diagonally to the left and down, we encounter the entries

$$1, 5, 15, 35,$$

which are, respectively,

$$\binom{4}{4}, \binom{5}{4}, \binom{6}{4}, \binom{7}{4}.$$

Now these numbers add up to 56, which is precisely the entry in the table that is immediately below and to the right of 35. In other words, we have discovered that

$$\binom{4}{4} + \binom{5}{4} + \binom{6}{4} + \binom{7}{4} = \binom{8}{5}. \tag{1.34}$$

This is no accident but the special case $n = 7, k = 4$ of the general identity

$$\binom{k}{k} + \binom{k+1}{k} + \cdots + \binom{n}{k} = \binom{n+1}{k+1}. \tag{1.35}$$

It turns out that the path interpretation of the binomial coefficients yields this identity almost immediately. Indeed, as we have seen, $\binom{8}{5}$ gives the number of lattice paths joining $(0, 0)$ to $(5, 3)$. Now, a lattice path from $(0, 0)$ to $(5, 3)$ must necessarily have as its last horizontal edge, one of the edges, a, b, c, and d of Figure 1.3.

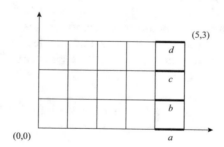

Fig. 1.3 Classification of lattice paths from the origin to $(5, 3)$ by the last horizontal edge.

Note, for instance, that there are precisely $\binom{6}{4}$ such paths whose last horizontal edge is c. The reason for this is that there are as many of these as there are lattice paths from $(0, 0)$ to $(4, 2)$. Similarly, we derive that there are $\binom{4}{4}, \binom{5}{4}$, and $\binom{7}{4}$ paths from $(0, 0)$ to $(5, 3)$ whose last horizontal edge is a, b, and d, respectively. Adding these counts must give the total number of paths from $(0, 0)$ to $(5, 3)$, and this is precisely (1.34).

Clearly, we can prove the identity in (1.35) in an entirely analogous manner.

Proceeding in the same vein, we may observe that any lattice path from $(0, 0)$ to $(4, 4)$ must pass through one of the points

$$A, B, C, D, E$$

of Figure 1.4.

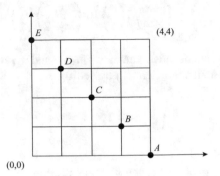

Fig. 1.4 A lattice path from the origin to $(4, 4)$ must pass through one of A, B, C, D, E.

We see that the number of lattice paths from $(0, 0)$ to $(4, 4)$ passing through C is $\binom{4}{2}^2$, since there are $\binom{4}{2}$ from $(0, 0)$ to C and $\binom{4}{2}$ from C to $(4, 4)$.
Similarly, we see that

$$\binom{4}{4}^2, \binom{4}{3}^2, \binom{4}{1}^2, \text{ and } \binom{4}{0}^2$$

of them pass through $A, B, D,$ and E, respectively.
Since these counts must add up to the total number of paths from $(0, 0)$ to $(4, 4)$, we must have the identity

$$\binom{4}{4}^2 + \binom{4}{3}^2 + \binom{4}{2}^2 + \binom{4}{1}^2 + \binom{4}{0}^2 = \binom{8}{4}. \tag{1.36}$$

We have thus discovered that the sum of the squares of the elements of the fourth row of the table in (1.29) must add up to the central element of the 8th row. That is, we must have

$$1 + 4^2 + 6^2 + 4^2 + 1 = 70,$$

which is indeed the case.
The general form of (1.36) is

$$\binom{n}{0}^2 + \binom{n}{1}^2 + \binom{n}{2}^2 + \cdots + \binom{n}{n}^2 = \binom{2n}{n}, \tag{1.37}$$

and this may be proved in the same manner by working with the $n \times n$ square grid with corners $(0, 0)$ and (n, n).

A more general identity involving 3 free parameters may be stated as follows:

$$\binom{m+n}{k} = \binom{m}{k}\binom{n}{0} + \binom{m}{k-1}\binom{n}{1} + \binom{m}{k-2}\binom{n}{2} + \cdots + \binom{m}{0}\binom{n}{k}.$$
(1.38)

This is valid for any nonnegative values of m, n, and k provided we extend the definition of the binomial coefficient $\binom{a}{b}$ by setting

$$\binom{a}{b} = 0 \text{ for } b > a .$$

We see that (1.38) reduces to (1.37) upon setting $m = k = n$.

The special case $m = 8$, $n = 6$, and $k = 4$ of (1.38) is

$$\binom{14}{4} = \binom{8}{4}\binom{6}{0} + \binom{8}{3}\binom{6}{1} + \binom{8}{2}\binom{6}{2} + \binom{8}{1}\binom{6}{3} + \binom{8}{0}\binom{6}{4} .$$
(1.39)

This is proved by observing that any lattice path from $(0, 0)$ to $(10, 4)$ must pass through one of the points, A, B, C, D, E of Figure 1.5.

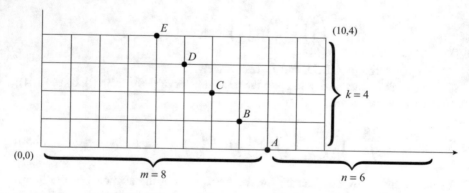

Fig. 1.5 Classification of lattice paths from the origin to $(10, 4)$.

A similar observation concerning lattice paths from $(0, 0)$ to the point $(m + n - k, k)$ yields the identity in (1.38) in full generality.

Note that from (1.38) and from the table in (1.39), we deduce that

$$\binom{10}{5} = \binom{8}{5}\binom{2}{0} + \binom{8}{4}\binom{2}{1} + \binom{8}{3}\binom{2}{2} = 56 + 70 \cdot 2 + 56 = 552 .$$

Similarly, we get that

$$\binom{10}{4} = 70 + 56 \cdot 2 + 28 = 210 .$$

Thus we see in particular that (1.38) may be used to extend the usefulness of the table in (1.29) by enabling us to calculate from it values of $\binom{n}{k}$ that do not appear in the table itself.

Finally, we should point out the 3-parameter identity

$$\sum_{i=0}^{k} \binom{m+i}{m}\binom{n+k-i}{n} = \binom{m+n+k+1}{m+n+1},\qquad (1.40)$$

which may be proved by working with vertical arrangements of *crossing edges* rather than diagonal arrangements of crossing points as we did for (1.39).

For instance, the case $m = 3, n = 5$, and $k = 4$ of (1.40) is obtained by observing that any lattice path from $(0, 0)$ to $(m+n+1, k)$ must pass through one of the edges, a, b, c, d, e of Figure 1.6.

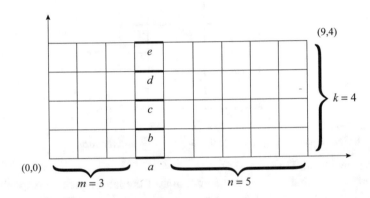

Fig. 1.6 Crossing edges for lattice paths from the origin to $(9, 4)$.

1.4   The Dyck Language, Ballot Sequences, Nested Parentheses, and 2-Rowed Young Tableaux

An $\{a, b\}$-word w is said to be a Dyck *word* (after the German mathematician Walter von Dyck) or briefly a D-*word* if and only if

(a) *w has an equal number of a' s and b' s,*

(b) *every initial segment of w has at least as many a' s as b' s.*

Thus, for instance,

$$w = aababbaabb \qquad (1.41)$$

is a D-word.

As it turns out, D-words can be used to represent a wide variety of combinatorial objects. Thus the study of D-words assumes particular importance.

Our first goal is to count them. To this end, note that D-words have always an even number of letters. Thus, for convenience, the number of D-words of length $2n$ will be denoted by D_n. In point of fact, the lattice path interpretation of $\{a, b\}$-words yields a very simple formula for D_n in terms of binomial coefficients.

To see how this comes about, note that the lattice path corresponding to the D-word in (1.41) lies entirely below the diagonal line d joining $(0, 0)$ to $(5, 5)$ (see Figure 1.7).

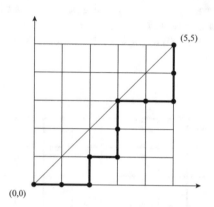

Fig. 1.7 The lattice path corresponding to the D-word $w = aababbaabb$.

This is true for all D-words.

Indeed, we see that if (i, j) is the endpoint of the lattice path corresponding to the initial segment w_1 of w, then w_1 has i a's and j b's. The property (b) of D-words is then that we must have $i \geq j$, and this is precisely what is required for the point (i, j) to be below d.

Thus, to calculate our number D_n, we are led to count the paths having this property. For convenience, let us say that a lattice path from $(0, 0)$ to (n, n) is a *strictly lower* path if it lies below d and touches d only at its endpoints. Paths that lie below d but touch it at some points (such is, for instance, the path of Figure 1.7) will be called *weakly lower* paths.

Note that we can *detach* a weakly lower path from d by adding to it an initial horizontal segment. For instance, carrying out this operation on the path of Figure 1.7, we obtain the path of Figure 1.8, which is a strictly lower path.

Clearly, this operation sends weakly lower paths from $(0, 0)$ to (n, n) into strictly lower paths from $(0, 0)$ to $(n+1, n)$ in a one-to-one fashion. Thus these two families of paths are equinumerous. We may thus obtain our number D_n by counting strictly lower paths from $(0, 0)$ to $(n + 1, n)$.

As a matter of fact, the latter paths are somewhat easier to count.

This follows from a rather clever idea, which is best explained in a special case. For instance, let us calculate the number D_5 by counting the strictly lower paths from $(0, 0)$ to $(6, 5)$.

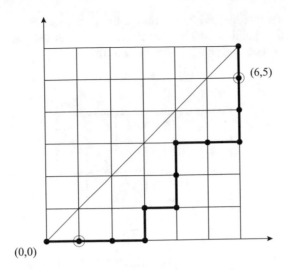

Fig. 1.8 Detaching a weakly lower path from the diagonal d.

We see from Figure 1.8 that all these paths pass through $(1, 0)$ as well. The total number of lattice paths from $(1, 0)$ to $(6, 5)$ is of course

$$\binom{10}{5}.$$

However, this number is larger than D_5, since quite a few of these paths, say I_5 of them, do touch or even cross the line d. Such is, for instance, the path P of Figure 1.9.

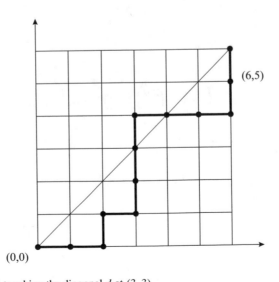

Fig. 1.9 A path P touching the diagonal d at $(3, 3)$.

However, note that if we reflect across d the portion of P which lies between $(1, 0)$ and the first point at which P intersects d, we get the path of Figure 1.10, which may be viewed as a path joining $(0, 1)$ to $(6, 5)$.

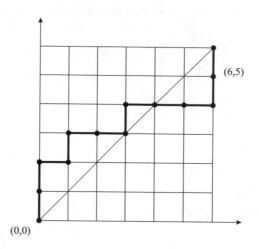

Fig. 1.10 P after reflection.

In fact, we see that there is a one-to-one correspondence between paths from $(1, 0)$ to $(6, 5)$ which meet d and unrestricted lattice paths from $(0, 1)$ to $(6, 5)$. We simply go from one kind of path to the other by reflecting across d the portion of the path that lies between the initial point and the first point of intersection with d.

This observation yields that I_5 is equal to the number of lattice paths from $(0, 1)$ to $(6, 5)$, that is,

$$I_5 = \binom{10}{4} .$$

We thus get that

$$D_5 = \binom{10}{5} - \binom{10}{4} = 252 - 210 = 42 .$$

In general we see that the number D_n may be written in the form

$$D_n = \binom{2n}{n} - I_n ,$$

where $\binom{2n}{n}$ counts the lattice paths from $(1, 0)$ to $(n+1, n)$ and I_n counts the subset of them that meet d.

The same reflection argument then yields that I_n is equal to the number of unrestricted lattice paths from $(0, 1)$ to $(n + 1, n)$. That is,

$$I_n = \binom{2n}{n - 1}.$$

We may thus state our result in the following form.

Theorem 1.4.1 *The number D_n of Dyck words of length $2n$ is given by the formula*

$$D_n = \binom{2n}{n} - \binom{2n}{n - 1}. \tag{1.42}$$

The numbers D_n are known as *Catalan* numbers after the Belgian mathematician Eugène Charles Catalan [2]. In the literature, they are most commonly denoted by the symbol C_n instead of D_n that we are using here. The first few of them are reproduced below:

$$D_0 = 1, \; D_1 = 1, \; D_2 = 2, \; D_3 = 5, \; D_4 = 14, \; D_5 = 42, \; D_6 = 132, \; D_7 = 429, \ldots$$

More than 200 different combinatorial structures and interpretations have been related to Dyck words [25]. We can only mention the most remarkable ones here.

To begin with, there is a colorful interpretation of Dyck words as ballot sequences. More precisely, let there be two candidates a and b at a local election, and suppose that there is a single voting booth. The sequence of ballots arranged in the order they have been cast may be represented by an $\{a, b\}$-word. For instance, we may use the word

aababba

to represent a sequence of 7 ballots, in which the 1st, 2nd, 4th, and 7th are for candidate a and the 3rd, 5th, and 6th are for candidate b.

We can easily see that elections represented by Dyck words are those in which both candidates receive an equal number of votes and one of the candidates, say a, never falls behind throughout the voting. For convenience, we shall refer to these as *Dull* elections.

This interpretation leads us naturally to another structure which can be represented by Dyck words. Indeed, note that we may represent the sequence of ballots given in (1.41) by the table in Figure 1.11, meaning that the 1st, 2nd, 4th, 7th, and 8th voters voted for candidate a and the 3rd, 5th, 6th, 9th, and 10th voted for b.

Putting it in another way, the bottom row in Figure 1.11 gives the positions of the a's in the word

aababbaabb

3	5	6	9	10
1	2	4	7	8

Fig. 1.11 Representation of the sequence of ballots in a Dull election.

and the top row gives the positions of the b's.

Note that in a Dull election, as we fill the table, there will never be more entries in the top row than in the bottom row. This implies that in the resulting table each entry in the top row is larger than the entry immediately below it. For instance, the entries in the fourth column increase as we go up (7 in the bottom row and 9 in the top row as we see in Figure 1.11), since candidate a got his/her first 4 votes before candidate b did.

A table such as that in Figure 1.11 with n_1 squares in the top row and n_2 squares in the bottom row (with $n_1 \leq n_2$), filled with the integers

$$1, 2, \ldots, n_1 + n_2,$$

which increase along the rows from left to right and along the columns from bottom up, is called a *Young tableau* of shape (n_1, n_2). These combinatorial objects were introduced by the British mathematician Alfred Young in 1900 [29].

Thus, in Figures 1.11 and 1.12, we have Young tableaux of shapes $(5, 5)$ and $(5, 8)$, respectively.

3	5	9	10	12			
1	2	4	6	7	8	11	13

Fig. 1.12 Young tableau of shape $(5, 8)$ corresponding to the word *aababaaabbaba*.

Here the Young tableau in Figure 1.12 corresponds to the word

$$aababaaabbaba . \tag{1.43}$$

Clearly, Dull elections involving $2n$ voters can just as well be represented by D-words of length $2n$ or by Young tableaux of shape (n, n). Thus we derive that $2n$-letter Dyck words and Young tableaux of shape (n, n) are equinumerous; in particular, the number of these tableaux is also given by the formula in (1.42).

The latter viewpoint suggests that we should represent Young tableaux of arbitrary shapes (n_1, n_2) by $\{a, b\}$-words. That is, given a tableau T, let $W(T)$ be the $\{a, b\}$-word whose ith letter is a or b according as i is in the bottom or top row of T.

Note that if we erase all the entries greater than i from a Young tableau T, we will end up erasing more entries from the top row than from the bottom row. The result will then be a subtableau T_i, which is also a Young tableau.

This given, we see that the initial segment consisting of the first i letters of $W(T)$ will be precisely the word $W(T_i)$ corresponding to T_i and as such will have more a's than b's. For instance, if we erase all entries greater than 6 in Figure 1.12, we obtain the subtableau in Figure 1.13, and this corresponds to the initial segment

$$aababa$$

of the word in (1.43).

$$T = \begin{array}{|c|c|c|c|} \hline 3 & 5 \\ \hline 1 & 2 & 4 & 6 \\ \hline \end{array}$$

Fig. 1.13 Subtableau corresponding to the initial segment $aababa$ of $aababaaabbaba$.

This shows that Young tableaux of shape (n_1, n_2) correspond in this manner to $\{a, b\}$-words with n_2 a's and n_1 b's having the property (b) given in the definition of D-words.

This fact leads us to the following result.

Theorem 1.4.2 *The number D_{n_1,n_2} of Young tableaux of shape (n_1, n_2) may be expressed in the form*

$$D_{n_1,n_2} = \binom{n_1 + n_2}{n_1} - \binom{n_1 + n_2}{n_1 - 1}. \tag{1.44}$$

Proof We illustrate the argument for the case $n_1 = 3, n_2 = 5$. Let T be the Young tableau in Figure 1.14.

$$T = \begin{array}{|c|c|c|c|c|} \hline 3 & 6 & 7 \\ \hline 1 & 2 & 4 & 5 & 8 \\ \hline \end{array}$$

Fig. 1.14 A Young tableau T of shape $(3, 5)$.

Then,

$$W(T) = aabaabba .$$

The lattice path corresponding to $W(T)$ is shown in Figure 1.15. Since T has 3 squares in the top row and 5 squares in the bottom row, the path P will end up at the point $(5, 3)$. Moreover, from our previous observations, we deduce that each initial segment of $W(T)$ will have at least as many a's as b's. This forces the lattice path corresponding to $W(T)$ to lie weakly below the diagonal d.

Thus we may conclude that we have a bijection between Young tableaux of shape $(3, 5)$ and weakly lower paths from $(0, 0)$ to $(5, 3)$.

As we have observed previously, we can transform these paths into strictly lower paths from $(1, 0)$ to $(6, 3)$ by the addition of an initial horizontal segment. Thus

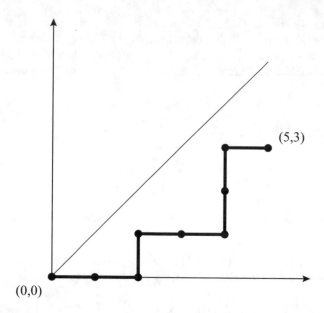

Fig. 1.15 The lattice path corresponding to $W(T)$.

our desired number can be obtained by counting strictly lower paths from $(1, 0)$ to $(6, 3)$. However, the number of these paths may be expressed in the form

$$\binom{3+5}{3} - J,$$

where $\binom{3+5}{3}$ counts all the lattice paths from $(1, 0)$ to $(6, 3)$ and J counts the subset of these paths that intersect d. However, the same reflection argument we used to obtain our formula for D_n yields that J is equal to the number of unrestricted paths from $(0, 1)$ to $(6, 3)$ (see Figure 1.16). Thus $J = \binom{3+5}{2}$ and the number $D_{3,5}$ of Young tableaux of shape $(3, 5)$ may be expressed in the form

$$D_{3,5} = \binom{3+5}{3} - \binom{3+5}{2},$$

which is the case $n_1 = 3$, $n_2 = 5$ of (1.44).

It is clear that the same argument can be carried out for any values of n_1 and n_2, and thus the formula (1.44) does hold in full generality.   □

The next structures we can relate to Dyck words are *correctly nested* strings of parentheses. To see how these strings arise, let us be given quantities

$$a_1, a_2, \ldots, a_n,$$

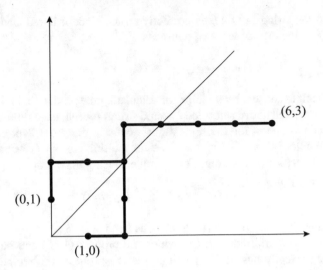

Fig. 1.16 Path from (0, 1) to (6, 3) after reflection.

which for convenience may be thought of as real numbers. We can view the expression

$$(((a_1 \cdot a_2) \cdot a_3) \cdot (a_4 \cdot a_5)) \cdot a_6 \qquad (1.45)$$

as a recipe for calculating the product

$$a_1 \cdot a_2 \cdot a_3 \cdot a_4 \cdot a_5 \cdot a_6 . \qquad (1.46)$$

That is, (1.45) may be interpreted as saying that we should first multiply a_1 and a_2 and then multiply the result with a_3. This done, the result is to be multiplied first by the product of a_4 and a_5, and finally, the product of a_1 through a_5 we have computed is to be multiplied by a_6.

The question then arises: how many different recipes are there for computing the product in (1.46)?

This is actually an important question, for unlike real numbers, our quantities $a_1, a_2, \ldots, a_6$ could be varying in an algebraic structure with a non-associative multiplication, and the product in (1.46) could take as many different values as there are recipes of type (1.45). However, at this point, we shall simply study the {(,)}-string

$$((())()) \qquad (1.47)$$

that we obtain when we remove the a's from the recipe in (1.45).

Now, a string of parentheses (and) is said to be *correctly nested* if it can be reduced to the null string by successively removing adjacent pairs ().

Note that the string in (1.47) is correctly nested since it is reduced to the null string by the following sequence of removals:

$$((())()) \rightarrow (()()) \rightarrow (()) \rightarrow () \rightarrow \epsilon \,.$$

It is not difficult to see that each recipe for calculating the product in (1.46) yields in the manner indicated above a correctly nested string. Indeed, each removal of a pair () corresponds to one of the multiplications we are to carry out. Notice, however, that it is possible for two different recipes for multiplying out (1.46) to yield the same correctly nested {(,)}-string. For example, erasing the a's in

$$a_1 \cdot (((a_2 \cdot a_3) \cdot a_4) \cdot (a_5 \cdot a_6)) \tag{1.48}$$

yields the same correctly nested {(,)}-string as (1.45).

Remarkably, the different ways in which the product (1.46) can be calculated and also the correctly nested {(,)}-strings that arise out of the recipes for these calculations themselves are both related to Dyck words.

Observe that if we replace each "(" by an a and each ")" by a b, we get

$$aaabbbabb,$$

which is seen to be a D-word. This is a completely general fact, which can be stated as follows.

Theorem 1.4.3 *A* {(,)}-*string is correctly nested if and only if the* {a, b}-*word corresponding to it by the replacements* ($\rightarrow a,$) $\rightarrow b$ *is a D-word.*

To prove this result, we need a few preliminary observations. Let us note first that we have the following two basic properties of Dyck words:

1. *Every D-word may be reduced to the empty word by the successive removal of adjacent pairs* ab.
2. *Removing or adding an adjacent pair* ab *anywhere in a D-word produces another D-word.*

It is worthwhile spending a bit of time here to see this clearly. Again, the lattice path interpretation is very helpful. However, it is more convenient to represent our D-words by a slightly different kind of path; namely, one in which

    each a *is replaced by a* 45° *edge* $(i, i) \rightarrow (i + 1, i + 1)$,
    each b *is replaced by a* −45° *edge* $(i, i) \rightarrow (i + 1, i - 1)$.

For instance, the D-word $aababbbaabb$ given in (1.41) is represented in this for by the lattice path in Figure 1.17.

We shall refer to these paths briefly as *diagonal paths*.

We can easily see that the path in Figure 1.17 may be obtained from the one in Figure 1.7 by reflecting across d, rotating −45° and stretching by a factor of $\sqrt{2}$. From this fact, we may conclude that the diagonal paths corresponding to D-words

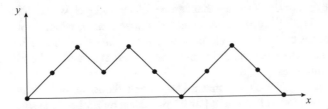

Fig. 1.17 Representation of the D-word *aababbaabb* by a lattice path with diagonal steps.

are now paths that start at $(0, 0)$, end up at a point $(2n, 0)$ for some integer n, and stay weakly above the x-axis.

For this reason, we shall refer to diagonal paths having these properties as Dyck paths.

This given, we see that an adjacent pair ab in a word w now produces a *peak* on the diagonal path corresponding to w. This simple observation makes property *1.* of Dyck words above entirely obvious. Indeed, it corresponds to the fact that

each Dyck*path is reduced to the empty path by the successive removal of the leftmost peak.*

For instance, the path of Figure 1.17 is reduced to the empty path by the sequence of steps in Figure 1.18.

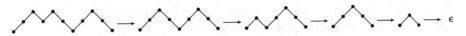

Fig. 1.18 Successive removal of the leftmost peak of a Dyck path.

As for property *2.* of the Dyck words mentioned, we just observe that it translates into

inserting or removing a peak anywhere in a Dyck *path produces another* Dyck *path,*

and this is also quite obvious geometrically.

We are now in a position to establish Theorem 1.4.3.

Proof We need to only show that the substitutions $a \leftrightarrow ($, $b \leftrightarrow)$ give a bijection between D-words and correctly nested strings. Now clearly, property *1.* of the Dyck words yields that D-words go into correctly nested strings.

To prove the converse, note that a correctly nested string S, by definition, can be built up from the empty string by successive insertions of adjacent pairs (). Now property *2.* of Dyck words guarantees that the successive strings thus obtained, including S itself, will all have to correspond to a D-word by the substitutions $(\rightarrow a,) \rightarrow b$. Thus our proof is complete. $\square$

Another collection of strings that can be related to Dyck words are the sequences of curly brackets "{" (open brace) and "}" (close brace), which in many programming languages such as C++ are used to define a block of code. Blocks appearing inside other blocks are usually indented with their start and end curly brackets for readability purposes.

More precisely, let us say that a C++ block sequence is indicated by a string over the alphabet with the letters { and } produced by proper *indenting* of the block, perhaps using tabs. An example of such a sequence is given below:

```
{
}
{
    {
        {
        }
    }
    {
    }
}
```

However, a glimpse at this figure should easily reveal that proper curly bracket indenting is none other than another geometric way of representing D-words. In fact, the D-word obtained by the replacements {$\to a$, } $\to b$ corresponds to the Dyck path of Figure 1.19, which is clearly seen to be essentially a 90° rotation of the indented curly bracket sequence given above.

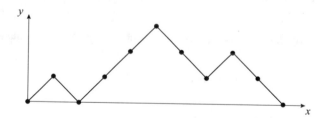

Fig. 1.19 The Dyck path corresponding to the indentation sequence of curly brackets.

The results of this section can then be summarized as follows.

Theorem 1.4.4 *There are natural one-to-one correspondences between any two of the following classes of objects:*

1. *D-words of length 2n,*
2. *Young tableaux of shape (n, n),*
3. *correctly nested strings of 2n parentheses,*

4. Dyck *paths of length* 2n,
5. *properly indented C++ curly bracket sequences of length* 2n.

Their common cardinality is given by the expression

$$D_n = \binom{2n}{n} - \binom{2n}{n-1}.$$

We end this section with one more result.

The collection of all D-words including the empty word will be here and after referred to as the *2-letter Dyck language*. For convenience the listing series of the Dyck language will be denoted by $\mathcal{D}$.

We find out that our representation of D-words by Dyck paths yields the following remarkable fact.

Theorem 1.4.5 *The listing series of the* Dyck *language satisfies the following quadratic equations:*

1. $\mathcal{D} = \epsilon + a\mathcal{D}b\mathcal{D}$ *and*
2. $\mathcal{D} = \epsilon + \mathcal{D}a\mathcal{D}b$.

Proof We simply observe that a nonempty Dyck path P starts immediately up to level one and sooner or later must return to touch the x-axis for the first time. Of course, the portion of P after that first return is another Dyck path.

It is easy to see that these geometrical facts yield us that each nonempty D-word w can be uniquely factored in the form

$$w = aw_1bw_2,$$

where aw_1b gives the portion of the corresponding Dyck path up to the first return to the x-axis and w_2 is a D-word that gives the remaining portion. This gives the first equation. The second equation may be proved in an entirely analogous manner. □

1.5 Injective Words

Here and in the following, we shall denote the general n-letter alphabet by the symbol X_n. For small values of n (that is, for $n \le 26$), it is sometimes convenient to let X_n be the appropriate section of the customary alphabet. That is, we take

$$X_2 = \{a, b\},$$
$$X_3 = \{a, b, c\},$$
$$X_4 = \{a, b, c, d\}, \tag{1.49}$$

$$\vdots$$

Of course, we can also take

$$X_2 = \{1, 2\},$$
$$X_3 = \{1, 2, 3\},$$
$$X_4 = \{1, 2, 3, 4\},$$
$$\vdots$$

The latter choices may lead to confusion, however, when we need more than ten letters. For instance, faced with the expression "21" we might not know whether it represents a one-letter word *(twenty-one)* or a two-letter word *(two one)*.

For these reasons, when n is large, we shall invariably set

$$X_n = \{x_1, x_2, \ldots, x_n\}.$$

With this alphabet the above ambiguities are removed, since *twenty-one* is represented by x_{21} and *two one* by $x_2 x_1$.

A word is said to be *injective* if it is not the null word and all its letters are different. For instance,

$$cda, \quad rfzn, \quad x_2 x_{20} x_5 x_{10} x_{22}$$

are injective words.

The language consisting of all the injective words over the alphabet $\mathcal{A}$ will be denoted by $J(\mathcal{A})$. The collection of all k-letter words in $J(\mathcal{A})$ will be denoted by $J_k(\mathcal{A})$.

Thus, for instance, with this notation, we have

$$s J(a, b) = a + b + ab + ba, \tag{1.50}$$
$$s J_2(a, b, c) = ab + ba + ac + ca + bc + cb.$$

Our first goal here is to count the number of k-letter injective words over an n-letter alphabet. More precisely, we shall obtain a formula for the number of words in $J_k(X_n)$.

Note that $J_4(a, b, c)$ is empty since there is no way we can produce a 4-letter word with a, b, c without using at least twice one of the letters. More generally, we see that $J_k(X_n)$ is empty for $k > n$. This given, we shall tacitly assume when working with $J_k(X_n)$ that $k \leq n$.

We have the following relation:

$$J_3(a, b, c, d) = a J_2(b, c, d) + b J_2(a, c, d) + c J_2(a, b, d) + d J_2(a, b, c). \tag{1.51}$$

This follows from the simple observation that a word in $J_3(a, b, c, d)$ must start with $a, b, c,$ or d and what follows after must be an injective word in the remaining letters.

Since clearly $J_2(b, c, d)$, $J_2(a, c, d)$, $J_2(a, b, d)$, and $J_2(a, b, c)$ all have the same number of elements, we deduce from (1.50) and (1.51) that

$$\# J_3(a, b, c, d) = 4 \cdot \# J_2(a, b, c) = 4 \cdot 6 = 24 . \tag{1.52}$$

Similarly, we obtain the relation

$$J_2(a, b, c) = a J_1(b, c) + b J_1(a, c) + c J_1(a, b)$$
$$= a(b + c) + b(a + c) + c(a + b) . \tag{1.53}$$

Thus, as we already know from the second part of (1.50), we have

$$\# J_2(a, b, c) = 3 \cdot 2 . \tag{1.54}$$

The same reasoning can be carried out in full generality, and we can see that (1.51) and (1.53) are just special cases of the identity

$$J_k(x_1, x_2, \ldots, x_n) = x_1 J_{k-1}(x_2, x_3, \ldots, x_n) +$$
$$x_2 J_{k-1}(x_1, x_3, \ldots, x_n) + \cdots + \tag{1.55}$$
$$x_n J_{k-1}(x_1, x_2, \ldots, x_{n-1}) .$$

It is worthwhile mentioning that this identity embodies an algorithm for constructing all injective k-letter words in a given alphabet. Indeed, it is not difficult to translate (1.55) into a recursive procedure for listing the words in $J_n(x_1, x_2, \ldots, x_n)$ in lexicographic order.

Proceeding as before, we note that the collections

$$J_{k-1}(x_2, x_3, \ldots, x_n), \ J_{k-1}(x_1, x_3, \ldots, x_n), \ldots, \ J_{k-1}(x_1, x_2, \ldots, x_{n-1})$$

have all the same number of elements. Thus from (1.55), we deduce that

$$\# J_k(x_1, x_2, \ldots, x_n) = n \cdot \# J_{k-1}(x_1, x_2, \ldots, x_{n-1}) .$$

Setting for a moment

$$\# J_k(x_1, x_2, \ldots, x_n) = M_{n,k} ,$$

this may be rewritten as

$$M_{n,k} = n \cdot M_{n-1,k-1} . \tag{1.56}$$

This recursion solves our enumeration problem.

For instance, from (1.56), we successively deduce that

$$M_{8,3} = 8 \cdot M_{7,2} = 8 \cdot 7 \cdot M_{6,1},$$

and since $M_{6,1}$ is clearly equal to 6, we get

$$M_{8,3} = 8 \cdot 7 \cdot 6 = 336 \,.$$

Proceeding more generally, we see that successive applications of (1.56) yield

$$
\begin{aligned}
M_{n,6} &= n \cdot M_{n-1,5} \\
&= n(n-1)M_{n-2,4} \\
&= n(n-1)(n-2)M_{n-3,3} \\
&= n(n-1)(n-2)(n-3)M_{n-4,2} \\
&= n(n-1)(n-2)(n-3)(n-4)M_{n-5,1} \\
&= n(n-1)(n-2)(n-3)(n-4)(n-5) \,.
\end{aligned}
$$

From these observations, we can easily see that the general result may be stated as follows.

Theorem 1.5.1 *The number $M_{n,k}$ of k-letter injective words over an n-letter alphabet is given by the expression*

$$n(n-1)(n-2)\cdots(n-k+1) \,. \tag{1.57}$$

Remark 1.5.1 The expressions

$$x(x-1)(x-2)\cdots(x-k+1) \tag{1.58}$$

and

$$x(x+1)(x+2)\cdots(x+k-1) \tag{1.59}$$

will often occur in our treatment. Note that when we do carry out the multiplications in (1.58) and (1.59), we obtain in both cases polynomials of degree k in x. For this reason, (1.58) and (1.59) are, respectively, referred to as the *lower factorial* and the *upper factorial* polynomials. Because of the ubiquity of these polynomials, it is convenient to introduce a shorthand notation. A symbol that is often used is $(x)_k$. Unfortunately, the same symbol is used equally often in the literature to represent either of the polynomials in (1.58) and (1.59). To make things worse, the upper factorial polynomials in (1.59) are also referred to in the literature as the *rising factorial* or the *shifted factorial* polynomials.

In our treatment, we shall use $(x)_k$ only to denote the polynomial in (1.58). When both (1.58) and (1.59) occur in our presentation, it is possible to use the more refined notation

$$x(x-1)(x-2)\cdots(x-k+1) = (x)_{\downarrow k} \tag{1.60}$$
$$x(x+1)(x+2)\cdots(x+k-1) = (x)_{\uparrow k} .$$

Injective words may be used to represent linear arrangements of distinct objects. In fact, such an arrangement may be viewed as a word in an alphabet whose symbols are the objects themselves.

Linear arrangements of k distinct objects are often called k-*permutations*. Thus Theorem 1.5.1 may be found stated as follows.

The number of k-permutations of k out of n given objects is $(n)_k$.

A linear arrangement of n objects out of n given ones is usually referred to as a *permutation* of the given objects. Thus the case $k = n$ of Theorem 1.5.1 can be stated as follows:

Theorem 1.5.2 *The number of permutations of n objects is given by the expression*

$$n(n-1)(n-2)\cdots 2\cdot 1 .$$

This expression is usually referred to as n *factorial* and is denoted by the symbol $n!$.

Note that we have

$$M_{8,3} = 8\cdot 7\cdot 6 = \frac{8!}{5!} ,$$

and in general (for $k \leq n$), we can write

$$M_{n,k} = \frac{n!}{(n-k)!} . \tag{1.61}$$

It develops that we can relate injective words and 2-letter words and thereby obtain an explicit formula for our binomial coefficients.

The idea is quite simple. For instance, suppose we wish to put together all 4-letter injective words in the alphabet $\{a, b, c, d, e, f, g, h, i\}$. One way to do this is to choose first which 4 letters will be used in the word (there are $\binom{9}{4}$ ways of making this choice). After this is done, we arrange the chosen 4 letters in all possible ways, and by Theorem 1.5.2, there are 4! permutations of these 4 letters possible.

Thus all together, we produce $\binom{9}{4}\cdot 4!$ different words, and we are led to the identity

$$9\cdot 8\cdot 7\cdot 6 = (9)_4 = M_{9,4} = \binom{9}{4}\cdot 4! .$$

Solving for $\binom{9}{4}$ gives

$$\binom{9}{4} = \frac{9 \cdot 8 \cdot 7 \cdot 6}{4!} = \frac{9!}{4!5!} .$$

The same idea, in the general case, yields the relation

$$M_{n,k} = (n)_k = \binom{n}{k} k! .$$

Thus, combining with (1.61), we obtain the following remarkable fact.

Theorem 1.5.3 *The binomial coefficient can be written in the form*

$$\binom{n}{k} = \frac{(n)_k}{k!} = \frac{n!}{k!(n-k)!} . \tag{1.62}$$

Remark 1.5.2 It should be noted that it is by no means obvious that either of the expressions in (1.62) evaluates to an integer. However, our combinatorial interpretation automatically yields this result. We shall see that this simple fact has remarkable consequences. In particular, the fact that $\binom{2n}{n}$ is an integer has astonishing implications in the theory of prime numbers.

Theorem 1.5.3 has a further surprising implication. Namely, we have the following result.

Theorem 1.5.4 *The number D_n of Dyck words with 2n letters can be written in the form*

$$D_n = \frac{1}{n+1}\binom{2n}{n} . \tag{1.63}$$

Proof Formulas (1.42) and (1.62) combined give

$$D_n = \frac{(2n)!}{n!n!} - \frac{(2n)!}{(n-1)!(n+1)!}$$

$$= (2n)! \left(\frac{1}{n!n!} - \frac{1}{(n-1)!(n+1)!} \right)$$

$$= \frac{(2n)!}{n!(n+1)!}(n+1-n)$$

$$= \frac{1}{(n+1)} \frac{(2n)!}{n!n!} ,$$

which is another way of writing (1.63).   □

There is a more symmetric and instructive way to derive formula (1.62). We illustrate the idea again in the case $n = 9, k = 4$. To this end, we use the alphabets

$$\mathcal{A} = \{a_1, a_2, a_3, a_4\}, \quad \mathcal{B} = \{b_1, b_2, b_3, b_4, b_5\}.$$

Note that there are $(4 + 5)! = 9!$ permutations of the letters

$$a_1, a_2, a_3, a_4, b_1, b_2, b_3, b_4, b_5. \tag{1.64}$$

Putting it in another way, there are $9!$ injective 9-letter words over the alphabet $\mathcal{A} + \mathcal{B}$. Now each of these words reduces to an $\{a, b\}$-word upon removal of the subscripts.

For instance,

$$a_3 a_2 b_1 b_4 a_1 b_3 b_2 b_5 a_4$$

reduces in this manner to

$$aabbabbba. \tag{1.65}$$

There are of course many different permutations of the letters in (1.64) that lead to the word in (1.65). However, we see that each of them is obtained by placing the subscripts $1, 2, 3, 4$ to the a's and the subscripts $1, 2, 3, 4, 5$ to the b's in (1.65).

Now each way of subscripting the a's may be viewed as a permutation of $1, 2, 3, 4$ and each way of subscripting the b's may be viewed as a permutation of $1, 2, 3, 4, 5$. Thus we may conclude that there is a total of

$$4! \cdot 5!$$

different permutations of the letters in (1.64), which reduce to the word in (1.65) upon removal of subscripts.

Now clearly, regardless of which $\{a, b\}$-word we work with, as long as it has 4 a's and 5 b's, we are led to the same count.

We can thus decompose the $9!$ permutations of the letters in (1.64) into groups each containing $4!5!$ permutations, the permutations of a given group reducing to the same $\{a, b\}$-word upon removal of subscripts.

Since we have as many groups as there are $\{a, b\}$-words with 4 a's and 5 b's, we conclude that we must have

$$9! = \binom{9}{4} 4! 5!, \tag{1.66}$$

which is (1.62) in the special case $n = 9, k = 4$.

It is not difficult to see that the general case can be treated in exactly the same manner.

Remark 1.5.3 We may refer to the basic step that yielded us (1.66) as the *shepherd principle*. The reason for this terminology goes back to a silly joke that is as follows:

Two passengers in a train are traveling through Basque Land. Through the window, they observe a large flock of sheep lying on their side in the hot summer day. One of the passengers, a shepherd (evidently a mathematically inclined one) says:

There are 12344 sheep in this flock!

The other passenger wonders:

How did you count them so fast?

To which the shepherd replies:

Simple! I counted the legs and divided by four.

Formula (1.66) may be viewed as an application of this principle in the following manner. We interpret each $\{a, b\}$-word w with 4 a's and 5 b's as a sheep. The permutations of $\mathcal{A}+\mathcal{B}$ which reduce to w upon removal of subscripts are interpreted as its *legs*.

The total number of legs in this strange flock is of course 9!. By the previous reasoning, each sheep w has 4!5! legs. Thus the number of sheep by the shepherd principle is given by the formula

$$\#\text{sheep} = \frac{\#\text{legs}}{\#\text{legs in a sheep}} = \frac{9!}{4!5!} \, .$$

1.6   Increasing Words

A word

$$w = x_{i_1} x_{i_2} \cdots x_{i_n}$$

over the alphabet $X_k = \{x_1, x_2, \ldots, x_k\}$ is said to be *weakly increasing* or *alphabetic* if

$$i_1 \leq i_2 \leq \cdots \leq i_n$$

and *strictly increasing* if

$$i_1 < i_2 < \cdots < i_n \, .$$

The languages of weakly and strictly increasing words will be denoted by *WI* and *SI*, respectively. Note that over the 2-letter alphabet $X = \{a, b\}$, the weakly increasing words are all of the form

$$w = aa \cdots abb \cdots b = a^h b^k \, .$$

That is, the words of $WI(a, b)$ are obtained by concatenating a term from the series

$$1 + a + a^2 + \cdots = \frac{1}{1 - a}$$

with a term from the series

$$1 + b + b^2 + \cdots = \frac{1}{1 - b} .$$

Thus we may write

$$WI(a, b) = \frac{1}{(1 - a)} \frac{1}{(1 - b)} .$$

Similarly, we deduce that

$$WI(x_1, x_2, \ldots, x_k) = \frac{1}{(1 - x_1)} \frac{1}{(1 - x_2)} \cdots \frac{1}{(1 - x_k)} .$$

We can proceed in the same manner for the language SI . Indeed, we see that

$$SI(a, b) = 1 + a + b + ab = (1 + a)(1 + b) .$$

Clearly, since in a strictly increasing word, each letter can only occur once, to build up $SI(x_1, x_2, \ldots, x_k)$, we are to choose one term from each of the factors $(1 + x_i)$ $(i = 1, 2, \ldots, k)$ in that order. Thus

$$SI(x_1, x_2, \ldots, x_k) = (1 + x_1)(1 + x_2) \cdots (1 + x_k) .$$

We also see that strictly increasing words are yet another way of representing subsets, where the word

$$x_{i_1} x_{i_2} \cdots x_{i_n}$$

represents the subset

$$S = \{i_1, i_2, \ldots, i_n\} .$$

Consequently, the number of n-letter words over $SI(x_1, x_2, \ldots, x_k)$ is simply given by the binomial coefficient

$$\binom{k}{n} .$$

The situation for *WI* is slightly more subtle. For convenience, let WI_n denote the collection of n-letter words in *WI* and set

$$W_{n,k} = \#WI_n(x_1, x_2, \ldots, x_k) .$$

Note that the elements of *WI* may also be written in the form

$$w = x_1^{p_1} x_2^{p_2} \cdots x_k^{p_k} , \tag{1.67}$$

where the integer p_i gives the number of occurrences of x_i in w. Thus for $w \in WI_n$, we must have

$$p_1 + p_2 + \cdots + p_k = n \qquad (p_i \geq 0) . \tag{1.68}$$

It often happens (for instance, in the study of roots of polynomials) that we are to consider collections of points each occurring with a certain multiplicity. Such collections are referred to as *multisets*. Thus, the words of $WI_n(x_1, x_2, \ldots, x_k)$ may also be used to represent multisubsets of the k-element set $X = \{x_1, x_2, \ldots, x_k\}$. The exponent p_i in (1.67) can then be referred to as the *multiplicity* of x_i. In this vein, the sum in (1.68), which is the length of w, gives the *cardinality* of the corresponding multiset.

To calculate $W_{n,k}$, it is convenient to resort to yet another way of representing multisets. The process by which we arrive at it is amusing. We shall start with an example. The multiset corresponding to the word

$$x_1^3 x_2^4 x_3^0 x_4^1$$

can be depicted by the *balls in boxes* arrangement shown in Figure 1.20.

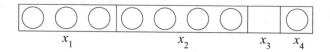

Fig. 1.20 A *balls in boxes* arrangement.

Reducing this picture to the minimum needed to recover it, we are led to the diagram in Figure 1.21.

Fig. 1.21 Succinct representation of the *balls in boxes* arrangement of Figure 1.20.

From this diagram, we ultimately arrive at the *word*

$$00010000110$$

over the 2-letter alphabet $X = \{0, 1\}$, where the balls are represented by 0 and the bars by 1. This word is then a rather clever coding of the arrangement of balls in boxes given in Figure 1.20.

It is easy to see that every 11-letter word with 8 occurrences of 0 and 3 occurrences of 1 can be converted, by reversing the above procedure, into an arrangement of 8 balls in four boxes, and it may thus be used to represent a word

$$w = x_1^{p_1} x_2^{p_2} x_3^{p_3} x_4^{p_4}$$

with

$$p_1 + p_2 + p_3 + p_4 = 8 \,.$$

Of course, the number of 11-letter words with 8 occurrences of 0 and 3 occurrences of 1 is given by the binomial coefficient

$$\binom{8 + 3}{3} \,.$$

We have thus proved that

$$W_{8,4} = \binom{11}{3} = 165 \,.$$

Similarly, we can use a word such as in (1.67), with the exponents satisfying (1.68), as a representation of an arrangement of n balls in k boxes. The number of balls in the ith box is made equal to the multiplicity of x_i. As before, we can then convert the picture into an $(n + k - 1)$-letter word with n occurrences of 0 and $k - 1$ occurrences of 1, the former representing the balls and the latter representing the walls separating the k boxes. We can thus finally conclude in the general case we have

$$W_{n,k} = \binom{n + k - 1}{k - 1} \,.$$

In some literature, the symbol

$$\left(\!\!\binom{k}{n}\!\!\right)$$

is used to represent $W_{n,k}$. We shall not do this here and use the binomial coefficient as we did above.

We should point out that properly speaking what is depicted in Figure 1.20 is a configuration of *indistinguishable balls in distinguishable boxes* .

Indeed, the ith box is labeled x_i, while the balls have no individuality at all since they are only used to depict quantity (the exponents p_i of the x_i).

In the next section, we shall consider the reverse situation, that is, we shall study configurations of

distinguishable balls in indistinguishable boxes .

These are used to represent *set partitions* that are a basic combinatorial tool as we shall see in more detail in Chapter 6.

1.7 Set Partitions and Restricted Growth Words

A decomposition of a set Ω into nonempty disjoint subsets $A_1, A_2, \ldots, A_k$ is briefly referred to as a *partition* of Ω. Formally we have

$$a) \quad \Omega = A_1 \cup A_2 \cup \cdots \cup A_k$$
$$b) \quad A_i \cap A_j = \emptyset \ \text{ for } i \neq j \, .$$

It will be convenient here and after to condense these two conditions into the single expression

$$\Omega = A_1 + A_2 + \cdots + A_k,$$

the "+" sign denoting *disjoint union*.

The subsets $A_1, A_2, \ldots, A_k$ are referred to as the *parts* of the partition. A partition of a set into k parts will sometimes be referred to as a *k-partition*. For example,

$$A_1 = \{9\}, \ \ A_2 = \{3, 5, 7\}, \ \ A_3 = \{8, 10, 11, 12\}, \ \ A_4 = \{1, 2, 4, 6\} \tag{1.69}$$

is a 4-partition of the set

$$\Omega = \{1, 2, 3, \ldots, 12\} \, .$$

It is important to note that the term *set partition* is always used in a context where the order of the parts is immaterial. That is, a set partition simply represents a way of splitting a set into parts. When the order of the parts matters, the term *ordered set partition* is used instead. This given, the convention is made to display set partitions with the parts written out in order of increasing least elements. Thus, for instance, for the partition in (1.69), the smallest elements (within the parts) are 1, 3, 8, 9, and accordingly it should be written in the form

$$A_1 = \{1, 2, 4, 6\}, \ \ A_2 = \{3, 5, 7\}, \ \ A_3 = \{8, 10, 11, 12\}, \ \ A_4 = \{9\} \, . \tag{1.70}$$

Set partitions are depicted as configurations of distinguishable balls in indistinguishable boxes (the balls representing the integers 1, 2, 3, etc. and the boxes representing the parts). It will be convenient to briefly refer to these as *DBIB* configurations. In drawing them, we again follow the same convention. That is, the boxes are drawn in order of increasing least elements, and the elements within the boxes are placed in increasing order of labels.

In other contexts, even the order in which the balls are placed within the boxes counts, but then we are dealing with something which is more appropriately called a configuration of *balls in tubes*, which will be studied in Chapter 6.

With these conventions, the partition in (1.70) is depicted as shown in Figure 1.22.

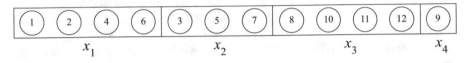

Fig. 1.22 A *DBIB* configuration.

This manner of visualization leads us to a very useful formal language representation of set partitions. To see how this comes about, we shall work on the example above and show how the picture is transformed into a word.

The process is quite simple. We interpret the balls as *positions* and boxes (in the order given by the picture) as *letters*. In position i of the word, we place the letter x_j if the ball labeled "i" is in the jth box. Doing this yields the word

$$x_1 \, x_1 \, x_2 \, x_1 \, x_2 \, x_1 \, x_2 \, x_3 \, x_4 \, x_3 \, x_3 \, x_3 \, .$$

Next, we need to find out what kind of words we do get in this manner from *DBIB* configurations. Now, remarkably, these words have a very simple description.

Note for instance that in the above example, since the smallest element in the 3rd part is 8, it is not until the 8th position in the word that we encounter an x_3. Moreover, since the parts are displayed in order of increasing least elements, the letter x_2 must have occurred in the word before x_3. Indeed that letter did occur in the position given by the smallest label in the 2nd box (which is smaller than 8 by our convention). We thus see that every time a new letter occurs, the letter just preceding it (in the alphabet) must already have occurred before it in the word. Strings satisfying these requirements are referred to as *restricted growth* words. The collection of all restricted growth words will be referred to here as the $\mathcal{RG}$-language. To be precise, this language is defined as follows.

A word

$$w = x_{i_1} x_{i_2} \cdots x_{i_n}$$

over the alphabet $X = \{x_1, x_2, \ldots, x_k\}$ belongs to $\mathcal{RG}$ if and only if for each $s = 2, 3, \ldots, n$,

$$i_s \leq 1 + \max\{i_1, i_2, \ldots, i_{s-1}\} . \tag{1.71}$$

Let us denote by $\mathcal{RG}_k$ the language of all words in $\mathcal{RG}$ in which the largest letter appearing is x_k. The experience gained in the previous sections, and a moment's thought, should tell us that the language $\mathcal{RG}_4$ is given by the expression

$$\mathcal{RG}_4 = x_1 \frac{1}{1-x_1} x_2 \frac{1}{1-x_1-x_2} x_3 \frac{1}{1-x_1-x_2-x_3} x_4 \frac{1}{1-x_1-x_2-x_3-x_4} .$$

For clearly, the first letter can only be x_1. Thereafter, before the first occurrence of x_2, any number of x_1's can occur; moreover, after the first occurrence of x_2 and before the first occurrence of x_3, only x_1's and x_2's can occur. Similarly, after the first occurrence of x_3 and before the first occurrence of x_4, only x_1's, x_2's, x_3's can occur, etc.

1.8 Stirling Numbers of the Second Kind

For convenience, let us set here

$$\Omega_n = \{1, 2, \ldots, n\}, \tag{1.72}$$

and let the collection of all k-partitions of Ω_n be denoted by $\Pi_{n,k}$. It is customary to set

$$S_{n,k} = \#\Pi_{n,k}$$

and refer to this as the *Stirling number of the second kind*. We shall consider the $S_{n,k}$ as filling an infinite matrix

$$[S_{n,k}] \qquad (n, k \geq 1)$$

the agreement being that

$$S_{n,k} = 0 \text{ if } k > n . \tag{1.73}$$

It is clear that we have

$$S_{n,1} = 1 = S_{n,n} ,$$

so that the first column and the diagonal entries of this matrix are 1's. The remaining entries can be calculated by means of the following recursion.

Theorem 1.8.1 *For all $n \geq 1, k \geq 2$, we have*

$$S_{n+1,k} = S_{n,k-1} + kS_{n,k} . \tag{1.74}$$

Proof The result is an immediate consequence of the fact that a partition of Ω_{n+1} has one of the following properties:

> (a) $n + 1$ is in a part all by itself, (1.75)
>
> (b) the part that contains $n + 1$ has some other elements.

Indeed, we can use this fact to put together an algorithm for listing all the partitions of Ω_{n+1} when those of Ω_n are given.

To this end, note that we can obtain the elements of $\Pi_{n+1,k}$ having property (a) in (1.75) taking an element of $\Pi_{n,k-1}$ and adding to it the singleton part

$$\{n + 1\} .$$

Clearly, these partitions are enumerated by the first term on the right hand side of (1.74). On the other hand, those having property (b) in (1.75) can be obtained from an element of $\Pi_{n,k}$ by adding $n + 1$ to one of the parts. Since this can be done in k different ways, each time obtaining a different partition, we can see that the number of elements of $\Pi_{n+1,k}$ having property (b) is given by the second term on the right hand side of (1.74). Thus (1.74) must hold as asserted. □

Using this recurrence, we can very quickly construct the entries in Table 1.1 where the blank entries are taken to be zero in accordance with (1.73).

Table 1.1 The Stirling numbers of the second kind $S_{n,k}$.

$n \backslash k$	1	2	3	4	5	6	7
1	1						
2	1	1					
3	1	3	1				
4	1	7	6	1			
5	1	15	25	10	1		
6	1	31	90	65	15	1	
7	1	63	301	350	140	21	1

An element of the kth column here (following the recursion in (1.74)) is obtained by multiplying the element directly above it by k and adding the result to the element directly above and to the left. Thus, for instance, we have

$$S_{5,3} = 7 + 3 \times 6 = 25 .$$

Remark 1.8.1 We should note that the recursion in (1.74) can also be obtained by working with the language $\mathcal{RG}_k$. Indeed, if we look at a word of length $n + 1$ in $\mathcal{RG}_k$, there are two possibilities for the letter in the $(n + 1)$th position:

>   (a) *It is the first x_k in the word.*

>   (b) *It is not the first x_k .*

By removing this letter, we get a word of length n, which in case (a) is in $\mathcal{RG}_{k-1}$ and in case (b) is still in $\mathcal{RG}_k$. Now, the number of words falling in case (a) is thus given by $S_{n,k-1}$, while clearly, for each word of $\mathcal{RG}_k$, there are exactly k words of length $n + 1$ in $\mathcal{RG}_k$ that yield it upon removal of the $(n + 1)$th letter. This gives (1.74) again.

Remark 1.8.2 Later on we will be studying the polynomials $\sigma_n(x)$ formed by the entries in the nth row of Table 1.1. These are traditionally called the *exponential polynomials* and are defined by setting

$$\sigma_n(x) = \sum_{k=1}^{n} S_{n,k} \, x^k . \tag{1.76}$$

From Table 1.1, the first few of them are as follows:

$$\sigma_1(x) = x$$
$$\sigma_2(x) = x + x^2$$
$$\sigma_3(x) = x + 3x^2 + x^3$$
$$\sigma_4(x) = x + 7x^2 + 6x^3 + x^4 .$$

The *total* number of partitions of the set Ω_n is usually referred to as the *Bell number* and is denoted by B_n. We clearly have

$$B_n = \sum_{k=1}^{n} S_{n,k} = \sigma_n(1) .$$

In other words, these numbers are given by the successive row sums of Table 1.1, thus

$$B_1 = 1, \ B_2 = 2, \ B_3 = 5, \ B_4 = 15, \ B_5 = 52, \ B_6 = 203, \ B_7 = 877, \ldots$$

The growth of this sequence is very rapid, indeed B_{19} ($= 5832742205057$) has already 13 digits. The B_n may be computed by the following curious recursion:

Theorem 1.8.2

$$B_{n+1} = \sum_{k=0}^{n} \binom{n}{k} B_{n-k} . \tag{1.77}$$

Proof A partition of the set $\{1, 2, \ldots, n + 1\}$ can be constructed by first putting together the part that contains $n + 1$ and then partitioning the remaining set of elements. Now the part that contains $n + 1$ may have k of the elements $1, 2, \ldots, n$. The number of ways to choose these k elements is $\binom{n}{k}$. Having chosen them, the remaining $n - k$ elements can, of course, be partitioned in B_{n-k} different ways. Combining these observations, we immediately derive that (1.77) must hold as asserted. □

We close this section by establishing an important result involving the $S_{n,k}$.

Theorem 1.8.3

$$x^n = \sum_{k=1}^{n} S_{n,k} (x)_k . \tag{1.78}$$

Proof Recall that the notation $(x)_k$ that appears in (1.78) is our shorthand for the lower factorial polynomial

$$x(x - 1) \cdots (x - k + 1) .$$

Let $\mathcal{L}_n(x)$ for a moment denote the language consisting of all the words of length n over an alphabet with x letters. If the alphabet is

$$\mathcal{A} = \{a_1, a_2, \ldots, a_x\},$$

then

$$\mathcal{L}_n(x) = (a_1 + a_2 + \cdots + a_x)^n . \tag{1.79}$$

The identity in (1.78) is obtained by counting the words of $\mathcal{L}_n(x)$ in two different ways. Firstly, from (1.79), we immediately get that

$$\#\mathcal{L}_n(x) = x^n . \tag{1.80}$$

The second count is derived from the following observations.

Let w be a word of $\mathcal{L}_n(x)$, and let $k \leq n$ be the number of distinct letters that appear in w. Looking at the positions that each of these letters occupies leads to a partition of the set of positions. More precisely, we place i and j in the same part if and only if, in w, positions i and j both are occupied by the same letter. This breaks up the set Ω_n into k parts. Thus each w in $\mathcal{L}_n(x)$ is associated with an element of $\Pi_{n,k}$. Clearly, the words of $\mathcal{L}_n(x)$ which are associated with a given fixed partition

$$\pi = \{A_1, A_2, \ldots, A_k\}$$

are obtained by choosing an injective $\mathcal{A}$-word

$$a_{i_1} a_{i_2} \cdots a_{i_k}$$

and placing a_{i_1} in each of the positions corresponding to A_1, placing a_{i_2} in each of the positions corresponding to A_2, etc., and finally placing a_{i_k} in each of the positions corresponding to the elements of A_k. This fact, combined with Theorem 1.5.1, gives that the number of words of $\mathcal{L}_n(x)$ which are associated with π is given by the lower factorial polynomial

$$(x)_k = x(x-1)\cdots(x-k+1) .$$

Since the number of π in $\Pi_{n,k}$ is $S_{n,k}$, we see that the term

$$S_{n,k} \, (x)_k$$

counts the number of words of $\mathcal{L}_n(x)$ which lead to a k-partition of the set of positions. We must then conclude that

$$\sum_{k=1}^{n} S_{n,k} \, (x)_k = \#\mathcal{L}_n(x) .$$

Comparing with (1.80) yields our desired relation (1.77).   □

Remark 1.8.3 A relation between polynomials of degree n that is valid at more than n distinct values of the argument must be valid for all values. Thus, although (1.78) was established only for integer values of x, it must remain valid for all x. Thus Theorem 1.8.3 may be interpreted as saying that the Stirling numbers of the second kind are simply the coefficients in the expansion of x^n in terms the lower factorial polynomials. This is precisely the context in which the $S_{n,k}$ were first encountered.

1.9   Permutations and Stirling Numbers of the First Kind

A permutation σ of the numbers $1, 2, \ldots, n$ written in the two-line notation

$$\begin{pmatrix} 1 & 2 & \ldots & n \\ \sigma_1 & \sigma_2 & \ldots & \sigma_n \end{pmatrix} \tag{1.81}$$

can be interpreted as a rule for mapping the set $[n] = \{1, 2, \ldots, n\}$ onto itself. That is, we interpret the symbol in (1.81) as saying that σ maps the element i onto the element σ_i.

This viewpoint leads us to a very useful graphical representation of a permutation. The idea is to plot $1, 2, \ldots, n$ as points in the plane and draw an arrow between a point i and its image σ_i.

For instance, if we apply this construction to the permutation

$$\sigma = \begin{pmatrix} 1\,2\,3\,4\,5\,6\,7\,8 \\ 4\,2\,8\,7\,1\,6\,5\,3 \end{pmatrix},$$

we obtain the directed graph in Figure 1.23.

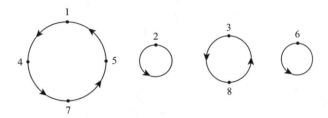

Fig. 1.23 Cycle diagram of the permutation 4 2 8 7 1 6 5 3.

We can easily see that whatever permutation we start with, we end up with a graph consisting of a certain number of disjoint cyclical connected components. These components are usually referred to as the *cycles* of the permutation.

It is customary to represent a cycle by the symbol obtained by writing its vertices, enclosed in parentheses, in the order they appear in the cycle, starting from the smallest element. Thus, for the example in Figure 1.23, we obtain the symbols

$$(1, 4, 7, 5)(2)(3, 8)(6) . \tag{1.82}$$

We usually refer to this construction as the *cycle factorization* of a permutation and write

$$\sigma = (1, 4, 7, 5)(2)(3, 8)(6) .$$

We shall also refer to the graph in Figure 1.23 as the *cycle diagram* of the permutation.

Let us now denote by $P_{n,k}$ the collection of permutations of $[n]$ which factorize into a total of k cycles, and let $p_{n,k}$ denote the number of such permutations. We remark that in some treatments, the notation $c(n, k)$ is used for our $p_{n,k}$.

We have the following analog of Theorem 1.8.1.

Theorem 1.9.1 *For all $n \geq 1, k \geq 2$, we have*

$$p_{n+1,k} = p_{n,k-1} + np_{n,k} . \tag{1.83}$$

Proof Just as we did for $\Pi_{n+1,k}$, we note that an element of $P_{n+1,k}$ has one of the following properties:

a) $n+1$ is in a cycle all by itself,   (1.84)

b)   the cycle containing $n+1$ has other elements.

This fact can be used to devise a recursive algorithm for constructing the elements of $P_{n+1,k}$ from those of $P_{n,k-1}$ and $P_{n,k}$. Indeed, an element of $P_{n+1,k}$ having property (a) of (1.84) has a cycle diagram that is obtained by adding the cycle in Figure 1.24 to the cycle diagram of an element of $P_{n,k-1}$.

Fig. 1.24 Cycle mapping $n+1$ to itself.

Thus the elements of $P_{n+1,k}$ having property (a) contribute the first term on the right hand side of (1.83). On the other hand, an element of $P_{n+1,k}$ having property $b)$ of (1.84) has a cycle diagram that can be constructed from the cycle diagram of an element of $P_{n,k}$ by prescribing where in this diagram $n+1$ should be inserted. This can be done by specifying which of the vertices $1, 2, \ldots, n$ is to be the endpoint of the arrow coming out of $n+1$.

For instance, consider the permutation

$$\begin{pmatrix} 1 & 2 & 3 & 4 & 5 & 6 & 7 & 8 & 9 & 10 & 11 & 12 & 13 \\ 7 & 11 & 12 & 5 & 10 & 6 & 13 & 1 & 8 & 4 & 3 & 2 & 9 \end{pmatrix}.$$

From its cycle diagram given in Figure 1.25, we can get 13 elements of $P_{14,4}$ by inserting 14 in the middle of any one of the 13 arrows.

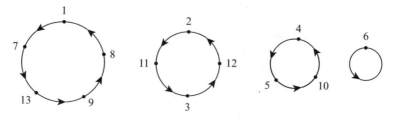

Fig. 1.25 The cycle diagram of the permutation 7 11 12 5 10 6 13 1 8 4 3 2 9 in one-line notation.

In particular, if the arrow coming out of 14 is to have 8 as an endpoint, we obtain the cycle diagram in Figure 1.26.

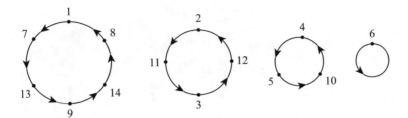

Fig. 1.26 The cycle diagram if the image of 14 is 8.

On the other hand, if the endpoint of the arrow coming out of 14 is to be 6, then the resulting cycle diagram is the one in Figure 1.27.

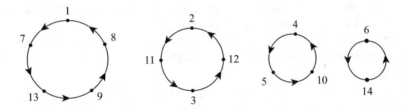

Fig. 1.27 The cycle diagram if the image of 14 is 6.

We must thus conclude that the elements of $P_{n+1,k}$ which have property (b) of (1.84) are counted by the second term on the right hand side of (1.83). This establishes formula (1.83).   □

The *Stirling numbers of the first kind,* usually denoted by the symbol $s_{n,k}$, are defined by setting

$$s_{n,k} = (-1)^{n+k} P_{n,k} .\tag{1.85}$$

It is customary to consider the $s_{n,k}$ as filling an infinite matrix

$$[s_{n,k}] \qquad (n, k \geq 1)\tag{1.86}$$

with the agreement that

$$s_{n,k} = 0 \text{ if } k > n .$$

Our recurrence in (1.83) yields the following recurrence for $s_{n,k}$.

Theorem 1.9.2 *For all $n \geq 1$, $k \geq 2$, we have*

$$s_{n+1,k} = s_{n,k-1} - n s_{n,k} .$$ (1.87)

Proof Making the replacements

$$p_{n+1,k} = (-1)^{n+1+k} s_{n+1,k} ,$$
$$p_{n,k-1} = (-1)^{n+k-1} s_{n,k-1} ,$$
$$p_{n,k} = (-1)^{n+k} s_{n,k} ,$$

the relation in (1.83) becomes

$$(-1)^{n+1+k} s_{n+1,k} = (-1)^{n+k-1} s_{n,k-1} + n(-1)^{n+k} s_{n,k},$$

and (1.87) follows immediately upon dividing this by $(-1)^{n+1-k}$.   □

To construct a table for the $s_{n,k}$, rather than using the recurrence in (1.87), we can construct the table for the $p_{n,k}$ first and then change signs in a checkerboard fashion according to the definition in (1.85).

For example, from (1.83), we quickly construct Table 1.2 with the convention that the blank entries in this table are equal to zero.

Here we should note that the number of permutations of n elements whose cycle diagram consists of a single cycle is $(n-1)!$, thus

$$p_{n,1} = (n-1)! .$$

On the other hand, there is only one permutation whose cycle diagram consists of n cycles, namely the permutation $(1)(2) \cdots (n)$, so that

$$p_{n,n} = 1 .$$

This gives us the first column and the diagonal entries of the table. Finally, the remaining terms are obtained according to the recursion in (1.83). More precisely, a term in the $(n+1)$th row is obtained by multiplying the term just above it by n and adding the result to the term just above and to the left. Thus, for instance, we get

Table 1.2 $p_{n,k}$, the number of permutations of $[n]$ with a total of k cycles.

$n\backslash k$	1	2	3	4	5	6	7
1	1						
2	1	1					
3	2	3	1				
4	6	11	6	1			
5	24	50	35	10	1		
6	120	274	225	85	15	1	
7	720	1764	1624	735	175	21	1

Table 1.3 The Stirling numbers of the first kind $s_{n,k}$.

$n\backslash k$	1	2	3	4	5	6	7
1	1						
2	-1	1					
3	2	-3	1				
4	-6	11	-6	1			
5	24	-50	35	-10	1		
6	-120	274	-225	85	-15	1	
7	720	-1764	1624	-735	175	-21	1

$$p_{6,3} = 50 + 5 \times 35 = 225 .$$

Changing signs in Table 1.2 according to formula (1.85), we obtain Table 1.3, the table of Stirling numbers of the first kind, again using the convention that the blank entries in this table are equal to zero.

The construction we have used to prove the recurrence in (1.83) yields also an interesting formal language representation of permutations. To see how this comes about, let us work on the example in Figure 1.25. Starting with the highest label (which is 13 in our case), we remove it and record the fact that the arrow coming out of 13 pointed to 9 by placing an x_9 at the end of our word. We now have a cycle diagram of a permutation of $1, 2, \ldots, 12$. We remove 12 and place an x_2 right before the x_9. We are left with the cycle diagram of a permutation of $1, 2, \ldots, 11$. We remove 11 and place an x_3 before $x_2 x_9$. Removing in this manner $10, 9, 8, 7$, we produce the successive strings

$$x_4 x_3 x_2 x_9 \rightarrow x_8 x_4 x_3 x_2 x_9 \rightarrow x_1 x_8 x_4 x_3 x_2 x_9 \rightarrow x_1 x_1 x_8 x_4 x_3 x_2 x_9 .$$

At this point, we are left with the cycle diagram in Figure 1.28.

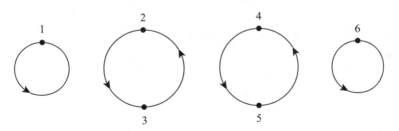

Fig. 1.28 Intermediate cycle diagram in which the largest label (6 in this case) is a cycle by itself.

Note now that removing the largest label 6 causes the disappearance of a cycle. We record this by appending an x_0 to the previous string, obtaining $x_0 x_1 x_1 x_8 x_4 x_3 x_2 x_9$. Continuing in this manner, remembering to place an x_0 every time a cycle is lost, we end up with the word

$$x_0\, x_0\, x_2\, x_0\, x_4\, x_0\, x_1\, x_1\, x_8\, x_4\, x_3\, x_2\, x_9 \ .$$

It is easy to see that in this manner we obtain a bijection between the cycle diagrams of permutations of [13] and words in the language

$$\mathcal{P}_{13} = x_0(x_0+x_1)(x_0+x_1+x_2)(x_0+x_1+x_2+x_3) \cdots (x_0+x_1+x_2+x_3+\cdots+x_{12}) \ .$$

We can generalize this construction and derive that the cycle diagrams of permutations of [n] can be represented by words of the language

$$\mathcal{P}_n = x_0(x_0 + x_1)(x_0 + x_1 + x_2) \cdots (x_0 + x_1 + x_2 + x_3 + \cdots + x_{n-1}) \ . \qquad (1.88)$$

This given, we can derive a rather surprising consequence.

Theorem 1.9.3 *The numbers* $p_{n,k}$ *are the coefficients of the upper factorial polynomial. More precisely, we have*

$$\sum_{k=1}^{n} p_{n,k}\, x^k = x(x+1)(x+2)\cdots(x+n-1) \ . \qquad (1.89)$$

Proof If we replace the letter x_0 by an x in (1.88) and each x_i $(i \geq 1)$ by a 1, the right hand side of (1.88) becomes the right hand side of (1.89). On the other hand, in the left hand side of (1.88), each word with k occurrences of x_0 will contribute an x^k. However, these words, by our construction, are precisely those that correspond to permutations with k cycles. Thus, under this substitution, the left hand side of (1.88) reduces to the left hand side of (1.89). This establishes Theorem 1.9.3. $\square$

Upon replacing x by $-x$ in (1.89) and multiplying both sides by $(-1)^n$, the right hand side of (1.89) becomes the lower factorial polynomial

$$x(x-1)\cdots(x-n+1) = (x)_n \ ,$$

and we immediately derive the following result.

Theorem 1.9.4

$$\sum_{k=1}^{n} s_{n,k}\, x^k = (x)_n \ . \qquad (1.90)$$

Comparing (1.90) with the identity in (1.78), we can see that the Stirling numbers of the first kind *undo* what the Stirling numbers of the second kind do. That is, they are the coefficients in the expansion of the lower factorial polynomial in terms of the polynomials more familiarly expressed in the *power* basis. From this fact, it must necessarily follow that the matrices $[S_{n,k}]$ and $[s_{n,k}]$ are inverses of each other.

1.10 Exercises for Chapter 1

1.10.1 Let $X = \{a, b, c\}$. Give a verbal description of the languages generated by the following expressions:

(a) $(a + b + c)^5$,
(b) aX^*,
(c) $a^*(b + c)^*$,
(d) $1 + aX^* + bX^* + cX^*$,
(e) $aX^+ + bX^+ + cX^+$,
(f) $a^*(b + c)^3$.

1.10.2 Show that

$$(a + b + c)^* = 1 + (a + b + c) + (a + b + c)^2(a + b + c)^*.$$

1.10.3 Let $X = \{a, b\}$. Give a verbal description of the language

$$1 + b + (a + b)^2 X^*.$$

1.10.4 Describe the collection of words generated by the following expressions:

(a) $\frac{1}{1-(x_2+x_3)} x_1 \frac{1}{1-(x_2+x_3)}$,
(b) $\frac{1}{1-(x_2+x_3)} \frac{x_1}{1-x_1} \frac{1}{1-(x_2+x_3)}$,
(c) $(x_2 + x_3)^n \frac{x_1}{1-x_1} (x_2 + x_3)^n$.

1.10.5 Let $X = \{x_1, x_2, \ldots, x_n, y_1, y_2, \ldots, y_m\}$. Describe the collection of words generated by the following expressions:

(a) $(x_1 + x_2 + \cdots + x_n)X^*(y_1 + y_2 + \cdots + y_m)$,
(b) $(x_1 + x_2 + \cdots + x_n)(y_1 + y_2 + \cdots + y_m)X^*$,
(c) $(x_1 + x_2 + \cdots + x_n)^N(y_1 + y_2 + \cdots + y_m)^M$,
(d) $\frac{1}{1-(x_1+x_2+\cdots+x_n)} \frac{1}{1-(y_1+y_2+\cdots+y_m)}$,
(e) $\frac{1}{1-(x_1+x_2+\cdots+x_n)(y_1+y_2+\cdots y_m)}$.

1.10.6 Let $X = \{x_1, x_2, y_1, y_2\}$ and $\mathcal{L}$ be the language consisting of all words in X^* where the x's and y's alternate.

(a) Show that the generating series of the words in $\mathcal{L}$ which start with an x is

$$\frac{1}{1 - (x_1 + x_2)(y_1 + y_2)}(x_1 + x_2)(1 + (y_1 + y_2)).$$

(b) Construct the generating series of $\mathcal{L}$.
(c) Calculate the number of words of length n in $\mathcal{L}$.

1.10.7 Construct the language $\mathcal{L} \subseteq \{a, b\}^*$ which is a solution of the equation

$$\mathcal{L} = a\mathcal{L} + b.$$

1.10.8 Let $w_1, w_2 \in X^*$. Show that

$$\mathcal{L} = w_1\mathcal{L} + w_2$$

has a unique solution if and only if $w_1 \neq 1$.

1.10.9 Let $X = \{a, b\}$. Show that there is no language $\mathcal{L} \subseteq X^*$ which satisfies

$$\mathcal{L} + ab\mathcal{L} = 1 + a\mathcal{L}.$$

1.10.10 Show that $\mathcal{L} = a^*b^*$ is the solution of the equation

$$\mathcal{L} + ba\mathcal{L} = 1 + a\mathcal{L} + b\mathcal{L}.$$

1.10.11 Let $X = \{a, b\}$ and $\mathcal{L}$ be the language of all words over X with an even number of a's. Argue that $\mathcal{L}$ satisfies the equation

$$\mathcal{L} = 1 + b\mathcal{L} + a(X^* - \mathcal{L}).$$

1.10.12 Let X be a finite alphabet and $w \in X^*$. Show that $x = w^2$ is the unique solution to the equation

$$x^2 = wxw$$

over X^*.

1.10.13 Let X be a finite alphabet and $w \in X^*$. Show that the equation

$$x^r = (wx)^s w$$

has no solutions over X^* unless $r - s$ divides $(s + 1)|w|$.

1.10.14 Show that the only solutions to the equation

$$xy = yx$$

over X^* are of the form $x = w^p$ and $y = w^q$, where $w \in X^*$ and $p, q \geq 0$.

1.10.15 Let $X = \{a, b\}$. Show that if $w \in aX^*b$, then the number of occurrences of ab in w plus the number of occurrences of ba in w is always an odd number.

1.10.16 Let $X = \{x_1, x_2, \ldots, x_n\}$. Given $w \in X^*$, calculate the number of solutions to

$$xy = w, \quad x, y \in X^*.$$

1.10.17 Calculate the number of solutions to

$$u_1 u_2 \cdots u_k = w,$$

where $w \in X^*$ is a fixed word and $u_1, u_2, \ldots, u_k \in X^*$.

1.10.18 Let X be an alphabet and $\mathcal{L} \subseteq X^*$. Given an integer k, denote by $k\mathcal{L}$ the formal series

$$\sum_{w \in \mathcal{L}} kw.$$

Show that

$$(1 + X_1^* + X_2^* + \cdots)^2 = 1 + 2X_1^* + 3X_2^* + 4X_3^* + \cdots$$

1.10.19 In general, show that if X is an alphabet, then

$$(X^*)^n = 1 + \binom{n}{1} X_1^* + \binom{n+1}{2} X_2^* + \cdots + \binom{n+k-1}{k} X_k^* + \cdots$$

1.10.20 Let $\mathcal{L}$ be the language of words over $\{a, b, c\}$ where each occurrence of a is followed by the letter b and each occurrence of c is followed by an a.

(a) Show that $\mathcal{L}$ satisfies

$$\mathcal{L} = 1 + ab\mathcal{L} + b\mathcal{L} + cab\mathcal{L}.$$

(b) Conclude that the generating series for $\mathcal{L}$ is

$$\frac{1}{1 - (ab + b + cab)}.$$

1.10.21 Let $X = \{x_1, x_2, x_3, x_4\}$ and $\mathcal{L}$ be the language of words over X in which each occurrence of x_i is followed by x_{i+1} for $i = 1, 2, 3$.

(a) Show that $\mathcal{L}$ satisfies

$$\mathcal{L} = 1 + x_4\mathcal{L} + x_3 x_4 \mathcal{L} + x_2 x_3 x_4 \mathcal{L} + x_1 x_2 x_3 x_4 \mathcal{L}.$$

(b) Construct the generating series of $\mathcal{L}$.

1.10.22 Let X be as in problem 1.10.21 and $\mathcal{L}$ be the language of words over X in which each x_1 is followed by an x_4 and each x_2 is followed by an x_3. Construct the generating series of $\mathcal{L}$.

1.10.23 Suppose $X = \{a, b\}$. Let $\mathcal{L}_1$ be the language over X consisting of all words in which aa does not appear, and let $\mathcal{L}_2$ be the language where aa appears at most once.

(a) Show that $\mathcal{L}_1$ and $\mathcal{L}_2$ are related by

$$\mathcal{L}_2 = \mathcal{L}_1 + aa + aab\mathcal{L}_1 + \mathcal{L}_1 baa + \mathcal{L}_1 baab\mathcal{L}_1.$$

(b) Show that

$$\mathcal{L}_1 = 1 + a + ab\mathcal{L}_1 + b\mathcal{L}_1.$$

(c) Construct the generating series of $\mathcal{L}_1$ and $\mathcal{L}_2$.

1.10.24 Let $X = \{x_1, x_2, x_3\}$ and $\mathcal{L}$ be the language of all words over X in which $x_1 x_2 x_3$ does not appear.

(a) Show that

$$\mathcal{L} = 1 + x_1 + x_1 x_2 + x_1^2 \mathcal{L} + x_1 x_3 \mathcal{L} + x_2 \mathcal{L} + x_3 \mathcal{L}.$$

(b) Construct the generating series of $\mathcal{L}$.

1.10.25 Let $X = \{a, b\}$ and $\mathcal{L}$ consist of words over X in which aab does not appear. Show that

$$\mathcal{L} = 1 + a + a^2 + ab\mathcal{L} + b\mathcal{L},$$

and construct the generating series of $\mathcal{L}$.

1.10.26 Let $X = \{a, b\}$. Construct the generating series of the words where the given word does not appear:

(a) ab
(b) abb
(c) aba
(d) $baba$.

1.10.27 Palindromes over $\{a, b\}$ are words that read the same forwards and backwards. Show that the language $\mathcal{P}$ of palindromes satisfies

$$\mathcal{P} = 1 + a + b + a\mathcal{P}a + b\mathcal{P}b.$$

What is the number of palindromes of length n?

1.10.28 Let $D_i \subseteq \{0, 1, 2, \ldots\}$ for $i = 1, 2, \ldots, n$. Show that the generating series of increasing words in $x_1, x_2, \ldots, x_n$ where x_i does not appear j times for each $j \in D_i$ is

$$\left(\frac{1}{1 - x_1} - \sum_{j \in D_1} x_1^j \right) \left(\frac{1}{1 - x_2} - \sum_{j \in D_2} x_2^j \right) \cdots \left(\frac{1}{1 - x_n} - \sum_{j \in D_n} x_n^j \right).$$

1.10.29 Let $\mathcal{L}$ denote the language of words over $\{a, b, c\}$ where none of the strings aa, bb, and cc appears. Let $\mathcal{L}_a, \mathcal{L}_b$, and $\mathcal{L}_c$, respectively, denote the words in $\mathcal{L}$ that start with a, b, and c and put

$$f(x, y, z) = \sum_{w \in \mathcal{L}} x^{|w|_a} y^{|w|_b} z^{|w|_c},$$

where $|w|_a, |w|_b$, and $|w|_c$ denote the number of occurrences of a, b, and c in w, respectively.

(a) Show that the languages $\mathcal{L}, \mathcal{L}_a, \mathcal{L}_b$, and $\mathcal{L}_c$ satisfy the following equations:

$$\mathcal{L} = 1 + \mathcal{L}_a + \mathcal{L}_b + \mathcal{L}_c$$
$$\mathcal{L} = 1 + (a + b + c)\mathcal{L} - a\mathcal{L}_a - b\mathcal{L}_b - c\mathcal{L}_c$$
$$\mathcal{L}_a = a(\mathcal{L} - \mathcal{L}_a)$$
$$\mathcal{L}_b = b(\mathcal{L} - \mathcal{L}_b)$$
$$\mathcal{L}_c = c(\mathcal{L} - \mathcal{L}_c).$$

(b) Show that the generating series of $\mathcal{L}$ can be written in the form

$$\mathcal{L} = \frac{1}{1 + \sum_{n \geq 1}(-1)^n (a^n + b^n + c^n)}.$$

(c) Verify that

$$f(x, y, z) = \frac{(1 + x)(1 + y)(1 + z)}{1 - xy - xz - yz - 2xyz}.$$

1.10.30 Suppose $\mathcal{L} \subseteq \{x_1, x_2, x_3\}^*$ is the language of all words in which $x_1 x_2$ and $x_2 x_3$ do not appear. Let $\mathcal{L}_1, \mathcal{L}_2$, and $\mathcal{L}_3$ denote the collection of words in $\mathcal{L}$ with initial letters x_1, x_2, and x_3, respectively.

(a) Show that

$$\mathcal{L}_1 = x_1 + x_1 \mathcal{L}_1 + x_1 \mathcal{L}_3$$
$$\mathcal{L}_2 = x_2 + x_2 \mathcal{L}_1 + x_2 \mathcal{L}_2$$
$$\mathcal{L}_3 = x_3 + x_3 \mathcal{L}_1 + x_3 \mathcal{L}_2 + x_3 \mathcal{L}_3.$$

(b) Put

$$f(t) = \sum_{w \in \mathcal{L}} t^{|w|} \quad \text{and} \quad f_i(t) = \sum_{w \in \mathcal{L}_i} t^{|w|}$$

for $i = 1, 2, 3$. Verify the matrix equation

$$
\begin{bmatrix} -t \\ -t \\ -t \end{bmatrix} = \begin{bmatrix} t-1 & 0 & t \\ t & t-1 & 0 \\ t & t & t-1 \end{bmatrix} \begin{bmatrix} f_1(t) \\ f_2(t) \\ f_3(t) \end{bmatrix}.
$$

(c) Use Cramer's rule to show that

$$
f_1(t) = \frac{\det \begin{bmatrix} -t & 0 & t \\ -t & t-1 & 0 \\ -t & t & t-1 \end{bmatrix}}{\det \begin{bmatrix} t-1 & 0 & t \\ t & t-1 & 0 \\ t & t & t-1 \end{bmatrix}} = \frac{-t^3 + t^2 - t}{t^3 - 2t^2 + 3t - 1}.
$$

Similarly, calculate $f_2(t)$ and $f_3(t)$.

(d) Show that

$$
f(t) = \frac{1}{1 - 3t + 2t^2 - t^3}.
$$

1.10.31 Let $X = \{x_1, x_2, \ldots, x_m\}$. Calculate the number of words in X_n^* with

(a) exactly one x_1,
(b) at least one x_1,
(c) at most one x_1,
(d) exactly k x_1's,
(e) exactly one x_1 and two x_2's,
(f) x_1 appearing an even number of times,
(g) x_1 appearing an odd number of times,
(h) x_1 and x_2 both appearing an even number of times.

1.10.32 Let $X = \{x_1, x_2, \ldots, x_m\}$. Calculate the number of words in X_n^* where

(a) the first letter is one of $x_1, x_2, \ldots, x_k$,
(b) the first two letters are from the set $\{x_1, x_2, \ldots, x_k\}$,
(c) the first and the last two letters are from the set $\{x_1, x_2, \ldots, x_k\}$,
(d) no adjacent letters are the same,
(e) no adjacent letters are different,
(f) each occurrence of x_1 is followed by an x_2.

1.10.33 Over the alphabet $X = \{x_1, x_2, x_3, y_1, y_2, y_3, y_4\}$, how many words are there which use each letter exactly

(a) once,
(b) twice,
(c) k times.

1.10.34 Let X be the alphabet of problem 1.10.33. Calculate the number of words in X^* in which

(a) each letter is used at most once,
(b) each letter is used exactly once and the x's appear before the y's,
(c) each letter is used exactly twice and the x's appear before the y's.

1.10.35 Let X be the alphabet of problem 1.10.33. Put together the generating series of all words in X^* where the x's appear before the y's.

1.10.36 Let X be the alphabet of problem 1.10.33. What is the number of injective words of length k in which exactly two y's appear?

1.10.37 Calculate the number of increasing words of length m over the alphabet $\{x_1, x_2, \ldots, x_n\}$ in which no letter appears exactly once.

1.10.38 Calculate the number of words of length n over the English alphabet in which the vowels and consonants alternate.

1.10.39 Calculate the number of 4-digit even integers.

1.10.40

(a) What is the number of strings of length n over the alphabet $\{a, b, c\}$ which contain exactly k c's?
(b) Justify the identity

$$\sum_{k=0}^{n} \binom{n}{k} 2^{n-k} = 3^n$$

by a counting argument based on your answer to part (a).

1.10.41 Calculate the number of rearrangements of the word *ABRACADABRA* which start and end with an A.

1.10.42 Calculate the number of words of length n over the English alphabet which start with a vowel and end with a consonant.

1.10.43 How many arrangements of the word *ARRANGEMENTS* are there?

1.10.44 How many ways are there to select k elements from the set $[n]$ such that

(a) no two consecutive numbers are selected?
(b) the numbers selected differ by at least three?

1.10.45 Find the number of ways of selecting k elements from $[n]$ such that the numbers selected differ by at least d.

1.10.46 Calculate the number of subsets $A \subseteq [n]$ with

$$|A| \leq \max\{i \mid i \in A\}.$$

1.10.47 Calculate the number of subsets $A \subseteq [n]$ with

$$|A| = \max\{i \mid i \in A\}.$$

1.10.48 We select six letters from the English alphabet at random, one by one, with possible repetitions. What is the probability that

(a) all letters are different?
(b) in the six-letter word obtained the number of vowels is the same as the number of consonants?
(c) the first and the last letters are the same?

1.10.49 In how many ways can one distribute 20 objects among 50 people

(a) if the objects are distinguishable?
(b) if the objects are indistinguishable?
(c) if the objects are distinguishable but no person can get more than one object?
(d) if the objects are indistinguishable but no person can get more than one object?

1.10.50 How many rearrangements of the word *PARALLEL* are there in which *AA* and *LLL* do not appear?

1.10.51 We select one letter at random from the word *SELECTIONS* and one from *RADIATION*. What is the probability that

(a) the two letters are neighbors in the English alphabet?
(b) the two letters are different?

1.10.52 Call a nonempty subset $A \subseteq [n]$ *good* if

$$|A| \geq \min\{i \mid i \in A\}$$

and *perfect* if

$$|A| = \min\{i \mid i \in A\}.$$

(For instance, if $n = 9$, the subsets $\{2, 5\}$ and $\{3, 4, 5, 8\}$ are good, but $\{5, 7, 9\}$ is bad. $\{2, 5\}$ is also perfect.)

(a) Show that if both A and B are good, then so is $A \cup B$.
(b) Suppose both A and B are perfect. When is $A \cup B$ perfect? When is $A \cap B$ perfect?
(c) Show that A and $\overline{A}$ both cannot be bad.
(d) How many good subsets of $[n]$ are there?
(e) How many perfect subsets of $[n]$ are there?

1.10.53 Find a correspondence between the ways of placing 8 nonattacking rooks on an 8×8 chessboard and the collection of words $J_8(x_1, x_2, \ldots, x_8)$.

1.10.54 Calculate the number of ways of placing 8 white and 8 black pawns on an 8 × 8 chessboard.

1.10.55 Calculate the number of ways of placing 8 white and 8 black pawns on an 8 × 8 chessboard such that

(a) the black pawns are on the lower 4 × 8 half of the board and the white ones are on the upper half;
(b) the white ones are placed below the main diagonal of the board;
(c) the white pawns are below the main diagonal and the black pawns are above the main diagonal of the board.

1.10.56 Calculate the number of ways of placing 8 white and 8 black pawns on an 8 × 8 chessboard with

(a) white pawns on white squares and the black ones on black squares, and
(b) white pawns are on the lower 4 × 8 half of the board and the black ones are placed symmetrically on the upper half.

1.10.57 Calculate the number of ways of placing 8 different pieces on an 8 × 8 chessboard.

1.10.58 How many different ways can all 32 chess pieces be arranged on an 8 × 8 chessboard?

1.10.59 Use the first equation in Theorem 1.4.5 to construct all Dyck words of length at most 6 in lexicographic order.

1.10.60 From Theorem 1.4.5, deduce that the numbers D_n satisfy the recursion

$$D_{n+1} = \sum_{i=0}^{n} D_i D_{n-i} \ .$$

1.10.61 Show that

$$D_{n+1} = \frac{2(2n + 1)}{n + 2} D_n \ .$$

1.10.62 Put

$$f(t, z) = \sum_{w \in \mathcal{D}} t^{|w|_a} z^{|w|_b} \ ,$$

where $|w|_a$ and $|w|_b$ denote the number of occurrences of a and b in w, respectively. Show that $f(t, z)$ satisfies the equation

$$f(t, z) = 1 + tzf(t, z)^2.$$

1.10.63 Call a Dyck path from $(0, 0)$ to $(2n, 0)$ *primitive* if it touches the x-axis only at these two points. Find a formula for the number of primitive Dyck paths.

1.10.64 What is the number of strictly lower paths from $(0, 0)$ to the point $(7, 6)$?

1.10.65 Find a formula for the number of weakly lower paths from $(0, 0)$ to $(n, n - 2)$.

1.10.66

(a) Show that the number of weakly lower paths from $(0, 0)$ to (n, n) in which the first i steps are horizontal is

$$\frac{i}{2n - i} \binom{2n - i}{n}.$$

(b) Count the number of weakly lower paths from $(0, 0)$ to (n, n) in which the first i steps are horizontal and the last j steps are vertical.

1.10.67 Suppose c is a nonzero integer and (n, m) is a lattice point with $n > m + c$. Show that the number of paths from the origin to (n, m) which do not touch the line $y = x - c$ is given by

$$\binom{n + m}{n} - \binom{n + m}{n - c}.$$

1.10.68 Calculate the number of Dyck paths from $(0, 0)$ to $(2n, 0)$ which stay weakly below the line $y = k$.

1.10.69 Show that D_n is odd if and only if $n = 2^r - 1$ for some positive integer r.

1.10.70

(a) Calculate the number of Young tableaux of the following shapes:

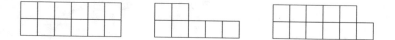

(b) Construct the Young tableaux corresponding to the following words:

$$aababba, \ ababaabaa, \ aaabbb.$$

1.10.71 Consider the complete parenthesization of the expressions in (1.45) and (1.48) obtained by adding a "(" to the beginning and a ")" to the end, resulting in

$$((((a_1 \cdot a_2) \cdot a_3) \cdot (a_4 \cdot a_5)) \cdot a_6), \quad (a_1 \cdot (((a_2 \cdot a_3) \cdot a_4) \cdot (a_5 \cdot a_6))).$$

Now make the replacements $(\to a, a_1, a_2, \ldots, a_{n-1} \to b, a_n \to \epsilon,$ and $) \to \epsilon,$ where ϵ denotes the null word. This results in the words

$$aaaabbbabb, \ abaaabbbab.$$

(a) Show that this construction defines a bijection between different recipes for computing the product $a_1 a_2 \cdots a_{n+1}$ and D-words of length $2n$.
(b) Construct the parenthesization of $a_1 a_2 a_3 a_4 a_5 a_6$ that corresponds to the Dyck path in Figure 1.17.

1.10.72 A *stack* is a data structure in which items can be piled up on top of one another by the *push* operation, and the top item can be retrieved by the *pop* operation. If the input items are the integers $1, 2, \ldots, n$ pushed onto the stack in that order, then the sequence of items retrieved by the pop operations results in a word that is a permutation of $1, 2, \ldots, n$. Such permutations are called *stack sortable*. For example, the following sequence of stack operations

$$push(1) \to push(2) \to pop(2) \to push(3) \to push(4) \to pop(4) \to pop(3) \to pop(1)$$

transforms 1234 into 2431, and thus 2431 is stack sortable.

(a) Construct a bijection between stack sortable permutations and Dyck words.
(b) Construct the stack sortable permutation that corresponds to the Dyck path in Figure 1.17.

1.10.73 Consider the ways of connecting $2n$ points that lie on a circle in pairs such that no two chords intersect. As an example, when $n = 2$, there are only two ways of doing this:

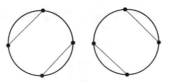

And for $n = 3$, there are 5 ways as shown below:

(a) Starting from the top point in the circle and proceeding clockwise, first mark the two endpoints of each chord with an a and a b, and then read the resulting labels clockwise. Show that this defines a bijection between such pairings and Dyck words.

(b) Construct the chords that correspond to the Dyck path in Figure 1.17.

1.10.74 A partition of $[n]$ is *noncrossing* if its parts do not interleave, i.e., if a and b belong to one part and x and y to another, then they are not arranged in the order of $a < x < b < y$. Show that the number of noncrossing partitions of $[n]$ is D_n.

1.10.75 Consider the collection of all lattice paths from the origin to the point (n, m) where we allow x colors to color the horizontal steps and y colors to color the vertical steps. Let $\pi(n, m)$ denote the collection of all such xy-colored paths.

(a) Show that there is a one-to-one correspondence between $\pi(n, m)$ and the words from the alphabet

$$\{h_1, h_2, \ldots, h_x, v_1, v_2, \ldots, v_y\}$$

with n occurrences of h and m occurrences of v.

(b) Construct a bijection between the words in the language

$$(h_1 + h_2 + \cdots + h_x + v_1 + v_2 + \cdots + v_y)^M$$

and the collection of all xy-colored lattice paths of length M that start at the origin.

(c) Use the above observations to give another proof of the binomial theorem

$$\sum_{n+m=M} \binom{M}{n} x^n y^m = (x + y)^M.$$

1.10.76 Examine the symmetry of the lattice paths from the origin to the point (n, n) to prove that

$$\binom{2n}{n}$$

is even for $n > 0$.

1.10.77 Classify the lattice paths from the origin to the point (n, m) according to the point they first touch the line $y = m$ to prove the identity

$$\binom{n+m}{n} = \sum_{k=0}^{n} \binom{m+k-1}{k}.$$

1.10.78 Let L be the vertical line $x = r$, $1 \le r \le n$. Classify the lattice paths from the origin to the point (n, m) according to their first contact point with the line L. Show that this yields

$$\binom{n+m}{n} = \sum_{k=0}^{m} \binom{r+k-1}{k}\binom{m-k+n-r}{m-k}.$$

1.10.79 Show that

$$\binom{n}{1}^2 + 2\binom{n}{2}^2 + 3\binom{n}{3}^2 + \cdots + n\binom{n}{n}^2 = n\binom{2n-1}{n}.$$

1.10.80 Calculate the coefficient of $a^2b^5c^3d$ in the expansion of $(a+b+c+d)^{11}$.

1.10.81 Find the sum of all fractions of the form

$$\frac{1}{u!v!w!},$$

where u, v, and w are nonnegative integers whose sum is 15.

1.10.82 Find the coefficient of t^n in the expansion of $(x+yt)^N$.

1.10.83 Calculate the number of lattice paths from the origin to the point $(6,6)$ which do not pass through the point $(3,3)$.

1.10.84 Calculate the number of lattice paths from A to B in the grids G_1 and G_2.

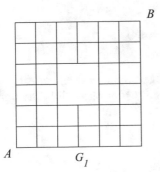

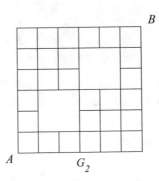

1.10.85 Calculate the number of lattice paths from the origin to the point $(6,6)$ which do not pass through either of the points $(1,1)$ and $(5,5)$.

1.10.86 Calculate the number of lattice paths from the origin to the point $(9,6)$ which pass through the points $(3,2)$ and $(6,4)$.

1.10.87 Calculate the number of lattice paths from A to B in the grids G_3 and G_4.

1.10.88 Show that

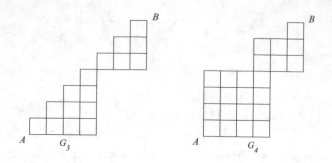

$$\lim_{n\to\infty}\frac{1}{n^k}\binom{n}{k}=\frac{1}{k!}.$$

1.10.89 Suppose a particle starts at the origin at time $t = 0$ and at each interval of time stays in its position with probability p or moves one unit to the right with probability $q = 1 - p$. Show that the probability $P_n(k)$ that the particle is at the point $(k, 0)$ at the end of the nth time interval is given by

$$\binom{n}{k}q^k p^{n-k}.$$

Thus,

$$\sum_{k\geq 0} P_n(k)x^k = (p+qx)^n.$$

1.10.90 Suppose a particle starts at the origin at time $t = 0$ and at each interval of time moves one unit to the left with probability p or one unit to the right with probability $q = 1 - p$.

(a) Show that

$$\sum_{-n\leq k\leq n} P_n(k)x^k = \left(px^{-1}+qx\right)^n,$$

where $P_n(k)$ is defined as in problem 1.10.89.
(b) What is the probability that the particle is back at the origin at time $t = 6$?
(c) Find a formula for $P_n(k)$.

1.10.91 Suppose a particle starts at the origin at time $t = 0$ and at each interval of time moves one unit to the right, left, up, or down with probabilities p_1, p_2, p_3, and p_4, respectively. Let $P_n(k, l)$ denote the probability that the particle is at the point (k, l) at the end of the nth time interval. Show that

$$\sum_{k,l} P_n(k, l) x^k y^l = \left(p_1 x + p_2 x^{-1} + p_3 y + p_4 y^{-1} \right)^n .$$

1.10.92 Construct the next row of Stirling numbers in Table 1.1.

1.10.93 List all partitions of $\{1, 2, 3, 4\}$.

1.10.94 List all partitions of the set $\{1, 2, 3, 4\}$ into 3 parts.

1.10.95 List all partitions of the set $\{1, 2, 3, 4\}$ where each part has

(a) an odd number of elements,
(b) an even number of elements.

1.10.96 Give a verbal description of the following languages in terms of set partitions:

(a) $x_1 x_2 (x_1 + x_2)^* x_3 (x_1 + x_2 + x_3)^*$,
(b) $x_1^+ x_2 (x_1 + x_2)^* x_3 \cdots x_k (x_1 + x_2 + \cdots + x_k)^* (x_2 + x_3 + \cdots + x_k)$,
(c) $x_1^+ x_2^+ x_3^+$

1.10.97 Interpret the following languages in terms of set partitions:

(a) $x_1^+ x_2 (x_1 + x_2)^* x_3 x_4 (x_1 + x_2 + x_3 + x_4)^*$,
(b) $x_1^+ x_2 (x_1 + x_2)^+ x_3 (x_1 + x_2 + x_3)^*$,
(c) $x_1 x_2 (x_1 + x_2)^* x_3 x_4 (x_1 + x_2 + x_3 + x_4)^*$,
(d) $x_1 x_2 \cdots x_k (x_1 + x_2 + \cdots + x_k)^*$.

1.10.98 Give a verbal description of the language

$$x_1^+ + x_1^+ x_2 (x_1 + x_2)^* + x_1^+ x_2 (x_1 + x_2)^* + x_3 (x_1 + x_2 + x_3)^* .$$

1.10.99 How many 7-letter words can be constructed using three A's, two B's, and two C's such that the first A occurs before the first B, and the first B occurs before the first C? (Examples: $ABAACCB$, $ABBACAC$.)

1.10.100 Calculate the number of rearrangements of the word

$$a_1 a_1 a_2 a_2 a_3 a_3 a_4 a_4$$

in which the first occurrence of a_i is before the first occurrence of a_{i+1} for $i = 1, 2, 3$.

1.10.101 Calculate the number of rearrangements of the word

$$a_1 a_1 a_1 a_2 a_2 a_2 a_3 a_3 a_3$$

in which the first occurrence of a_i is before the first occurrence of a_{i+1} for $i = 1, 2$.

1.10.102 Use formula (1.78) to derive the identity

$$(-1)^n = \sum_{k=1}^{n}(-1)^k k! S_{n,k} \ .$$

1.10.103 Let $\mathcal{L}_s$ denote the language of words over the alphabet $\{x_1, x_2, \ldots, x_s\}$ in which x_i^2 does not appear for $i = 1, 2, \ldots, s$. Let $\mathcal{L}_s^i$ denote the words in $\mathcal{L}_s$ which start with the letter x_i and put

$$f_s(t) = \sum_{w \in \mathcal{L}_s} t^{|w|}, \quad f_s^i(t) = \sum_{w \in \mathcal{L}_s^i} t^{|w|} \ .$$

(a) Show that

$$\mathcal{L}_s = 1 + (x_1 + x_2 + \cdots + x_k)\mathcal{L}_s - x_1\mathcal{L}_s^1 - x_2\mathcal{L}_s^2 - \cdots - x_s\mathcal{L}_s^s \ .$$

(b) Conclude that

$$f_s^i(t) = \frac{t}{1 - (s-1)t} \quad (i = 1, 2, \ldots, s).$$

(c) Show that the language of partitions into k parts where no part contains consecutive integers is

$$x_1 x_2 (1 + \mathcal{L}_2^1) x_3 (1 + \mathcal{L}_3^1 + \mathcal{L}_3^2) x_4 \cdots x_k (1 + \mathcal{L}_k^1 + \mathcal{L}_k^2 + \cdots + \mathcal{L}_k^{k-1}) \ .$$

(d) Show that the number of partitions of $[n]$ into k parts in which no part contains consecutive integers is $S_{n-1,k-1}$.

1.10.104 Construct a bijection between unrestricted partitions of $[n-1]$ into $k-1$ parts and partitions of $[n]$ into k parts in which no part contains consecutive integers.

1.10.105

(a) Construct the language of partitions into k parts in which 1 and 2 are in different parts.

(b) Show that the number of partitions of $[n]$ into k parts in which 1 and 2 are in different parts is

$$S_{n,k} - S_{n-1,k} \ .$$

1.10.106 Show that the number of partitions of $[n]$ into k parts in which i is in the ith part for $i = 1, 2, \ldots, k$ is k^{n-k}.

1.10.107 Show that the number of partitions of $[n]$ into k parts in which 2 is in the first part is $S_{n-1,k}$.

1.10.108 Use the identity (1.55) to describe a procedure to generate all injective k-letter words over an alphabet.

1.10.109 Construct the eighth row of Table 1.3.

1.10.110 Show that for $n > 1$, the entries in the nth row of Table 1.3 sum to zero.

1.10.111

(a) List all permutations of $\{1, 2, 3, 4\}$ whose cycle factorization consists of cycles of even length only.
(b) List all permutations of $\{1, 2, 3, 4\}$ whose cycle factorization contains precisely two cycles of length 1.

1.10.112 Write the following permutations in two-row notation:

(a) $(132)\,(47)\,(56)$,
(b) $(1234)\,(567)$,
(c) $(1)\,(2)\,(35)\,(46)$.

1.10.113 Construct the cycle factorizations and the corresponding words in $\mathcal{P}_8$ of the permutations

$$\sigma = \begin{pmatrix} 1\,2\,3\,4\,5\,6\,7\,8 \\ 3\,6\,4\,1\,5\,2\,8\,7 \end{pmatrix}, \quad \tau = \begin{pmatrix} 1\,2\,3\,4\,5\,6\,7\,8 \\ 8\,3\,7\,4\,6\,1\,2\,5 \end{pmatrix}.$$

1.10.114 Give a direct counting argument to show that for every $n, k \geq 1$,

$$S_{n,k} \leq p_{n,k}\,.$$

Verify this by comparing the entries in Tables 1.1 and 1.2.

1.10.115 Construct the word in $\mathcal{P}_{14}$ that corresponds to the permutation

$$\sigma = \begin{pmatrix} 1\,2\,3\ \ 4\ \ 5\ \ 6\,7\,8\,9\ \ 10\ 11\ 12\ 13\ 14 \\ 7\,6\ 14\ 13\ 11\ 2\ 1\,9\,8\ 12\ \ 5\ \ 10\ \ 4\ \ \ 3 \end{pmatrix}.$$

1.10.116 Construct the cycle diagram of the permutation whose word is

$$x_0 x_1 x_2 x_2 x_0 x_3 x_1 x_6 x_0 x_7 x_9 x_0 x_2 x_8\,.$$

1.10.117 Find the number of words in $\mathcal{P}_n$

(a) that contain exactly one x_0,
(b) that end with the letter x_0,
(c) where x_0 appears exactly in the odd-numbered indices.

1.10.118 Describe the collection of permutations that correspond to the words in problem 1.10.117.

1.10.119 Describe the set of permutations

(a) whose words in the language $\mathcal{P}_n$ are injective,
(b) that correspond to the words in $\mathcal{P}_n \cap x_0(x_0 + x_1)^{n-1}$,
(c) that correspond to the words in $\mathcal{P}_{2n} \cap x_0^n J_n(x_1, x_2, \ldots, x_{2n-1})$.

1.10.120 Suppose we use the one-line notation for permutations. Thus the permutation whose cycle diagram appears in Figure 1.27 is represented as

$$\sigma = 7\ 11\ 12\ 5\ 10\ 6\ 13\ 1\ 8\ 4\ 3\ 2\ 9.$$

Starting with the highest label, remove it from σ and record the fact that it precedes six elements by writing down an x_6. Now remove 12 and record that 12 precedes nine letters in the resulting permutation by placing an x_9 before x_6. Continuing in this manner, we end up with a word

$$w(\sigma) = x_0 x_0 x_1 x_2 x_4 x_4 x_6 x_3 x_0 x_7 x_9 x_9 x_9 x_6 .$$

(a) Construct $w(\sigma)$, where σ is the permutation given in problem 1.10.115.
(b) Denote by P_n the collection of permutations of $\{1, 2, \ldots, n\}$. Show that the above construction defines a bijection between the collection of words $w(\sigma), \sigma \in P_n$, and the language $\mathcal{P}_n$, i.e.,

$$\sum_{\sigma \in P_n} w(\sigma) = x_0(x_0 + x_1)(x_0 + x_1 + x_2) \cdots (x_0 + x_1 + x_2 + x_3 + \cdots + x_{n-1}) = \mathcal{P}_n .$$

(c) A pair of letters σ_i, σ_j with $i < j$ is an *inversion* in σ if $\sigma_i > \sigma_j$. Let $\mathrm{inv}(\sigma)$ denote the number of inversions of σ. Show that

$$\sum_{\sigma \in P_n} x^{\mathrm{inv}(\sigma)} = 1(1 + x)(1 + x + x^2) \cdots (1 + x + \cdots + x^{n-1}).$$

(d) Using part (c), deduce that for $n > 1$, exactly half the permutations in P_n have an odd number of inversions.
(e) For $n > 2$, let $n_k = \#\{\sigma \in P_n \mid \mathrm{inv}(\sigma) \equiv k \pmod 3\}$ for $k = 0, 1, 2$. Show that $n_0 = n_1 = n_2 = \frac{n!}{3}$. What is the generalization of this?

1.10.121 Show that

$$\sum_{\sigma \in P_n} \mathrm{inv}(\sigma) = \frac{1}{2}\binom{n}{2} n! .$$

1.11   Sample Quiz for Chapter 1

1. Suppose $\mathcal{L}$ is a language over the alphabet $\{a, b\}$ that satisfies

$$\mathcal{L} = 1 + a^2\mathcal{L} + b^2\mathcal{L}.$$

   (a) Put together the listing series of $\mathcal{L}$.

   (b) Construct the words of length 6 in $\mathcal{L}$ in lexicographic order.

2. Calculate the number of lattice paths from A to B in the following grid:

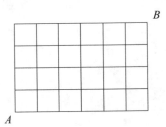

   How many of these paths pass through the point $(3, 2)$?

3. (a) Construct the Young tableau that corresponds to the following diagonal path:

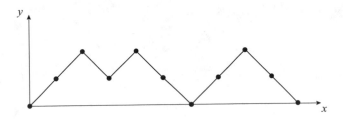

   (b) Calculate the number of diagonal paths from the origin to the point $(8, 2)$ that stay weakly above the x-axis.

4. Construct the listing series of weakly increasing words over the alphabet $\{x_1, x_2, x_3\}$ in which each letter appears at least once. How many such words of length 5 are there?

5. Construct the partition of the set $S = \{1, 2, \ldots, 7\}$ that corresponds to the word

$$x_1x_2x_1x_3x_3x_2x_1 \, .$$

   What is the number of partitions of S into 3 parts? What is the total number of partitions of S?

6. Construct the cycle diagram and the corresponding word of the Permutation:

$$\sigma = \begin{pmatrix} 1 & 2 & 3 & 4 & 5 & 6 & 7 & 8 & 9 & 10 & 11 & 12 & 13 & 14 \\ 7 & 11 & 12 & 5 & 10 & 6 & 13 & 1 & 14 & 4 & 3 & 2 & 9 & 8 \end{pmatrix}$$

Chapter 2
Partitions and Generating Functions

2.1 Ferrers Diagrams

Here we shall be concerned with manipulating objects built up from *rows of squares* as in Figure 2.1.

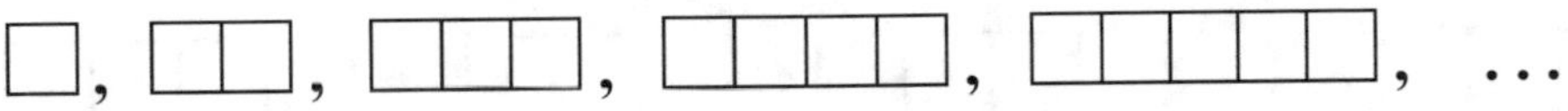

Fig. 2.1 Rows of squares.

It is convenient to add to this list the symbol 1, denoting the empty row, i.e., the one which has no squares. Thus our basic building blocks will be drawn from the set that we can list as shown in Figure 2.2.

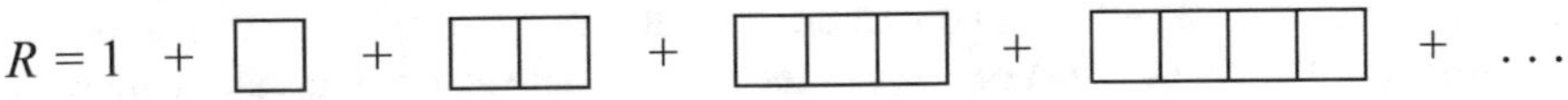

$$R = 1 + \;\square\; + \;\square\square\; + \;\square\square\square\; + \;\square\square\square\square\; + \;\cdots$$

Fig. 2.2 Listing the rows of squares with the empty row.

We can define a useful type of multiplication on R. Given two rows R_i and R_j, we shall understand by the product $R_i \times R_j$ the figure obtained by placing the shorter row on top of the longer one as indicated in Figure 2.3.

Fig. 2.3 Multiplying rows of squares.

© Springer Nature Switzerland AG 2021
Ö. Eğecioğlu, A. M. Garsia, *Lessons in Enumerative Combinatorics*, Graduate Texts in Mathematics 290, https://doi.org/10.1007/978-3-030-71250-1_2

Multiplying a row by the empty row 1 has no effect (see Figure 2.4).

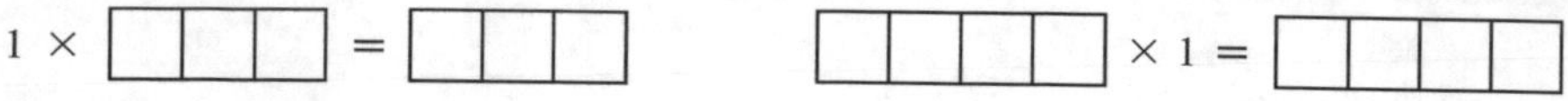

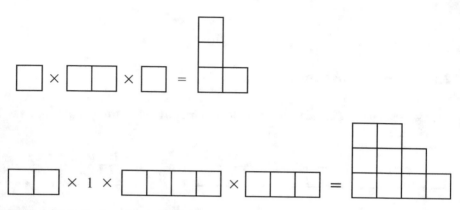

Fig. 2.4 Multiplication by the empty row.

We can multiply more than two rows as well by piling them on top of one another as shown in Figure 2.5.

Fig. 2.5 Multiplication of more than two rows of squares.

Thus the product of rows of lengths $\lambda_1 \geq \lambda_2 \geq \cdots \geq \lambda_r$ taken from R is the staircase-like figure obtained by piling up these rows in a left justified manner, starting with the longest one at the bottom. These configurations are called *Ferrers diagrams* (after the British mathematician Norman Ferrers) or *FD*'s for short. The rows of a FD are ordered from bottom up, so that we can refer to the longest one as the first row, the next longest one as the second row, and so on. We shall denote the collection of all FDs by $\mathcal{FD}$.

By using the above rule of multiplication, we can easily generate some special classes of FDs. For instance the collection of all FDs where each of the rows has length 2 is given by the infinite sum in Figure 2.6.

Fig. 2.6 FDs where each row is of length 2.

Those of rows of length 3 together with the empty diagram are represented by the formal sum in Figure 2.7.

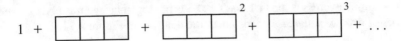

Fig. 2.7 The empty row together with FDs with rows of length 3.

Now it is very tempting to sum the expression in Figure 2.7 using the formula for the sum of a geometric series, that is

$$1 + x + x^2 + x^3 + \cdots = \frac{1}{1-x}.$$ (2.1)

Since we are not about to substitute any numerical values in any of these expressions here, the questions about convergence will never arise in our case. We will use the formalism in Figure 2.8.

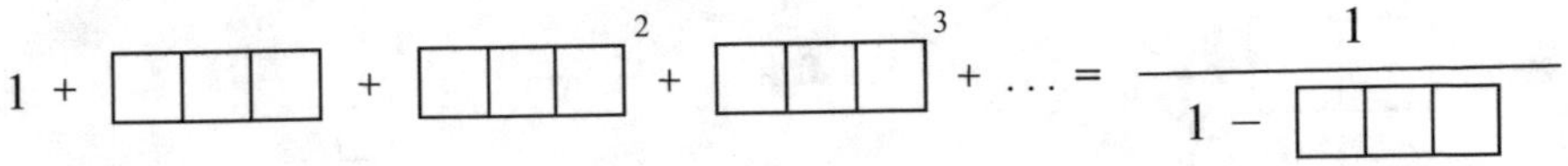

$$1 + \boxed{} + \boxed{}^2 + \boxed{}^3 + \cdots = \frac{1}{1 - \boxed{}}$$

Fig. 2.8 Compact representation of the diagrams in Figure 2.7.

This is with the understanding that the expression on the right hand side of Figure 2.8 is nothing but a shorthand that represents all FDs that appear on the left hand side. Similarly, all FDs with rows of length 2 given in Figure 2.6 can be represented by the symbol in Figure 2.9 in the same vein.

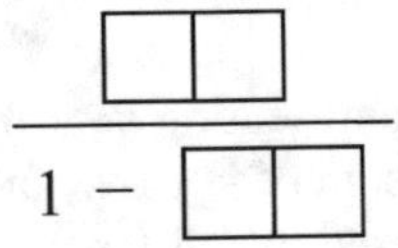

$$\frac{\boxed{}}{1 - \boxed{}}$$

Fig. 2.9 Compact representation of all FDs with rows of length 2.

We see that *all* FDs can be represented by the infinite product in Figure 2.10.

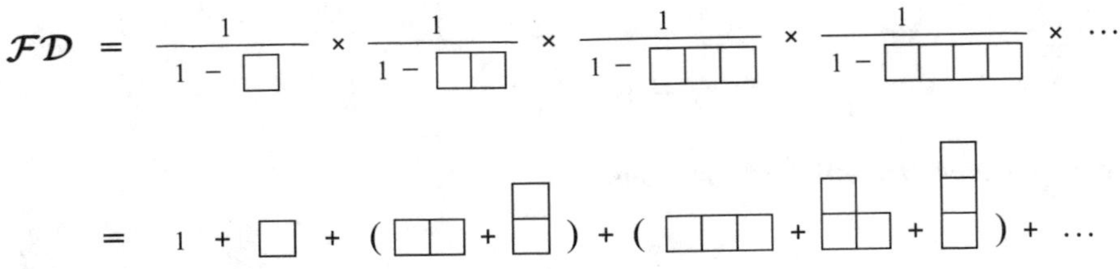

$$\mathcal{FD} = \frac{1}{1 - \square} \times \frac{1}{1 - \boxed{}} \times \frac{1}{1 - \boxed{}} \times \frac{1}{1 - \boxed{}} \times \cdots$$

$$= 1 + \square + (\boxed{} + \square) + (\boxed{} + \square + \square) + \cdots$$

Fig. 2.10 Representing all FDs.

This simply expresses the fact that a FD λ is put together by multiplying a certain number of rows of squares. For instance, the diagram in Figure 2.11

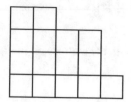

Fig. 2.11 A Ferrers diagram.

is obtained from the product above by picking the terms shown in Figure 2.12, and the term 1 from all the other factors.

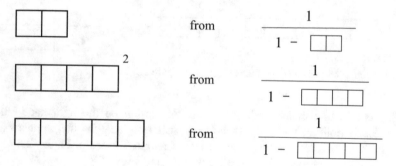

Fig. 2.12 Construction of the Ferrers diagram in Figure 2.11.

FDs are a geometric way of representing *integer partitions*. These are decompositions of a given integer n as a sum of positive integers called *parts*. For example, the number 4 can be so decomposed five different ways as listed below

$$
\begin{aligned}
4 &= 4 \\
&= 3 + 1 \\
&= 2 + 2 \\
&= 2 + 1 + 1 \\
&= 1 + 1 + 1 + 1,
\end{aligned}
\tag{2.2}
$$

and there are seven different partitions of 5

$$
\begin{aligned}
5 &= 5 \\
&= 4 + 1
\end{aligned}
$$

$$= 3 + 2$$

$$= 3 + 1 + 1 \qquad (2.3)$$

$$= 2 + 2 + 1$$

$$= 2 + 1 + 1 + 1$$

$$= 1 + 1 + 1 + 1 + 1.$$

Note that the order in which the parts of a partition are written down is not taken into account. Thus we might as well assume that the parts $\lambda_1, \lambda_2, \ldots, \lambda_k$ of a partition form a nonincreasing sequence, i.e., $\lambda_1 \geq \lambda_2 \geq \cdots \geq \lambda_k$.

Now we see that FDs and integer partitions are essentially two different ways of looking at the same object. The FDs in Figure 2.13, for instance, which are made up of exactly 4 cells are simply graphical representations of the five partitions of the integer 4 depicted in (2.2).

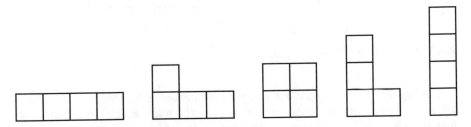

Fig. 2.13 Graphical representation of the five partitions of the integer 4.

Some facts about partitions with restrictions placed on the parts become totally transparent when viewed as FDs. As an example we have

Theorem 2.1.1 *The number of partitions of n where no part exceeds m is equal to the number of partitions of n into at most m parts.*

Proof First, as an example let us consider all partitions of $n = 7$ where no part exceeds 2. By inspection we see that there are four such partitions of 7:

$$2 + 2 + 2 + 1$$

$$2 + 2 + 1 + 1 + 1$$

$$2 + 1 + 1 + 1 + 1 + 1$$

$$1 + 1 + 1 + 1 + 1 + 1 + 1.$$

There are also four partitions of 7 into at most 2 parts as given below:

$$4 + 3$$

$$5 + 2$$

$$6 + 1$$

$$7.$$

Now the reason for this is very simple if we look at the corresponding FDs. In general if we trade the rows of a FD λ for its columns and vice versa, we obtain a new diagram called the *conjugate* of λ, denoted by λ'. Thus the conjugate is simply the reflection of the diagram along the line $y = x$. Conjugation preserves the number of cells of λ but the rows of λ become the columns of λ'. Clearly if we take the conjugate of λ' we get back λ itself (see Figure 2.14).

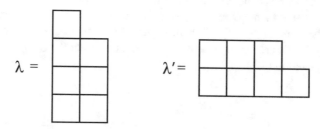

Fig. 2.14   Conjugating Ferrers diagrams.

Conjugation then sets up a one-to-one correspondence between all FDs with n cells where each row has no more than m cells and those which contain no more than m rows. This is the content of Theorem 2.1.1.   □

2.2   Generating Functions

Our next task is to try to enumerate FDs we can form using a fixed number n of cells, or equivalently the number of partitions of n. We shall denote this number by $P(n)$. For convenience, we set $P(0) = 1$. From the expansion Figure 2.10 and the two examples (2.2) and (2.3), we have the first few values of this sequence:

$$P(1) = 1$$

$$P(2) = 2$$

$$P(3) = 3$$

$$P(4) = 5$$

$$P(5) = 7.$$

Now it turns out that there really is no elementary closed formula of the kind we are accustomed to that expresses $P(n)$. But nevertheless, we can try to keep track of *all* of the numbers in the sequence $P(0)$, $P(1)$, $P(2)$, ... simultaneously in another way. This is done by passing to the *generating function* of this sequence from the expression in Figure 2.10.

Definition 2.2.1 The generating function of a sequence

$$a_0, a_1, a_2, a_3, \ldots$$

is the formal series

$$a_0 + a_1 q + a_2 q^2 + a_3 q^3 + \cdots = \sum_{m \geq 0} a_m q^m .$$

Remark 2.2.1 Note that we are using various powers of the symbol q merely as place holders, and from our point of view a generating function is not really a function but only a formal series. Nevertheless, this terminology is used for historical reasons.

In our particular case here where $a_n = P(n)$, the idea is to try to find a compact form for the generating function

$$F(q) = P(0) + P(1)q + P(2)q^2 + \cdots + P(n)q^n + \cdots$$
$$= \sum_{n \geq 0} P(n)q^n .$$

This can be accomplished by making use of a very simple fact from elementary algebra, namely the law of exponents:

$$q^i \cdot q^j = q^{i+j}$$

in conjunction with the way we multiply rows of squares.

Let us define the *weight* of a FD λ to be n if it is made up of n squares. Alternately, the weight of a partition is just the sum of its parts. We shall denote the weight of λ by $|\lambda|$.

Given two rows R_i and R_j we have

$$|R_i \times R_j| = |R_i| + |R_j| . \tag{2.4}$$

This simply expresses the fact that the total number of cells in the product $R_i \times R_j$ is the sum of the number of cells of R_i and of R_j. Using these weights as exponents, one has

$$q^{|R_i \times R_j|} = q^{|R_i|+|R_j|} = q^{|R_i|} \cdot q^{|R_j|}.$$

In case λ is the product of more than two rows of say lengths $\lambda_1, \lambda_2, \ldots, \lambda_r$, we similarly have

$$q^{|\lambda|} = q^{\lambda_1 + \lambda_2 + \cdots + \lambda_r} = q^{\lambda_1} \cdot q^{\lambda_2} \cdots q^{\lambda_r} \, .$$

But now this obvious fact allows us to calculate the generating function $F(q)$ easily from the generating series for $\mathcal{F}\mathcal{D}$.

We replace each row that appears in Figure 2.10 by q raised to the number of cells in the row. This gives

$$\sum_{\lambda \in \mathcal{F}\mathcal{D}} q^{|\lambda|} = \frac{1}{1-q} \times \frac{1}{1-q^2} \times \frac{1}{1-q^3} \times \frac{1}{1-q^4} \times \cdots . \tag{2.5}$$

Going back to the diagram λ illustrated in Figure 2.12, for instance, we see that $q^{|\lambda|}$ for this partition is the product of the terms

$$q^2 \quad \text{from} \quad \frac{1}{1-q^2}$$

$$q^8 \quad \text{from} \quad \frac{1}{1-q^4}$$

$$q^5 \quad \text{from} \quad \frac{1}{1-q^5}$$

and thus contributes

$$q^2 \cdot q^8 \cdot q^5 = q^{15}$$

to the left hand side of (2.5).

Now the crucial fact here is that in general each FD λ with n cells will contribute a term q^n to the sum

$$\sum_{\lambda \in \mathcal{F}\mathcal{D}} q^{|\lambda|} \, . \tag{2.6}$$

In other words, the number of times q^n appears in (2.5) is precisely $P(n)$. But this says that the sum (2.6) is the generating function $F(q)$. This observation together with the product in (2.5) then yield

Theorem 2.2.1 *The generating function of the numbers $P(n)$, $n = 0, 1, 2, \ldots$ is given by*

$$\sum_{n \geq 0} P(n) q^n = \prod_{m \geq 1} \frac{1}{1-q^m} \, . \tag{2.7}$$

With only minor modifications, the proof of Theorem 2.2.1 for the generating function of the sequence $P(n)$ also gives us the generating functions of many restricted classes of partitions. For example, suppose $S = \{s_1, s_2, s_3, \ldots\}$ is some subset of the positive integers, and we denote by

$$P_S(n)$$

the number of FDs with n cells, where the lengths of the rows are only allowed to be elements of S. For instance, if we are interested in FDs where the length of each row is an *odd* number, then picking

$$S = \{1, 3, 5, \ldots\}$$

expresses this requirement.

In analogy with the generating function of $P(n)$ we can construct the generating function of the sequence of numbers $P_S(n)$, $n = 1, 2, \ldots$.

$$F_S(q) = 1 + \sum_{n \geq 1} P_S(n)q^n . \tag{2.8}$$

The constant term 1 in (2.8) is included for convenience.

We have

Theorem 2.2.2 *The generating function of the sequence of numbers $P_S(n)$ is given by the product*

$$F_S(q) = \prod_{m \geq 1} \frac{1}{1 - q^{s_m}} .$$

Theorem 2.2.2 has numerous beautiful consequences in the theory of partitions. First of all, we are able to write down in a compact and alternate way many generating functions for restricted partitions by picking specific sets S for allowable rowlengths.

1. The generating function of number of FDs with odd rowlengths:

$$\prod_{m \geq 1} \frac{1}{1 - q^{2m-1}} .$$

2. The generating function of number of FDs with even rowlengths:

$$\prod_{m \geq 1} \frac{1}{1 - q^{2m}} .$$

3. The generating function of number of FDs with rowlengths congruent to 1 (mod 5):

$$\prod_{m \geq 0} \frac{1}{1 - q^{5m+1}} \cdot$$

4. The generating function of number of FDs with rowlengths congruent to 4 (mod 5):

$$\prod_{m \geq 0} \frac{1}{1 - q^{5m+4}} \cdot$$

5. The generating function of number of FDs with rowlengths congruent to 1 or 4 (mod 5):

$$\prod_{m \geq 0} \frac{1}{(1 - q^{5m+1})(1 - q^{5m+4})} \cdot$$

It is customary to use the term *parts* interchangeably with rowlengths when talking about partitions. So we could as well say *partitions with odd parts* for the partitions in Example 1 above and *partitions with parts congruent to 1 (mod 5)* for the partitions in Example 3.

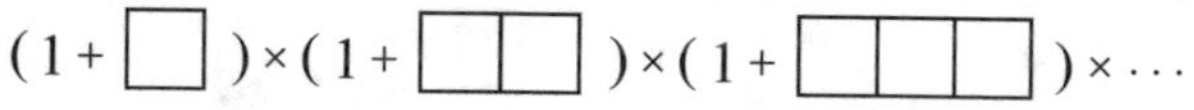

$$(1 + \boxed{}) \times (1 + \boxed{}) \times (1 + \boxed{}) \times \cdots$$

Fig. 2.15 Generating partitions with distinct parts.

Next we construct the generating function of FDs where the rowlengths are all *distinct* integers. In this case, Theorem 2.2.2 is no more applicable, but as soon as we can write down an expression that generates this class of partitions *themselves*, then the generating function can be obtained from this very easily. Since these partitions are generated by the expression in Figure 2.15, we have that the generating function of FDs with distinct parts is the product

$$\prod_{m \geq 1} (1 + q^m) . \tag{2.9}$$

We can use (2.9) for computational purposes as well. As an example, note that in generating the partitions of 5 into distinct parts, we do not have to go further than the fifth factor in Figure 2.15 since *no* partition of 5 can possibly contain a part larger than 5. This means that the coefficients of the terms

$$q, q^2, q^3, q^4, q^5$$

in the truncated product

$$(1+q)(1+q^2)(1+q^3)(1+q^4)(1+q^5) \tag{2.10}$$

should give us the number of partitions of $1, 2, 3, 4$, and 5, respectively, into distinct parts.

Indeed, multiplying out (2.10) gives

$$1 + q + q^2 + 2q^3 + 2q^4 + 3q^5 + \text{(higher degree terms)}.$$

We can check, for example, that 3 has two such partitions

$$3 \text{ and } 2+1,$$

and 5 has three of them

$$5, \ 4+1 \text{ and } 3+2$$

as expected.

Now the generating function (2.9) together with the generating function of partitions into odd parts (Example 1 above) gives us the following interesting partition identity.

Theorem 2.2.3 (Euler) *The number of partitions of n into distinct parts is equinumerous with the number of partitions of n into odd parts.*

Proof Suppose $f(q)$ is the generating function of the sequence $a_0, a_1, a_2, \ldots$ and $g(q)$ is the generating function of the sequence $b_0, b_1, b_2, \ldots$ where the a_i's and the b_i's are arbitrary numbers (or expressions). That is

$$f(q) = \sum_{n \geq 0} a_n q^n, \quad g(q) = \sum_{n \geq 0} b_n q^n.$$

Then $f(q) = g(q)$ simply means that for $n = 0, 1, 2, \ldots$, we have $a_n = b_n$.

In our case we know that the coefficient of the term q^n in

$$\prod_{m \geq 1} (1 + q^m)$$

is the number of partitions of n into distinct parts, and the coefficient of the term q^n in

$$\prod_{m \geq 1} \frac{1}{1 - q^{2m-1}}$$

is the number of partitions of n into odd parts.

By a simple calculation we have that for every $m \geq 1$

$$1 + q^m = \frac{1 - q^{2m}}{1 - q^m}$$

and hence

$$\prod_{m \geq 1}(1 + q^m) = \prod_{m \geq 1} \frac{(1 - q^{2m})}{(1 - q^m)}$$

$$= \frac{(1 - q^2)(1 - q^4)(1 - q^6)(1 - q^8)(1 - q^{10})(1 - q^{12})(1 - q^{14}) \cdots}{(1 - q)(1 - q^2)(1 - q^3)(1 - q^4)(1 - q^5)(1 - q^6)(1 - q^7) \cdots}$$

$$= \frac{1}{(1 - q)(1 - q^3)(1 - q^5)(1 - q^7) \cdots}$$

$$= \prod_{m \geq 1} \frac{1}{1 - q^{2m-1}} \, .$$

The equality of these two generating functions proves Theorem 2.2.3.   □

2.3   The Euler Recursion for the Partition Function

The number of partitions of n or equivalently, the number of FDs with n squares which we denote by $P(n)$ is usually referred to as the *partition function*.

By counting we have determined the first few values of $P(n)$ in the previous section. With a little more effort, these values can be extended. In fact the first 20 values of $P(n)$ starting with $P(0) = P(1) = 1$ are found to be as follows:

$$1, 1, 2, 3, 5, 7, 11, 15, 22, 30, 42, 56, 77, 101, 135, 176, 231, 297, 385, 490.$$
$$(2.11)$$

It was shown by Euler [6] that for all n

$$P(n) = P(n - 1) + P(n - 2)$$

$$- P(n - 5) - P(n - 7) \qquad\qquad (2.12)$$

$$+ P(n - 12) + P(n - 15)$$

$$\pm \cdots + (-1)^{m+1} \left(P(n - \tfrac{m(3m-1)}{2}) + P(n - \tfrac{m(3m+1)}{2}) \right) + \cdots .$$

We have already noted the convention that $P(0) = 1$. In (2.12) it is understood that $P(k) = 0$ if k is a negative number.

The pairs of numbers that we subtract off from n in this recursion are called *pentagonal* and *second pentagonal* numbers. Pentagonal numbers get their name from the fact that they enumerate the number of dots in the sequence of geometric figures shown in Figure 2.16.

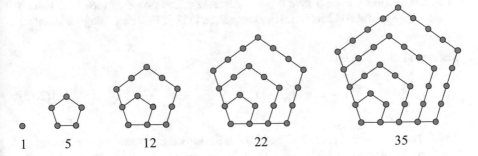

Fig. 2.16 Geometric representation of pentagonal numbers.

First few values of these sequences are as follows:

m	$\frac{m(3m-1)}{2}$	$\frac{m(3m+1)}{2}$
1	1	2
2	5	7
3	12	15
4	22	26
5	35	40
6	51	57
7	70	77

Using formula (2.12) we can quickly compute

$$P(2) = P(1) + P(0) = 2$$
$$P(3) = P(2) + P(1) = 2 + 1 = 3$$
$$P(4) = P(3) + P(2) = 3 + 2 = 5$$
$$P(5) = P(4) + P(3) - P(0) = 8 - 1 = 7$$
$$P(6) = P(5) + P(4) - P(1) = 12 - 1 = 11$$
$$P(7) = P(6) + P(5) - P(2) - P(0) = 18 - 3 = 15$$
$$P(8) = P(7) + P(6) - P(3) - P(1) = 26 - 4 = 22.$$

So we see that we can recursively recover the numbers in (2.11) and go further much quicker by means of (2.12) than by counting Ferrers diagrams or doing algebraic manipulations on the generating function.

In this section we shall give a proof of (2.12). The proof of based on Fabian Franklin's beautiful combinatorial demonstration [11] of Euler's pentagonal number theorem.

Theorem 2.3.1 (Euler)

$$\prod_{k \geq 1}(1 - q^k) = 1 + \sum_{m \geq 1}(-1)^m \left(q^{\frac{m(3m-1)}{2}} + q^{\frac{m(3m+1)}{2}}\right). \tag{2.13}$$

Proof The left hand side of (2.13) without the minus signs would be the generating function of partitions with distinct parts, as the actual FDs are generated by the product in Figure 2.15.

With the signs, every row used in building up a FD contributes a "−1" when we generate what we can call *signed* FDs with the formal product shown in Figure 2.17.

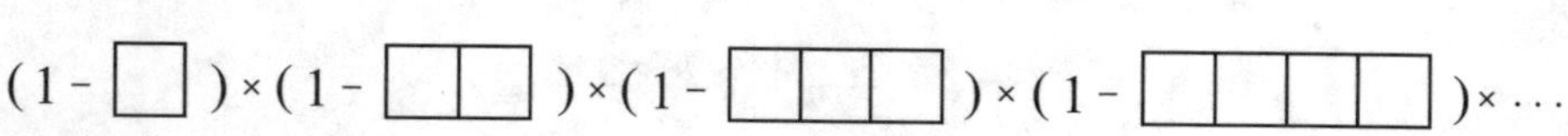

$$(1 - \boxed{}) \times (1 - \boxed{}) \times (1 - \boxed{}) \times (1 - \boxed{}) \times \cdots$$

Fig. 2.17 Product of signed rows.

Each FD λ with distinct parts $\lambda_1 > \lambda_2 > \cdots > \lambda_r$ that is produced by this expression contributes

$$q^{\lambda_1 + \lambda_2 + \cdots + \lambda_r} = q^{|\lambda|}$$

which simply keeps track of the number of squares in λ, and each such term comes with a sign

$$(-1)^r,$$

where $r = r(\lambda)$ is the number of parts of λ.

We see then the left hand side of (2.13) is the generating function of signed partitions into distinct parts.

Before we proceed with Franklin's proof, we need some notation.

We associate two numbers to a given partition λ with distinct parts, call them $m(\lambda)$ and $d(\lambda)$.

$m(\lambda)$ denotes the length of the smallest part of λ, i.e., $m(\lambda) = \lambda_r$. This is just the number of cells in the topmost row of the FD of λ, and it is defined whether or not λ has distinct parts.

$d(\lambda)$ denotes the length of the unit staircase that starts at the bottom right cell of the FD of λ and climbs north-west. Unit here means that each step is we take in this direction is a single square, or that the lengths of the steps of the staircase starting from the largest part are consecutive numbers. Examples of the cells that define $m(\lambda)$ and $d(\lambda)$ are shown in Figure 2.18.

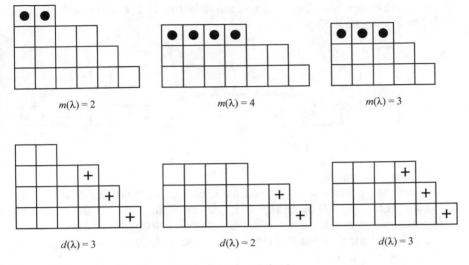

$$m(\lambda) = 2 \qquad\qquad m(\lambda) = 4 \qquad\qquad m(\lambda) = 3$$

$$d(\lambda) = 3 \qquad\qquad d(\lambda) = 2 \qquad\qquad d(\lambda) = 3$$

Fig. 2.18 Demonstration of how $m(\lambda)$ and $d(\lambda)$ are defined.

Examples:

The idea is to try to match a signed diagram λ that is produced by the expression in Figure 2.17 with a mate $\bar{\lambda}$ which is also one of the diagrams produced by Figure 2.17, in such a way that the weight stays the same, i.e., $|\lambda| = |\bar{\lambda}|$ but λ and $\bar{\lambda}$ have different signs. For such a pair then

$$(-1)^{r(\lambda)}q^{|\lambda|} + (-1)^{r(\bar{\lambda})}q^{|\bar{\lambda}|} = 0$$

and these terms cancel out in the generating function. The way we try to do this by considering $m(\lambda)$ and $d(\lambda)$ is quite simple. We look at the cases when $m(\lambda) > d(\lambda)$ and $m(\lambda) \leq d(\lambda)$ separately.

Case I: $m(\lambda) > d(\lambda)$

In this case we simply move the squares of the unit staircase up and make them the top row of the diagram. This new diagram is λ's mate $\bar{\lambda}$ as shown in Figure 2.19.

Fig. 2.19 The case when $m(\lambda) > d(\lambda)$.

Of course $\bar{\lambda}$ has one more row than λ.

Case II: $m(\lambda) \le d(\lambda)$

In this case we move the top row of λ like a slinky all the way to the bottom right following the boundary of λ to become the new staircase. The resulting diagram is $\bar{\lambda}$ as shown in Figure 2.20.

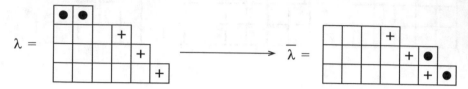

Fig. 2.20 The case when $m(\lambda) \le d(\lambda)$.

This time $\bar{\lambda}$ has one fewer part than λ.

Note that in each case $\bar{\lambda}$ has distinct parts, has the same number of cells as λ, but its sign is the opposite of the sign of λ because of the change in the number of rows. Moreover, if we were to start with the partitions $\bar{\lambda}$ in Figures 2.19 and 2.20, we would end up with the λ we started out with as $\bar{\lambda}$'s mate. So the terms corresponding to all such pairs cancel out.

What is left then? Certainly not everything cancels.

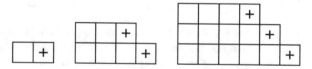

Fig. 2.21 Special FDs for Case I: $m(\lambda) > d(\lambda)$ but no move is possible.

If we look at our pairing a bit more closely, say in Case I when $m(\lambda) > d(\lambda)$, we see that there is actually a very special kind of partition for which we are unable to move the squares of the staircase and make them a new row at the top. This happens for the partitions that look like the ones in Figure 2.21. If we try to move the staircase squares, we get a legitimate FD, but the top two rows become identical and the parts are no longer distinct. It follows that these partitions have $d(\lambda) = m$ and $m(\lambda) = m + 1$ with weight

$$(m + 1) + (m + 2) + \cdots + 2m = \frac{m(3m + 1)}{2}$$

and sign $(-1)^m$.

If in turn we consider Case II more closely, we see that the change we make to the diagram of λ is not possible when $m(\lambda) = d(\lambda)$, since in this case moving the squares of the top row does not result in a FD. This is shown in Figure 2.22.

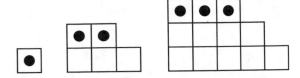

Fig. 2.22 Special FDs for Case II: $m(\lambda) = d(\lambda)$ but no move is possible.

If $m(\lambda) = m$, then such a λ has weight

$$m + (m + 1) + \cdots + (2m - 1) = \frac{m(3m - 1)}{2}$$

and sign $(-1)^m$. So these partitions are the only ones that we are unable to pair up with a mate, and it is the sum terms of the form

$$(-1)^m q^{\frac{m(3m-1)}{2}} \quad \text{and} \quad (-1)^m q^{\frac{m(3m+1)}{2}}$$

that make up the generating function. Therefore the only powers of q that remain have exponents that are pentagonal or second pentagonal numbers, with sign $(-1)^m$ for the corresponding m as indicated in (2.13). This is the content of Theorem 2.3.1.

□

From Theorem 2.3.1, it is just a small step to derive the recursion (2.12) for $P(n)$. The reciprocal of the product in (2.13) is the generating function of the number $P(n)$ of partitions of n:

$$\prod_{k \geq 0} \frac{1}{1 - q^k} = \sum_{m \geq 0} P(m)q^m .$$

Multiplying both sides of (2.13) by this, we get

$$1 = \left(1 + \sum_{m \geq 1}(-1)^m \left(q^{\frac{m(3m-1)}{2}} + q^{\frac{m(3m+1)}{2}}\right)\right)\left(\sum_{m \geq 0} P(m)q^m\right) . \qquad (2.14)$$

Now when we multiply two power series

$$f(q) = a_0 + a_1 q + a_2 q^2 + \cdots , \quad g(q) = b_0 + b_1 q + b_2 q^2 + \cdots$$

and collect like powers of q, we end up with

$$f(q)g(q) = a_0 b_0 + (a_0 b_1 + a_1 b_0)q + (a_0 b_2 + a_1 b_1 + a_2 b_0)q^2 + \cdots .$$

In other words if we write

$$f(q)g(q) = \sum_{n \geq 0} c_n q^n, v$$

then

$$c_n = \sum_{k=0}^{n} a_k b_{n-k} \ .$$

For the product in (2.14), we have

$$a_k = \begin{cases} (-1)^m & \text{if } k = \frac{m(3m-1)}{2} \text{ or } k = \frac{m(3m+1)}{2}, \\ 0 & \text{otherwise} \end{cases}$$

and $b_k = P(k)$.

Now (2.12) is a restatement of the fact that for $n \geq 1$,

$$P(n) + a_1 P(n-1) + a_2 P(n-2) + \cdots + a_n P(0) = 0 \ .$$

2.4   Inversion Generating Function for Permutations

A permutation

$$\sigma = \begin{pmatrix} 1 & 2 & \dots & n \\ \sigma_1 & \sigma_2 & \dots & \sigma_n \end{pmatrix}$$

written in 1-line notation is $\sigma_1 \sigma_2 \cdots \sigma_n$. A pair of indices $i < j$ is an *inversion* in σ if $\sigma_i > \sigma_j$. So an inversion is an out of order pair in the 1-line representation of σ. We define

$$\text{inv}(\sigma) = \sum_{i<j} \chi(\sigma_i > \sigma_j),$$

in other words $\text{inv}(\sigma)$ is the total number of inversions in σ. We use the notation $i(\sigma)$ interchangeably with $\text{inv}(\sigma)$.

There are two extreme cases we note: the permutation

$$\begin{pmatrix} 1 & 2 & 3 & \dots & n \\ 1 & 2 & 3 & \dots & n \end{pmatrix}$$

has no inversions. Its reverse

$$\begin{pmatrix} 1 & 2 & 3 & \dots & n \\ n & n-1 & n-2 & \dots & 1 \end{pmatrix}$$

has every pair of indices inverted, and there are $\binom{n}{2}$ of these. Therefore

$$\text{inv}\begin{pmatrix} 1\ 2\ 3\ \dots\ n \\ 1\ 2\ 3\ \dots\ n \end{pmatrix} = 0\ , \quad \text{inv}\begin{pmatrix} 1 & 2 & 3 & \dots\ n \\ n & n-1 & n-2 & \dots\ 1 \end{pmatrix} = \binom{n}{2}\ .$$

Listing all permutations of [3] we find

σ	$\text{inv}(\sigma)$
123	0
132	1
213	1
231	2
312	2
321	3

and in particular

$$\sum_{\sigma \in P_3} q^{\text{inv}(\sigma)} = 1 + 2q + 2q^2 + q^3$$

$$= (1+q)(1+q+q^2)\ .$$

A longer calculation with the 24 permutations in P_4 gives

$$\sum_{\sigma \in P_4} q^{\text{inv}(\sigma)} = 1 + 3q + 5q^2 + 6q^3 + 5q^4 + 3q^5 + q^6$$

$$= (1+q)(1+q+q^2)(1+q+q^2+q^3)\ .$$

We can call

$$\sum_{\sigma \in P_n} q^{\text{inv}(\sigma)}$$

the generating function of permutation *by inversions*. The coefficient of q^k here is the number of permutations in P_n with k inversions. To show that

$$\sum_{\sigma \in P_n} q^{\text{inv}(\sigma)} = (1+q)(1+q+q^2) \cdots (1+q+\cdots+q^{n-1}) \qquad (2.15)$$

for arbitrary n, we first consider *subexceedent functions*.

Here we will define a subexceedent function as one mapping $\{1, 2, \ldots, n\}$ to $\{0, 1, \ldots, n - 1\}$ such that for every i we have $f(i) < i$. So we necessarily have $f(1) = 0$. For $i = 2$, there are two choices: $f(2) = 0$ or $f(2) = 1$. For $i = 3$, $f(3) = 0$ or $f(3) = 1$ or $f(3) = 2$, etc. Let $\mathcal{R}_n$ denote subexceedent functions. It is easy to see that $|\mathcal{R}_n| = n!$. But we can actually say more. For $f \in \mathcal{R}_n$, define its weight by

$$w(f) = \sum_{k=1}^{n} f(k) .$$

Then

$$q^{w(f)} = q^{f(1)+f(2)+\cdots+f(n)} = q^{f(1)} \cdot q^{f(2)} \cdots q^{f(n)}.$$

But these are exactly the terms we collect when we multiply out

$$1(1 + q)(1 + q + q^2) \cdots (1 + q + \cdots + q^{n-1}) \tag{2.16}$$

(1 added to the front in (2.16) for emphasis). For example, for $n = 6$, for the subexceedent function

i	1	2	3	4	5	6
$f(i)$	0	1	2	0	2	4

(2.17)

We have $w(f) = 9$ and

$$q^{w(f)} = q^0 \cdot q^1 \cdot q^2 \cdot q^0 \cdot q^2 \cdot q^4$$

corresponds to picking the underlined terms when multiplying (2.16) out in the case $n = 6$:

$$\underline{1}(1+\underline{q})(1+q+\underline{q^2})(\underline{1}+q+q^2+q^3)(1+q+\underline{q^2}+q^3+q^4)(1+q+q^2+q^3+\underline{q^4}+q^5) .$$

We can clearly make the same correspondence for arbitrary n. Therefore

$$\sum_{f \in \mathcal{R}_n} q^{w(f)} = (1 + q)(1 + q + q^2) \cdots (1 + q + \cdots + q^{n-1}) . \tag{2.18}$$

Going back to permutations, we construct a pairing of subexceedent function $f \in \mathcal{R}_n$ and permutations $\sigma \in P_n$ such that

$$w(f) = \text{inv}(\sigma) .$$

This correspondence is pretty straightforward. We proceed with the example in (2.17) and construct a permutation in P_6 such that the number of inversions involving pairs $i < j$ for any j is the value $f(j)$. This is done in stages, where we construct a partial permutation for $j = 1, 2, \ldots 6$.

Starting with 1 initially, there are 2 positions where we can put 2: to the left of 1 or to the right of 1:

$$\_ \, 1 \, \_.$$

The right position creates no inversions and the left creates 1 inversion. Since $f(2) = 1$, we place 2 on the left to create an inversion. The current partial permutation becomes

$$\_ \, 2 \, \_ \, 1 \, \_.$$

The available positions for 3 from right to left create 0, 1, 2 inversions, respectively. Since $f(3) = 2$, we place it in the leftmost space to create 2 inversion to obtain the partial permutation

$$\_ \, 3 \, \_ \, 2 \, \_ \, 1 \, \_ \, .$$

Continuing in this fashion, guided by the value of $f(j)$ for where to insert j in the partial permutation of $\{1, 2, \ldots, j - 1\}$ that we have at hand, we obtain

$$1 \to 21 \to 321 \to 3214 \to 32514 \to 362514$$

with $w(f) = \mathrm{inv}(\sigma)$ where $\sigma = 362514$. We conclude that

$$\sum_{f \in \mathcal{R}_n} q^{w(f)} = \sum_{\sigma \in P_n} q^{\mathrm{inv}(\sigma)}$$

and this gives the expression (2.15) for the generating function of permutations by inversions.

The subexceedent function f that corresponds to a permutation σ is sometimes referred to its *inversion table*.

There is also a connection between subexceedent functions and ranking/unranking permutations that is worth mentioning at this point.

Given n, any integer r in the range $0 \le r < n!$ can be expressed uniquely in the so-called *factorial basis* in the form

$$d_0 0! + d_1 1! + \cdots + d_i i! + \cdots + d_{n-1}(n-1)! \qquad (2.19)$$

in which the factorial digits d_i satisfy $0 \le d_i \le i$ for $i = 0, 1, \ldots, n - 1$ ($d_0 1! = 0$ in front added in (2.19) for emphasis). This allows us to encode subexceedent functions, and therefore permutations as integers. We simply use the values $f(i)$

as the factorial digits by setting $d_{i-1} = f(i)$. By this process we can *rank* the permutations in P_n as the 0th permutation, 1st permutation, etc., all the way to $(n! - 1)$st permutation.

For $n = 6$, using the values of the function $f \in \mathcal{R}_6$ given in (2.17) as factorial digits, we obtain the number

$$0 \cdot 0! + 1 \cdot 1! + 2 \cdot 2! + 0 \cdot 3! + 2 \cdot 4! + 4 \cdot 5! = 533.$$

So among the numbers $0, 1, \ldots, 719 (= 6! - 1)$, the rank of f is 533. We can then use the correspondence between $f \in \mathcal{R}_n$ and $\sigma \in P_n$ to rank permutations as well. This gives the rank of the σ corresponding to f as $rank(362514) = 533$.

For $n = 3$, the complete table is as follows:

σ	$f(1)f(2)f(3)$			Factorial expansion	$rank(\sigma)$
123	0	0	0	$0 \cdot 0! + 0 \cdot 1! + 0 \cdot 2!$	0
132	0	0	1	$0 \cdot 0! + 0 \cdot 1! + 1 \cdot 2!$	2
213	0	1	0	$0 \cdot 0! + 1 \cdot 1! + 0 \cdot 2!$	1
231	0	1	1	$0 \cdot 0! + 1 \cdot 1! + 1 \cdot 2!$	3
312	0	0	2	$0 \cdot 0! + 0 \cdot 1! + 2 \cdot 2!$	4
321	0	1	2	$0 \cdot 0! + 1 \cdot 1! + 2 \cdot 2!$	5

Note that if we want to generate a random permutation in P_n, all we have to do is to generate a random integer r in the range $0 \le r < n!$, and *unrank* it by finding the factorial digits of r and then constructing the corresponding permutation through its inversion table.

2.5 Parity

Permutations are classified as *odd* or *even* depending on whether they have an odd or an even number of inversions. We assign a sign to σ by setting

$$\text{sign}(\sigma) = \begin{cases} -1 \text{ if } \text{inv}(\sigma) \text{ is odd,} \\ +1 \text{ if } \text{inv}(\sigma) \text{ is even.} \end{cases}$$

So the *parity* of σ is the parity of $\mathrm{inv}(\sigma)$. Evidently

$$\mathrm{sign}(\sigma) = (-1)^{\mathrm{inv}(\sigma)} .$$

Writing $\sigma = \sigma_1\sigma_2 \cdots \sigma_n$ in one-line notation, we observe that transposing (interchanging) two adjacent symbols $\sigma_i\sigma_{i+1}$ to $\sigma_{i+1}\sigma_i$ changes the parity of the permutation since the number $\mathrm{inv}(\sigma)$ goes up or down by one. Therefore making adjacent transpositions changes the parity. Actually we can say more. We can affect transposing two nonadjacent symbols σ_i and σ_{i+d} by a sequence of $2d - 1$ adjacent transpositions. It takes d adjacent transpositions to bring σ_i to where σ_{i+d} is, and then $d - 1$ additional adjacent transpositions to bring σ_{i+d} to index i, where σ_i was. Since $2d - 1$ is an odd number, we conclude that any transposition changes the parity of σ.

Consider now a permutation σ of $[n + 1]$ where $\sigma_{n+1} = n + 1$. Since it is the largest symbol and it is on the very right, $n + 1$ has no effect on the number of inversions. So the parity of such a σ is the same as the parity of the permutation

$$\begin{pmatrix} 1 & 2 & \dots & n \\ \sigma_1 & \sigma_2 & \dots & \sigma_n \end{pmatrix} . \tag{2.20}$$

Suppose now that the permutation in (2.20) is a cycle. If we insert $n + 1$ by breaking the arrow from i to σ_i in the cycle diagram, the new cycle of length $n + 1$ we obtain in one-line notation is

$$\sigma_1\sigma_2 \cdots \sigma_{i-1}\, n + 1\, \sigma_{i+1} \cdots \sigma_n\, \sigma_i .$$

Now this was obtained from the original σ by a single transposition interchanging σ_i and $\sigma_{n+1} = n + 1$. Therefore the $(n + 1)$-cycle we obtained by inserting $n + 1$ by breaking the arrow from i to σ_i, and the n-cycle we started out with have opposite parities.

Now we can use induction to show that an odd length cycle has parity $+1$ and an even length cycle has parity -1.

There is one more important property of the parity we will consider

$$\mathrm{sign}(\sigma) = \prod_{1 \le i < j \le n} \frac{\sigma_i - \sigma_j}{i - j} . \tag{2.21}$$

This is because σ is a permutation, and hence the pairs (σ_i, σ_j) in the numerator must account for all the pairs (i, j) that appear in the denominator, with the proviso that some pairs may be inverted. Each inverted pair contributes -1 to the product.

It can be shown that if we put together a permutation σ from two permutations τ and ρ defined on disjoint sets, then $\mathrm{sign}(\sigma) = \mathrm{sign}(\tau)\mathrm{sign}(\rho)$. In particular, the sign of a permutation is the product of the signs of the cycles in its cycle diagram.

We leave the formal proof of the following theorem to the exercises.

Theorem 2.5.1 *Suppose σ is a permutation on $[n]$ whose cycle diagram consists of $k = k(\sigma)$ cycles, of which c_i are of length i for $i = 1, 2, \ldots, n$. Then*

1. $\operatorname{sign}(\sigma) = (-1)^{c_2 + 2c_3 + \cdots + (n-1)c_n}$,
2. $\operatorname{sign}(\sigma) = -1$ *if and only σ has an odd number of even length cycles,*
3. $\operatorname{sign}(\sigma) = (-1)^{n - k(\sigma)}$.

2.6 Gaussian Polynomials

Knowing the generating function of permutations by inversions allows us to give a combinatorial interpretation of *Gaussian polynomials*, or the so-called *q-binomial coefficients*. These are defined as follows.

Denoting by the symbol $(q)_n$ the polynomial

$$(1 + q)(1 + q + q^2) \cdots (1 + q + \cdots + q^{n-1}) \tag{2.22}$$

we put

$$\begin{bmatrix} n + m \\ n \end{bmatrix} = \frac{(q)_{n+m}}{(q)_m (q)_n} . \tag{2.23}$$

It is clear that setting $q = 1$ in (2.22) one obtains $n!$, so with this substitution (2.23) reduces to

$$\frac{(n + m)!}{m! n!} = \binom{n + m}{n}$$

which is the ordinary binomial coefficient.

What is not so clear is that the right hand side of (2.23) is actually a polynomial in q, since the factors of the expressions in the denominator do not seem to appear among the factors of the numerator.

As a few examples we have

$$\begin{bmatrix} 5 \\ 2 \end{bmatrix} = \frac{(q)_5}{(q)_3 (q)_2}$$

$$= \frac{(1 + q)(1 + q + q^2)(1 + q + q^2 + q^3)(1 + q + q^2 + q^3 + q^4)}{(1 + q)(1 + q + q^2)(1 + q)}$$

$$= \frac{(1 + q + q^2 + q^3)(1 + q + q^2 + q^3 + q^4)}{(1 + q)}$$

$$= 1 + q + 2q^2 + 2q^3 + 2q^4 + q^5 + q^6 ,$$

after we carry out the algebraic operations.

Similarly, it turns out that

$$\begin{bmatrix} 8 \\ 3 \end{bmatrix} = 1 + q + 2q^2 + 2q^3 + 3q^4 + 3q^5 + 4q^6 + 3q^7 + 3q^8 + 2q^9 + 2q^{10} + q^{11} + q^{12}.$$
(2.24)

It may be worthwhile pointing out that our $(q)_n$ defined in (2.22) that appears in the definition of the q-binomial coefficients differs from the one most commonly found in the literature, namely $(q)_n = (1 - q)(1 - q^2) \cdots (1 - q^n)$ for $n > 0$ with $(q)_0 = 1$. However, it is easy to show that both definitions result in the same quotient in (2.23).

Now while (2.22) is the generating function of permutations of the set $\{1, 2, \ldots, n\}$ by inversions, as we have seen, i.e.,

$$\sum_{\sigma \in P_n} q^{\mathrm{inv}(\sigma)} = (1 + q)(1 + q + q^2) \cdots (1 + q + \cdots + q^{n-1})$$

the Gaussian polynomial is nothing but the generating function of *words* over the two-letter alphabet $\{1, 2\}$ with m 1's and n 2's by inversions. More precisely we set for such a word w

$$\mathrm{inv}(w) = \sum_{i < j} \chi\,(w_i > w_j)$$
(2.25)

and denote by $W_{m,n}$ the collection of all such words. What we will show is that

$$\sum_{w \in W_{m,n}} q^{\mathrm{inv}(w)} = \begin{bmatrix} n + m \\ n \end{bmatrix}.$$
(2.26)

Clearly, from (2.26) it will follow that the q-binomial coefficients are actually polynomials. Moreover we obtain more information than this. As an example, (2.26) also tells us that the coefficient of q^4 in (2.25), for instance, is the number of words with three 1's and five 2's with exactly 4 inversions. By inspection, we see that there are only three words in $W_{3,5}$ with 4 inversions as listed below

$$11222212, \quad 12211222, \quad 21121222\,.$$

This number is also the coefficient of the term q^4 in (2.24).

In general, we have that the coefficient of the term q^k in the polynomial

$$\begin{bmatrix} n + m \\ n \end{bmatrix}$$

is the number of words in $W_{m,n}$ which have k inversions.

To give a combinatorial proof of (2.26), we will show instead that

$$(q)_n (q)_m \sum_{w \in W_{m,n}} q^{\text{inv}(w)} = (q)_{m+n} \tag{2.27}$$

from which (2.26) follows by dividing both sides by $(q)_n (q)_m$.

The idea behind the proof of the identity (2.27) is best communicated by an example. To this end, let us consider the case $m = 3$ and $n = 5$. In this case, the left hand side of (2.27) is the weight generating function of triples

$$(\sigma, \tau, w), \tag{2.28}$$

where σ and τ are permutations of the sets $\{1, 2, 3\}$ and $\{1, 2, 3, 4, 5\}$, respectively, and w is a member of $W_{3,5}$. By the *weight* of such a triple, we mean the quantity

$$\text{inv}(\sigma) + \text{inv}(\tau) + \text{inv}(w).$$

For instance,

$$weight(312, 24153, 12211222) = 2 + 4 + 4 = 10. \tag{2.29}$$

Since the right hand side of (2.27) is simply the weight generating function of permutations of $\{1, 2, \ldots, 8\}$ by inversions, what is needed to prove (2.27) is a weight-preserving bijection between triples of the form (2.28) and these permutations. But the form of the triples in (2.29) already suggests what this bijection might be. We can construct a permutation of $\{1, 2, \ldots, 8\}$ from the given triple in (2.29) as follows: we first use the positions of the 1's in the word w to place the elements of σ in the given order:

$$
\begin{array}{cccccccc}
1 & 2 & 2 & 1 & 1 & 2 & 2 & 2 \\
3 & & & 1 & 2 & & &
\end{array}
$$

Now we place the symbols $4, 5, 6, 7, 8$ in the positions of the 2's in w using the order of the letters of τ as the recipe. This results in

$$
\begin{array}{cccccccc}
1 & 2 & 2 & 1 & 1 & 2 & 2 & 2 \\
3 & & & 1 & 2 & & & \\
& 5 & 7 & & & 4 & 8 & 6.
\end{array}
\tag{2.30}
$$

Together, these replacements yield the permutation

$$\pi = 3\,5\,7\,1\,2\,4\,8\,6.$$

We can check that

$$\mathrm{inv}(\pi) = 3 + 3 + 2 + 2 = 10 \,.$$

The reason for this is very simple. The inversions of π are of three types:

(A) Inversions among the symbols $1, 2, 3$,
(B) Inversions among the symbols $4, 5, 6, 7, 8$,
(C) Inversions where one symbol is from $1, 2, 3$, *and the other from* $4, 5, 6, 7, 8$.

By construction, type (A) and type (B) inversions are exactly the inversions of σ and τ, respectively. On the other hand, each symbol put in place of 2's in (2.30) is larger than every symbol placed in the positions of 1's in w. But this means that type (C) inversions in π are equinumerous with the inversions of w.

Note that this correspondence is also reversible. For instance, from the permutation

$$\rho = 35276148$$

we obtain the word

$$w = 12122122$$

by marking the positions of the symbols $1, 2, 3$ that appear in ρ by 1 and those of $4, 5, 6, 7, 8$ by 2's. Furthermore, we can read off the two permutations σ and τ from ρ as well: since the symbols $1, 2, 3$ appear in order 321 we have that $\sigma = 321$. The remaining letters are ordered as 57648 in ρ and this gives the permutation $\tau = 24315$. Finally, we have that ρ corresponds to the triple

$$(321, 57648, 12122122) \,.$$

Note that

$$\mathrm{inv}(\rho) = 2 + 1 + 5 + 3 = 11$$

and

$$\mathrm{inv}(\sigma) + \mathrm{inv}(\tau) + \mathrm{inv}(w) = 3 + 4 + 4 = 11 \,.$$

It is not difficult to see that this argument goes through in the case of general n and m as well.

This establishes (2.27).

Gaussian polynomials behave much the same way as the binomial coefficients, except for an occasional power of q that happens to enter into various identities. For instance, recall that we have the identity

$$\binom{n+m}{n} = \binom{n+m-1}{n-1} + \binom{n+m-1}{n}. \tag{2.31}$$

This reflects the fact that n-subsets of an $(n+m)$-set can be partitioned into two groups: those which contain the largest element $n+m$, which is counted by the first term on the right in (2.31), and those which avoid it, counted by the second binomial coefficient on the right.

By a similar partitioning argument based on words, we can show that the Gaussian polynomials satisfy

$$\begin{bmatrix} n+m \\ n \end{bmatrix} = \begin{bmatrix} n+m-1 \\ n-1 \end{bmatrix} + q^n \begin{bmatrix} n+m-1 \\ n \end{bmatrix}. \tag{2.32}$$

To do this, we separate the words in $W_{m,n}$ into two groups: those which end with the letter 2, and those which end with the letter 1. We can then write

$$W_{m,n} = W_{m,n-1} \cdot 2 + W_{m-1,n} \cdot 1.$$

Note that for $w \in W_{m,n-1}$ we have $\mathrm{inv}(w \cdot 2) = \mathrm{inv}(w)$. On the other hand, for any $w \in W_{m-1,n}$, we have $\mathrm{inv}(w \cdot 1) = \mathrm{inv}(w) + n$, since the last 1 causes an inverted pair with every 2 that appears in w and there are n of these. It follows that

$$\sum_{w \in W_{m,n}} q^{\mathrm{inv}(w)} = \sum_{w \in W_{m,n-1}} q^{\mathrm{inv}(w)} + \sum_{w \in W_{m-1,n}} q^{\mathrm{inv}(w)+n}$$

$$= \sum_{w \in W_{m,n-1}} q^{\mathrm{inv}(w)} + q^n \sum_{w \in W_{m-1,n}} q^{\mathrm{inv}(w)}$$

and this is just another way of writing (2.32) by (2.26).

Clearly, there is nothing special about the two-letter alphabet in the definition of inversions of a word, and (2.25) makes sense for an arbitrary word—as long as the alphabet is linearly ordered.

We leave it to the reader to prove the following general version of (2.26) by extending the combinatorial arguments presented in this section to *q-multinomial coefficients*, defined by

$$\begin{bmatrix} n_1 + n_2 + \cdots + n_k \\ n_1, n_2, \ldots, n_k \end{bmatrix} = \frac{(q)_{n_1+n_2+\cdots+n_k}}{(q)_{n_1}(q)_{n_2} \cdots (q)_{n_k}}.$$

Theorem 2.6.1 (MacMahon) *Let $W_{n_1,n_2,\ldots,n_k}$ denote the collection of words over the alphabet $\{1, 2, \ldots, k\}$ in which the letter i appears n_i times, $i = 1, 2, \ldots, k$. Then*

$$\sum_{w \in W_{n_1, n_2, \ldots, n_k}} q^{\mathrm{inv}(w)} = \begin{bmatrix} n_1 + n_2 + \cdots + n_k \\ n_1, n_2, \ldots, n_k \end{bmatrix}. \tag{2.33}$$

This theorem is due to Percy Alexander MacMahon [21].

We will conclude this section by presenting a different interpretation of the q-binomial coefficients as the weight generating function of restricted partitions. This dual point of view is useful, in that some identities involving Gaussian polynomials can be proved more easily by pictorial arguments using FDs than making use of words over a two-letter alphabet or permutation statistics.

We let $P_{m,n}$ denote the collection of Ferrers diagrams included in the $n \times m$ rectangle. Thus these are partitions with at most n parts, where no part is allowed to exceed m. The *weight* of an element $\lambda \in P_{m,n}$ is the sum of its parts, denoted by $|\lambda|$ as before. Thus, for instance, for the partition $\lambda \in P_{9,7}$ shown in Figure 2.23, we have that $|\lambda| = 26$. As another example, all the partitions that are included in the 2×3 rectangle are reproduced in Figure 2.24. There are clearly $\binom{2+3}{2}$ of these.

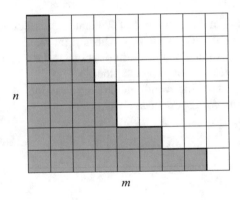

Fig. 2.23 A partition included in the 9×7 rectangle.

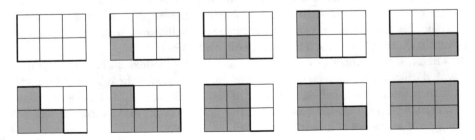

Fig. 2.24 Partitions included in the 2×3 rectangle.

From Figure 2.24 we get that the weight generating function of $P_{3,2}$ is given by

$$\sum_{\lambda \in P_{3,2}} q^{|\lambda|} = 1 + q + 2q^2 + 2q^3 + 2q^4 + q^5 + q^6 = \begin{bmatrix} 5 \\ 2 \end{bmatrix}.$$

Now what is going on here is that there is a simple coding of partitions in $P_{m,n}$ by words in $W_{m,n}$ in such a way that the weight of the given partition is equal to the number of inversions of the corresponding word. To see this, we traverse the boundary of λ starting from the north-west corner and moving a step at a time to the south-east corner. Each horizontal step we come across is coded as a 2 and each vertical step as a 1 as we proceed from left to right. For instance, this algorithm applied to the partition in example Figure 2.23 yields the word

$$w = 2112212112122212$$

with

$$\mathrm{inv}(w) = 1 + 1 + 3 + 4 + 4 + 5 + 8 = 26.$$

The reason for this is that the number of 2's that appear before a given 1 in w is precisely the number of cells in the row whose right hand vertical boundary was represented by this 1. Adding up these contributions is the weight of λ on one hand, and the number of inversions of w on the other.

Since we already know the weight generating function of $W_{m,n}$ by inversions, this weight-preserving correspondence gives

$$\sum_{\lambda \in P_{m,n}} q^{|\lambda|} = \begin{bmatrix} m+n \\ n \end{bmatrix} \tag{2.34}$$

by (2.26).

We can exploit what we know about Ferrers diagrams to obtain further results involving the Gaussian polynomials. We shall give a few such examples.

We have mentioned that q-binomial coefficients behave in much the same way as the ordinary binomial coefficients themselves. Thus we may wonder what the form of the identity analogous to the binomial expansion is when we consider q-binomial coefficients instead of the binomial coefficients. It turns out that the partition interpretation given above is a suitable tool to this end.

Recall that the formal expression in Figure 2.25 generates all Ferrers diagrams with distinct rows where no row has more than n cells. Now imagine that we code by an x the leftmost cell in each row of such a diagram λ, and assign another indeterminate q to all the other cells. The product of all these variables will be called the *monomial* of λ (see Figure 2.26).

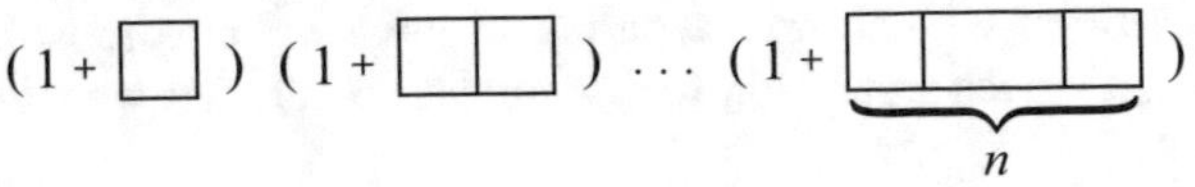

$$(1 + \boxed{}\,) \,(1 + \boxed{}\,) \cdots (1 + \underbrace{\boxed{}}_{n}\,)$$

Fig. 2.25 FDs with distinct rows where no row has more than n cells.

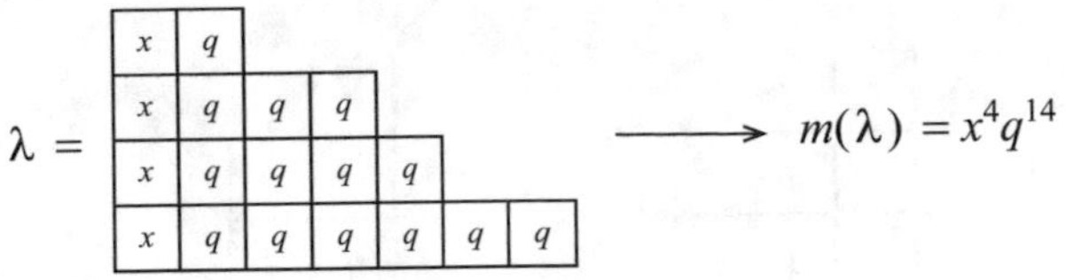

$$\lambda = \quad \longrightarrow \quad m(\lambda) = x^4 q^{14}$$

Fig. 2.26 The monomial of λ.

Clearly, the exponent of x in the monomial of λ keeps track of the number of rows of λ, and this exponent together with the exponent of q give the weight of λ. If we denote by $r(\lambda)$ the number of rows of λ, then

$$m(\lambda) = x^{r(\lambda)} q^{|\lambda| - r(\lambda)}.$$

From Figure 2.25 we have that

$$(1 + x)(1 + qx) \cdots (1 + q^{n-1}x) = \sum_{\lambda} m(\lambda), \qquad (2.35)$$

where the summation is over all such partitions.

However, the right hand side of (2.35) can be evaluated in another way. This can be done with the following alternate combinatorial interpretation.

A partition with k distinct parts can be thought of as a staircase with k rows, and a partition $\widetilde{\lambda}$ with *at most* k parts, not necessarily distinct. For instance, the partition in example given in Figure 2.26 can be broken up as indicated in Figure 2.27.

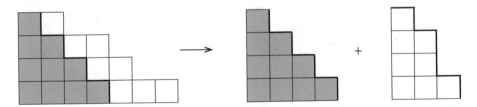

Fig. 2.27 Taking out a staircase from λ.

Note that the largest part of λ is reduced by k so that if the original partition has largest part at most n, the partition $\widetilde{\lambda}$ that is produced after we take out the staircase has largest part at most $n - k$. Furthermore this correspondence is reversible.

An arbitrary partition with at most k parts and largest part at most $n - k$ (i.e., an element of $P_{n-k,k}$) gives a partition with k *distinct* parts after we attach a staircase of height k to it.

We need one more observation: we shall view the weights x and q of the partition λ as distributed to the pair produced by the above trick as illustrated in Figure 2.28.

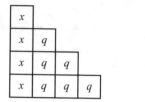 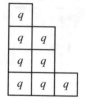

Fig. 2.28 The assignment of the weights.

Clearly

$$m(\lambda) = x^k q^{1+2+\cdots+k-1} q^{|\tilde{\lambda}|} .$$

If E_k denotes all partitions produced by the expression Figure 2.25 with exactly k rows, we then have

$$\sum_{\lambda \in E_k} m(\lambda) = x^k q^{\frac{k(k-1)}{2}} \sum_{\tilde{\lambda} \in P_{n-k,k}} q^{|\tilde{\lambda}|} = x^k q^{\frac{k(k-1)}{2}} \begin{bmatrix} n \\ k \end{bmatrix}, \tag{2.36}$$

where the last equality is a consequence of (2.34).

Now summing (2.36) for $k = 0, 1, \ldots, n$ and comparing with (2.35) we get

$$(1+x)(1+qx)\cdots(1+q^{n-1}x) = \sum_{k=0}^{n} \begin{bmatrix} n \\ k \end{bmatrix} x^k q^{\frac{k(k-1)}{2}} . \tag{2.37}$$

This is the *q-binomial theorem*.

Note that (2.37) reduces to the familiar binomial theorem

$$(1+x)^n = \sum_{k=0}^{n} \binom{n}{k} x^k$$

upon letting $q = 1$.

Next we consider the expansion of the product

$$\frac{1}{(1-x)(1-qx)\cdots(1-q^{n-1}x)} .$$ (2.38)

We see that this expression is the sum of terms of the form

$$x^{r(\lambda)}q^{|\lambda|-r(\lambda)}$$ (2.39)

over partitions λ which have rowlengths at most n and number of parts $r(\lambda)$. We classify these partitions according to the number of parts. If such a λ has k parts, then removing the first column from its FD, in which we imagine that each square carries the symbol x, we end up with a FD $\tilde{\lambda}$ which has at most k parts and whose largest part is at most $n-1$. These are exactly the partitions in the rectangle $P_{n-1,k}$. Thus summing the expressions in (2.39)

$$\sum_{r(\lambda)=k} x^{r(\lambda)}q^{|\lambda|-r(\lambda)} = x^k \sum_{\tilde{\lambda}\in P_{n-1,k}} q^{|\tilde{\lambda}|} = x^k \begin{bmatrix} n+k-1 \\ k \end{bmatrix} .$$

Adding the contributing over k, we obtain

$$\frac{1}{(1-x)(1-qx)\cdots(1-q^{n-1}x)} = \sum_{k\geq 0} \begin{bmatrix} n+k-1 \\ k \end{bmatrix} x^k ,$$ (2.40)

and this reduces to Newton's expansion

$$\frac{1}{(1-x)^n} = \sum_{k\geq 0} \binom{n+k-1}{k} x^k$$

for $q=1$.

An interesting consequence of the pair of formulas (2.37) and (2.40) is the following. We note that if we replace x by $-x$ in (2.40), we obtain the reciprocal of the left hand side of (2.37). In other words we must have

$$\left(\sum_{k=0}^{n} \begin{bmatrix} n \\ k \end{bmatrix} x^k q^{\frac{k(k-1)}{2}}\right) \left(\sum_{k\geq 0} \begin{bmatrix} n+k-1 \\ k \end{bmatrix} (-1)^k x^k\right) = 1.$$

This means that the coefficient of the term x^N for $N \geq 1$ must be zero on the left. For two generating functions

$$f(x) = a_0 + a_1 x + a_2 x^2 + \cdots , \qquad g(x) = b_0 + b_1 x + b_2 x^2 + \cdots ,$$

if

$$h(x) = f(x)g(x) = c_0 + c_1 x + c_2 x^2 + \cdots + c_N x^N + \cdots,$$

then

$$c_N = \sum_{k=0}^{N} a_k b_{N-k} . \tag{2.41}$$

Thus the product of the generating functions of two sequences $\{a_n\}$ and $\{b_n\}$ is the generating function of the sequence of their *convolution*, whose Nth term is given by (2.41).

For our example, we have

$$a_k = \begin{bmatrix} n \\ k \end{bmatrix} q^{\frac{k(k-1)}{2}}$$

$$b_k = \begin{bmatrix} n+k-1 \\ k \end{bmatrix} (-1)^k$$

so that we always have

$$\sum_{k=0}^{N} (-1)^k q^{\frac{k(k-1)}{2}} \begin{bmatrix} n \\ k \end{bmatrix} \begin{bmatrix} n+N-k-1 \\ N-k \end{bmatrix} = 0, \quad (n, N > 0) . \tag{2.42}$$

2.7   Miscellaneous Identities

We start with a very useful geometric consideration that allows us to prove a number of partition identities. The geometric nature of the decomposition makes some of the arguments intuitively evident.

2.7.1   Durfee Square

A partition λ contains a special square shaped partition in it called its *Durfee square*. This is best understood by considering Ferrers diagrams. The Durfee square of λ is the largest square that we can fit in the diagram of λ in such a way that it shares with λ the south-west corner cell of the diagram of λ.
Examples:

In Figure 2.29, we have three FDs with their corresponding Durfee squares of sizes 2×2, 1×1 and 3×3, respectively.

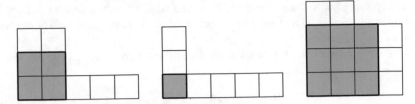

Fig. 2.29 Durfee squares.

Just as the FD itself is a very convenient visual tool to represent and manipulate integer partitions, using Durfee squares makes certain properties of partitions and generating functions transparent.

2.7.2 Hook Shapes

Let us start with the simplest case. A *hook shape* is a FD which literally looks like a hook or a right angle, much like the partition in the middle of Figure 2.29. We can write down all hooks with 3 cells (shown in Figure 2.30) and those with 4 cells (shown in Figure 2.31).

Fig. 2.30 All hooks with 3 cells.

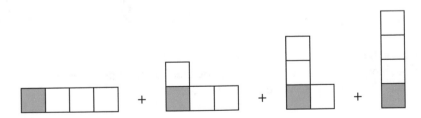

Fig. 2.31 All hooks with 4 cells.

Clearly there are n hook shaped FDs with a total of n cells, each obtained by bending a row of n cells like a slinky all possible ways. In particular all hooks have a 1×1 Durfee square.

If we denote by H the set of all hook shapes and denote by $|\lambda|$ the number of cells of λ, then

$$\sum_{\lambda \in H} q^{|\lambda|} = q + 2q^2 + 3q^3 + 4q^4 + \cdots \tag{2.43}$$

$$= \frac{q}{(1-q)^2},$$

which we find by calculus.

Let us obtain this generating function another way. We use the 1×1 Durfee square and build the rest of the hook in stages around it. To generate a hook shape, we start with a 1×1 square. Then to its right, we can add any one of the diagrams in Figure 2.32.

$$1 \;+\; \square \;+\; \square\square \;+\; \square\square\square \;+\; \square\square\square\square \;+\; \ldots$$

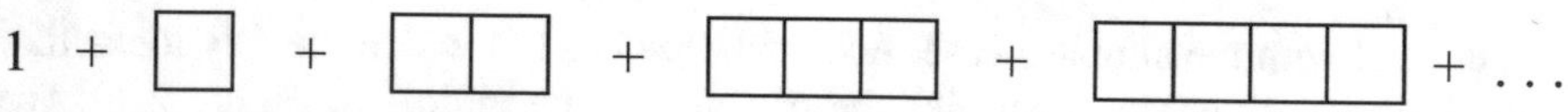

Fig. 2.32 Rows of squares.

After that, we can again use any one of the shapes in Figure 2.32 to add to the top of the Durfee square, but this time use the conjugate shape, i.e., as a column of cells. Clearly the generating function of the diagrams in Figure 2.32 is

$$\frac{1}{1-q}.$$

Using this twice, and remembering to multiply by q to account for our 1×1 Durfee square we have

$$\sum_{\lambda \in H} q^{|\lambda|} = \frac{q}{(1-q)^2}$$

as we found in (2.43).

Classification of families of partitions by their Durfee squares allows for many surprising identities. We will consider a few of them.

2.7.3 Classifying All Partitions

Let us start with the classification of all partitions by the size of their Durfee square. For convenience, we include 1 as the null partition with the 0×0 Durfee square.

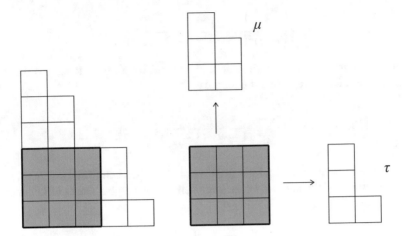

Fig. 2.33 Decomposition via the Durfee square.

Now if the Durfee square of λ is $m \times m$ for some $m \geq 1$, then it can be decomposed into three partitions as shown in Figure 2.33. Here we have m^2 cells contributed by the Durfee square. Of the other two parts, τ is a partition with at most m parts, and μ is a partition where no part exceeds m. By Theorem 2.1.1 we know these two families of partitions are equinumerous.

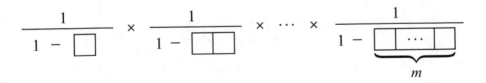

Fig. 2.34 Partitions with largest part $\leq m$.

Clearly, the μ's that are allowed are generated by the expression in Figure 2.34. The τ's that make up the part on the right are again generated by the same expression, except we do not use the partitions generated but their conjugates instead. The generating function we obtain from Figure 2.34 for the μ's (and similarly for the τ's) is then simply

$$\frac{1}{(1-q)(1-q^2)\cdots(1-q^m)}. \tag{2.44}$$

Therefore the generating function of all partitions which have an $m \times m$ Durfee square is

$$\frac{q^{m^2}}{(1 - q)^2(1 - q^2)^2 \cdots (1 - q^m)^2} \cdot$$

Adding the contributions over all m should account for all partitions, and therefore using Theorem 2.2.1 we obtain the identity

$$\prod_{m \geq 1} \frac{1}{1 - q^m} = 1 + \sum_{m \geq 1} \frac{q^{m^2}}{(1 - q)^2(1 - q^2)^2 \cdots (1 - q^m)^2} \cdot \qquad (2.45)$$

Using the same ideas and with a little more effort, we can prove a general result by keeping track of the number of parts as well as the weight. The coefficient of x^m in the product

$$\prod_{k \geq 1} \frac{1}{1 - xq^k} \qquad (2.46)$$

is the generating function of the number of partitions with exactly m parts. We can make use of this to develop two general identities at once.

First, we know the generating function of partitions with exactly m parts. This family is obtained from partitions with at most m parts by adding a column of m cells. It follows from (2.44) that the generating function of partitions with exactly m parts is

$$\frac{q^m}{(1 - q)(1 - q^2) \cdots (1 - q^m)} \cdot \qquad (2.47)$$

This then must be the coefficient of x^m in the expansion of (2.46). In other words

$$\prod_{m \geq 1} \frac{1}{1 - xq^m} = 1 + \sum_{m \geq 1} \frac{x^m q^m}{(1 - q)(1 - q^2) \cdots (1 - q^m)} \cdot$$

Secondly, we can generate all partitions according to the number of parts by classifying them according to the size of their Durfee square. We do this as follows.

From the decomposition in Figure 2.33, we can recover the number of parts of λ by adding to m the number of parts of μ. But the generating function for the μ's which keeps track of the number of parts as the exponent of x is

$$\frac{1}{(1 - xq)(1 - xq^2) \cdots (1 - xq^m)} \cdot$$

Clearly each part used to build up μ contributes 1 to the exponent of x in the expansion of this. We have no restriction on the τ's attached to the right of the Durfee square other than they should not have more than m parts. The generating function of the τ's is of course (2.44).

Putting these together and not forgetting that the Durfee square itself contributes

$$x^m q^{m^2} ,$$

we have

$$\prod_{m\geq 1} \frac{1}{1-xq^m} = 1 + \sum_{m\geq 1} \frac{x^m q^{m^2}}{(1-q)(1-q^2)\cdots(1-q^m)(1-xq)(1-xq^2)\cdots(1-xq^m)} ,$$

and (2.45) is the special case $x = 1$ of this identity.

2.7.4 Self-Conjugate Partitions

Another application of the Durfee square idea is the enumeration of *self-conjugate* partitions. These are partitions which are equal to their own conjugate, in other words partitions whose diagram is symmetric with respect to the diagonal line $y = x$.

We will construct the generating function of self-conjugate partitions two different ways.

Again consider the decomposition of a partition into the three pieces consisting of an $m \times m$ square and the two partitions τ and μ as we did in Figure 2.33.

However, now the τ and μ in decomposition of Figure 2.33 must be so that $\tau = \mu'$ since we need to fall back on the same partition after reflecting about the diagonal line (Fig. 2.35).

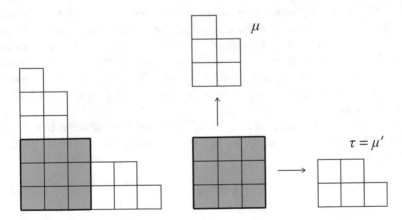

Fig. 2.35 Decomposition of a self-conjugate partition.

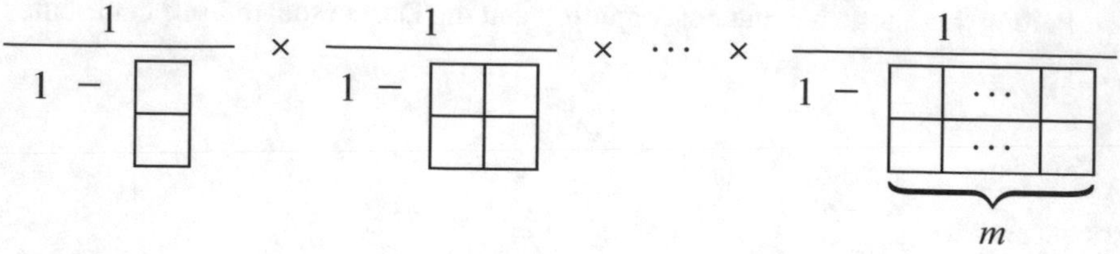

Fig. 2.36 Picking pairs of rows, each row $\leq m$.

This means that instead of using Figure 2.34 to generate the relevant partitions to add to the Durfee square, now we need to pick *pairs* of identical parts as generated symbolically by the expression in Figure 2.36. Here one of the double rows picked is used as a row of μ, while the other is used as a column of τ. We see that the generating function for self-conjugate partitions which have an $m \times m$ Durfee square is

$$\frac{q^{m^2}}{(1 - q^2)(1 - q^4)\cdots(1 - q^{2m})} .$$

Summing over all m, the generating function of self-conjugate partitions is given by

$$1 + \sum_{m \geq 1} \frac{q^{m^2}}{(1 - q^2)(1 - q^4)\cdots(1 - q^{2m})} . \tag{2.48}$$

Now it turns out that self-conjugate partitions can be put together another way in a completely original manner.

If a self-conjugate partition λ has an $m \times m$ Durfee square, then we can peel off m hook shapes from λ one after the other as in Figure 2.37. Notice that these hooks all have odd number of cells. This is easy to see in general since the row incident to the 1×1 Durfee square on the right and the column incident to it at the top have the same number of cells given that λ is self-conjugate. The hook shapes we peel off this way must also have distinct lengths for otherwise we would not have a legitimate Ferrers diagram.

It follows that the generating function of self-conjugate partitions is identical to the generating function of partitions with distinct odd parts. This latter generating function is

$$\prod_{m \geq 0} (1 + q^{2m+1}) . \tag{2.49}$$

So (2.49) and (2.48) are different expressions for the same generating function and we have proved the identity

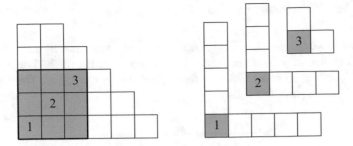

Fig. 2.37 Decomposition into hook shapes.

$$\prod_{m\geq 0}(1 + q^{2m+1}) = 1 + \sum_{m\geq 1} \frac{q^{m^2}}{(1 - q^2)(1 - q^4)\cdots(1 - q^{2m})}.$$

Again, with a little more effort we can prove a generalization of this identity. The idea is to keep track of the size of the Durfee square of a self-conjugate partition in addition to the weight of the partition. We do this by summing up terms of the form

$$q^{|\lambda|}x^m$$

over all self-conjugate partitions λ where m indicates that λ has an $m \times m$ Durfee square. Then we see from (2.48) that the sum of these terms over all self-conjugate partitions is given by

$$1 + \sum_{m\geq 1} \frac{x^m q^{m^2}}{(1 - q^2)(1 - q^4)\cdots(1 - q^{2m})}.$$

Let us go back to the construction of the same collection with an $m \times m$ Durfee square using the decomposition in Figure 2.37. The diagonal of the Durfee square has as many cells as the number of hooks we use to build the self-conjugate partition. Each of these hooks comes from a power of q from the product in (2.49). This means that we can keep track of the number of hooks used, which of course is the same as the length of the diagonal of the resulting Durfee square, by attaching an x to each power of q in (2.49). It follows that

$$\prod_{m\geq 0}(1 + xq^{2m+1}) = 1 + \sum_{m\geq 1} \frac{x^m q^{m^2}}{(1 - q^2)(1 - q^4)\cdots(1 - q^{2m})}. \qquad (2.50)$$

2.7.5   Partitions with Distinct Parts

We know that we can generate all partitions with largest part exactly m: first generate all partitions whose parts do not exceed m as shown in Figure 2.34, and then add a part of length m to make sure that the largest part is exactly m. The generating function for such partitions is then

$$\frac{q^m}{(1-q)(1-q^2)\cdots(1-q^m)} \tag{2.51}$$

as we have seen before.

We could just as well generate all partitions with exactly m parts by taking the conjugates of the partitions generated this way. So we see that the generating function of partitions with exactly m parts is also (2.51).

Now we can attach a staircase shape of height $m-1$ to such a partition to obtain another partition with exactly m parts, but the parts of the new partition have become all distinct as shown in Figure 2.38.

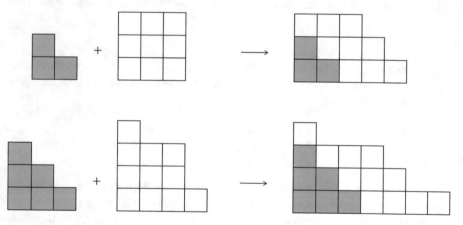

Fig. 2.38 Making parts distinct by attaching a staircase.

Examples:

This process is also reversible: given a partition with m distinct parts, we can subtract off a staircase of height $m-1$ from it to obtain a pair of partitions, a staircase shape with height $m-1$ and a partition with exactly m parts (but no condition of parts being distinct).

So the generating function for partitions with m distinct parts is

$$\frac{q^m \cdot q^{1+2+\cdots+m-1}}{(1-q)(1-q^2)\cdots(1-q^m)}. \tag{2.52}$$

Therefore the generating function for *all* partitions with distinct parts is the sum of (2.52) over all m. It follows that we have the following expression for this generating function:

$$1 + \sum_{m \geq 1} \frac{q^{\frac{m(m+1)}{2}}}{(1-q)(1-q^2)\cdots(1-q^m)} . \qquad (2.53)$$

On the other hand we already know how to generate partitions with distinct parts. These are given by the expression in Figure 2.15. In (2.53) then we then have another representation of this generating function and we have proved that the following identity holds:

$$\prod_{m \geq 1}(1+q^m) = 1 + \sum_{m \geq 1} \frac{q^{\frac{m(m+1)}{2}}}{(1-q)(1-q^2)\cdots(1-q^m)} . \qquad (2.54)$$

Notice that we could have obtained the identity (2.54) from (2.50) by taking $x = q$, and then replacing q by $\sqrt{q}$.

Either way, again a generalization is possible, and by keeping track of the number of parts as the exponent of a variable x, we can prove

$$\prod_{m \geq 1}(1+xq^m) = 1 + \sum_{m \geq 1} \frac{x^m q^{\frac{m(m+1)}{2}}}{(1-q)(1-q^2)\cdots(1-q^m)} ,$$

which reduces to (2.54) for $x = 1$.

2.7.6 Binary Expansions

Sometimes we can interpret a familiar identity in terms of properties of partitions as well. We recall Euler's theorem that the number of partitions of n into distinct parts is equinumerous with the number of partitions of n into odd parts. The proof of this property of partitions (Theorem 2.2.3) was obtained as a consequence of a generating function identity.

Here let us start with

$$\frac{1}{1-q} = \prod_{m \geq 0}(1+q^{2^m}) . \qquad (2.55)$$

This identity is a restatement of the well known property that any integer can be uniquely expressed in binary, that is it can be uniquely written as the sum of a finite subset of the numbers $1, 2, 2^2, 2^3, \ldots$..

Rewriting (2.55) in the form

$$1 - q = \frac{1}{(1+q)(1+q^2)(1+q^{2^2})(1+q^{2^3})\cdots} \tag{2.56}$$

we see that in the expansion

$$(1-q^1+q^{2\cdot1}-q^{3\cdot1}+\cdots)(1-q^2+q^{2\cdot2}-q^{3\cdot2}+\cdots)(1-q^{2^2}+q^{2\cdot2^2}-q^{3\cdot2^2}+\cdots)\cdots$$

the coefficient of the term q^n vanishes on the right for $n \geq 2$.

Now the right hand side of (2.56) can be interpreted as the generating function of certain signed FDs: each rowlength used is a power of 2, and the sign of a FD is simply the number of rows that is used to construct it. (2.56) says that after replacing each such λ by $q^{|\lambda|}$, the terms for $n \geq 2$ cancel out. In other words $n \geq 2$ has as many partitions into parts that are restricted to be powers of 2 with an even number of parts as it has with an odd number of them.

We can check this in the case $n = 7$. There are 3 partitions of 7 into parts that are powers of 2 where the number of parts is an odd number (the number of parts is 3, 5, and 7 below):

$$4 + 2 + 1$$
$$2 + 2 + 1 + 1 + 1$$
$$1 + 1 + 1 + 1 + 1 + 1 + 1$$

and also a matching number of partitions into parts that are powers of 2 and where the number of parts is even (the number of parts is 4, 4, and 6 below):

$$4 + 1 + 1 + 1$$
$$2 + 2 + 2 + 1$$
$$2 + 1 + 1 + 1 + 1 + 1 .$$

2.7.7   Rogers–Ramanujan Identities

Finally we consider partitions whose parts differ by at least 2. For example, 7 has three such partitions:

$$7$$
$$6 + 1 \tag{2.57}$$
$$5 + 2,$$

and 8 has four of them:

$$8$$
$$7+1$$
$$6+2 \qquad\qquad (2.58)$$
$$5+3\,.$$

We can obtain partitions with m parts in this class by starting with partitions λ with exactly m parts (with no other condition on the lengths of the parts), and adding to each such λ a *double staircase* of height $m-1$ and parts $2, 4, \ldots 2(m-1)$ from top to bottom (see Figure 2.39).

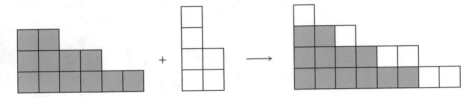

Fig. 2.39 Adding a double staircase.

This construction is also reversible. Any partition with m parts with the parts differing by at least 2 can be decomposed into a double staircase and a partition with exactly m parts as in Figure 2.39.

The generating function of partitions with exactly m parts is as given in (2.47). To obtain the generating function of partitions with m parts differing by at least 2, we have to add the weight of the double staircase used to construct it. Since this weight is

$$2+4+\cdots+2(m-1) = m(m-1)\,,$$

we need to multiply (2.47) by $q^{m(m-1)}$. Once we do this, the product we obtain is the generating function of the partitions λ with parts differing by at least 2 and having a total of m parts. To get the generating function of all λ with parts differing by at least 2, we need to add these over m. In other words the generating function of partitions whose parts differ by at least 2 is

$$1+\sum_{m\geq 1} \frac{q^{m^2}}{(1-q)(1-q^2)\ldots(1-q^m)}\,. \qquad (2.59)$$

Now it is interesting that this generating function is equal to the generating function of partitions whose parts are congruent to 1 or 4 modulo 5: i.e.,

$$1+\sum_{m\geq 1}\frac{q^{m^2}}{(1-q)(1-q^2)\dots(1-q^m)}=\prod_{m\geq 0}\frac{1}{(1-q^{5m+1})(1-q^{5m+4})}.\qquad (2.60)$$

In other words the number of partitions of n whose parts differ by at least 2 are equinumerous with the number of partitions of n whose parts are congruent to 1 or 4 modulo 5.

The identity (2.60) and its companion below

$$1+\sum_{m\geq 1}\frac{q^{m^2+m}}{(1-q)(1-q^2)\dots(1-q^m)}=\prod_{m\geq 0}\frac{1}{(1-q^{5m+2})(1-q^{5m+3})}\qquad (2.61)$$

are known as the Rogers-Ramanujan identities [1].

Incidentally, the partitions of 7 whose parts are congruent to 1 or 4 modulo 5 are

$$6+1$$
$$4+1+1+1$$
$$1+1+1+1+1+1+1,$$

and those of 8 are

$$6+1+1$$
$$4+4$$
$$4+1+1+1+1$$
$$1+1+1+1+1+1+1+1,$$

which are equinumerous with the partitions given in (2.57) and (2.58), respectively.

We should mention that the interpretation of these curious analytic identities combinatorially as partition identities have led to the development of a number of new combinatorial techniques and surprising results in this area [13].

2.8 Exercises for Chapter 2

2.8.1 Find the number of Ferrers diagrams which fit inside an $n \times m$ rectangle.

2.8.2 Calculate the number of Ferrers diagrams with exactly n parts that fit inside an $n \times m$ rectangle.

2.8.3 Use the fact that $(e^t)^2 = e^{2t}$ and compare coefficients of t^n on both sides to give another proof of the identity

$$\sum_{k=0}^{n} \binom{n}{k} = 2^n.$$

2.8.4 Derive the binomial theorem from the identity

$$e^{xt} \cdot e^{yt} = e^{(x+y)t}.$$

2.8.5 Show that the average number of inversions of permutations of $[n]$ is given by $\frac{1}{2}\binom{n}{2}$.

2.8.6

(a) Find the permutation $\sigma \in P_6$ whose rank is 100.

(b) Describe the permutation $\sigma \in P_n$ whose rank is 100 for a given $n > 6$.

2.8.7 For a permutation $\sigma \in P_n$ define

$$\mathrm{maj}(\sigma) = \sum_{i=1}^{n-1} i \, \chi \, (\sigma_i > \sigma_{i+1}).$$

This quantity is called the *major index* of σ. Show that

$$\sum_{\sigma \in P_n} q^{\mathrm{inv}(\sigma)} = \sum_{\sigma \in P_n} q^{\mathrm{maj}(\sigma)} \, .$$

2.8.8 Construct the permutation of P_7 that corresponds to the subexceedent function below:

i	1	2	3	4	5	6	7
$f(i)$	0	0	2	0	3	3	5

2.8.9 Starting with the generating function

$$f(q) = 1 + \frac{q}{1-q} + \frac{q^2}{(1-q)(1-q^2)} + \frac{q^3}{(1-q)(1-q^2)(1-q^3)} + \cdots$$

give a combinatorial interpretation to the string of identities

$$f(q) = \frac{1}{(1-q)} \left(1 + \frac{q^2}{1-q^2} + \frac{q^3}{(1-q^2)(1-q^3)} + \cdots \right)$$

$$= \frac{1}{(1-q)(1-q^2)} \left(1 + \frac{q^3}{1-q^3} + \frac{q^4}{(1-q^3)(1-q^4)} + \cdots \right)$$

$$\vdots$$

2.8.10 Starting with the generating function

$$g(q) = 1 + \frac{q}{1-q} + \frac{q^3}{(1-q)(1-q^3)} + \frac{q^5}{(1-q)(1-q^3)(1-q^5)} + \cdots$$

give a combinatorial interpretation to the string of identities

$$g(q) = \frac{1}{(1-q)}\left(1 + \frac{q^3}{1-q^3} + \frac{q^5}{(1-q^3)(1-q^5)} + \cdots\right)$$

$$= \frac{1}{(1-q)(1-q^3)}\left(1 + \frac{q^5}{1-q^5} + \frac{q^7}{(1-q^5)(1-q^7)} + \cdots\right)$$

$$\vdots$$

2.8.11 More generally, given a sequence $0 < s_1 < s_2 < \cdots$, with

$$s(q) = 1 + \frac{q^{s_1}}{1-q^{s_1}} + \frac{q^{s_2}}{(1-q^{s_1})(1-q^{s_2})} + \frac{q^{s_3}}{(1-q^{s_1})(1-q^{s_2})(1-q^{s_3})} + \cdots$$

give a combinatorial interpretation to the string of identities

$$s(q) = \frac{1}{(1-q^{s_1})}\left(1 + \frac{q^{s_2}}{1-q^{s_2}} + \frac{q^{s_3}}{(1-q^{s_2})(1-q^{s_3})} + \cdots\right)$$

$$= \frac{1}{(1-q^{s_1})(1-q^{s_2})}\left(1 + \frac{q^{s_3}}{1-q^{s_3}} + \frac{q^{s_4}}{(1-q^{s_3})(1-q^{s_4})} + \cdots\right)$$

$$\vdots$$

2.8.12 Starting with the generating function

$$h(q) = 1 + q + q^2(1+q) + q^3(1+q)(1+q^2) + q^4(1+q)(1+q^2)(1+q^3) + \cdots$$

give a combinatorial interpretation to the string of identities

$$g(q) = (1+q)\left(1 + q^2 + q^3(1+q^2) + q^4(1+q^2)(1+q^3) + \cdots\right)$$

$$= (1+q)(1+q^2)\left(1 + q^3 + q^4(1+q^3) + q^5(1+q^3)(1+q^4) + \cdots\right)$$

$$\vdots$$

2.8.13 More generally, given a sequence $0 < s_1 < s_2 < \cdots$, with

$$s(q) = 1 + q^{s_1} + q^{s_2}(1+q^{s_1}) + q^{s_3}(1+q^{s_1})(1+q^{s_2}) + q^{s_4}(1+q^{s_1})(1+q^{s_2})(1+q^{s_3}) + \cdots$$

give a combinatorial interpretation to the string of identities

$$s(q) = (1 + q^{s_1}) \left(1 + q^{s_2} + q^{s_3}(1 + q^{s_2}) + q^{s_4}(1 + q^{s_2})(1 + q^{s_3}) + \cdots \right)$$
$$= (1 + q^{s_1})(1 + q^{s_2}) \left(1 + q^{s_3} + q^{s_4}(1 + q^{s_3}) + q^{s_5}(1 + q^{s_3})(1 + q^{s_4}) + \cdots \right)$$

$$\vdots$$

2.8.14 Show that the number of partitions of n is equinumerous with the number of partitions of $2n$ into exactly n parts.

2.8.15 Verify that conjugation gives a bijection between the following two classes of partitions:

P_1 : no part appears exactly once,
P_2 : no two consecutive integers appear as parts and each part is > 1.

(a) Show that the generating function of P_1 is given by

$$\prod_{m \geq 1} \frac{1 + q^{3m}}{1 - q^{2m}}.$$

(b) Put

D : partitions with distinct parts all divisible by 3,
E : partitions with even parts.

Interpret your result in part (a) as a partition identity involving the classes of partitions P_1, D, and E.
(c) Construct a weight-preserving bijection between P_1 and $D \times E$.
(d) Show that

$$\prod_{m \geq 1} \frac{1 + q^{3m}}{1 - q^{2m}} = \prod_{m \geq 1} \frac{1}{(1 - q^{6m})(1 - q^{6m-2})(1 - q^{6m-3})(1 - q^{6m-4})}.$$

Interpret this equation as a partition identity.

2.8.16 Use Ferrers diagrams to give a combinatorial interpretation of the identity

$$\frac{q^{m^2}}{(1-q)(1-q^2)\cdots(1-q^m)} = (1+q)(1+q^2)\cdots(1+q^m)\frac{q^{m^2}}{(1-q^2)(1-q^4)\cdots(1-q^{2m})}.$$

2.8.17 Show that

$$\prod_{m \geq 0} \frac{1}{1 - q^{5m+1}} = 1 + \sum_{m \geq 1} \frac{q^m}{(1 - q^5)(1 - q^{10})\cdots(1 - q^{5m})}.$$

2.8.18 Show that

$$\prod_{m\geq 0}\frac{1}{1-q^{5m+4}} = 1 + \sum_{m\geq 1}\frac{q^{4m}}{(1-q^5)(1-q^{10})\cdots(1-q^{5m})}.$$

2.8.19 By classifying partitions according to the length of their longest hook, show that

$$\prod_{m\geq 1}\frac{1}{1-xq^m} = 1 + \sum_{r,s\geq 0}\begin{bmatrix} r+s \\ s \end{bmatrix} q^{r+s+1}x^{r+1}.$$

2.8.20 Given $d \geq 1$, construct the generating function of partitions whose parts differ by at least d.

2.8.21 Show that the number of partitions of n into odd parts each of which is strictly larger than 1 equals the number of partitions of n into distinct parts in which no part is a power of 2.

2.8.22

(a) Show that the generating function of partitions of n in which only the odd parts may be repeated is

$$\prod_{m\geq 1}\frac{1+q^{2m}}{1-q^{2m-1}}.$$

(b) Show that the generating function of partitions of n in which a part can appear at most three times is

$$\prod_{m\geq 1}\left(1+q^m+q^{2m}+q^{3m}\right).$$

(c) Show that these classes of partitions of n are equinumerous.

2.8.23 Show that the number of partitions of n in which each part appears at most three times is equinumerous with the number of partitions of n whose parts are congruent to 1 or 2 modulo 3.

2.8.24 Construct the generating function of partitions whose longest hook consists of m cells.

2.8.25 Show that

$$\frac{x(2x+1)}{(1-x)^3}$$

is the generating function of pentagonal numbers.

2.8.26 Show that the set of numbers of the form

$$\frac{r^2 - 1}{24},$$

where r runs over those integers relatively prime to 6 is the union of the set of pentagonal numbers and the set of second pentagonal numbers.

2.8.27 Set $w_0 = 1$ and define w_n as the difference between the number of partitions of n into an even number of parts, and the number of partitions of n into an odd number of parts. For example, $w_1 = -1$, $w_2 = 0$, $w_3 = -1$, $w_4 = 1$. Put together the generating function of $\{w_n\}$.

2.8.28 Show that the left side of the identity (2.61) is the generating function of partitions with smallest part equal to 2 and whose parts differ by at least two.

2.8.29 Jacobi's triple product identity is

$$\sum_{n=-\infty}^{\infty} z^n q^{n^2} = \prod_{n \geq 0} (1 - q^{2n+2})(1 + zq^{2n+1})(1 + z^{-1}q^{2n+1})$$

which holds for $z \neq 0$ and $|q| < 1$.

(a) Show that Euler's recursion for $P(n)$ follows from this identity.
(b) Use Jacobi's triple product identity to prove Gauss's theorem:

$$\sum_{n=-\infty}^{\infty} (-1)^n q^{n^2} = \prod_{m \geq 1} \frac{(1 - q^m)}{(1 + q^m)}.$$

2.8.30 Let $a(n)$ be the number of partitions of n into parts congruent to 2 modulo 4. So $a(n)$ is the number of partitions of $n/2$ into odd parts if n is even and zero if n is odd. Use the Jacobi triple product identity to show that

$$P(n) - P(n-1) - P(n-3) + P(n-6) + P(n-10) - P(n-15) - P(n-21) + \cdots = a(n),$$

where the indices $1, 6, 15, 28, \ldots$ that appear are the hexagonal numbers $m(2m-1)$ and $3, 10, 21, 26, \ldots$ are the second hexagonal numbers $m(2m+1)$. In particular for n odd

$$P(n) = \sum_{m \geq 1} (-1)^{m+1} \left(P(n - m(2m-1)) + P(n - m(2m+1)) \right)$$

2.8.31 Berger's identity is

$$\prod_{m\geq 1}(1-q^{rm})(1-q^{rm-1})(1-q^{r(m-1)+1}) = \sum_{n=-\infty}^{\infty}(-1)^n q^{\frac{r}{2}n(n-1)+n} .$$

(a) The exponents of q on the right are what are called polygonal numbers. What is the interpretation of this identity for $r = 3$?

(b) Divide through by the left hand side and compare coefficients to derive an identity satisfied by $P(n)$.

2.8.32 Heptagonal numbers are polygonal numbers of the form $m(5m-3)/2$. Show that

$$P(n) + \sum_{m\geq 1}(-1)^m \left(P(n - \tfrac{m(5m-3)}{2}) + P(n - \tfrac{m(5m+3)}{2})\right)$$

is equal to the number of partitions of n into parts that are congruent to 2 or 3 modulo 5.

2.8.33 Let

$$\prod_{m\geq 1}\frac{1}{1+q^m} = \sum_{n\geq 0} w_n q^n .$$

Show that for every $n \geq 1$, the number of partitions $P(n)$ satisfies the recursion

$$P(2n + 1) = -w_1 P(2n) - w_2 P(2n - 1) - \cdots - w_{2n+1} P(0),$$

$$P(2n) = P(n) - w_1 P(2n - 1) - w_2 P(2n - 2) - \cdots - w_{2n} P(0) .$$

2.8.34 A *composition* of n is an ordered partition of n. Consider compositions of n where each part is a nonzero pentagonal number, i.e., each part is of the form $\frac{m(3m-1)}{2}$ for $m = \pm 1, \pm 2, \ldots$. The sign of a part is defined to be $(-1)^{m+1}$, and the sign $sign(C)$ of a composition C is defined to be the product of the signs of its parts. For example, $n = 5$ has the following pentagonal compositions along with their signs:

$$-(5), +(2, 2, 1), +(2, 1, 2), +(1, 2, 2), +(2, 1, 1, 1),$$
$$+ (1, 2, 1, 1), +(1, 1, 2, 1), +(1, 1, 1, 2), +(1, 1, 1, 1, 1) .$$

Show that

$$P(n) = \sum_{C} sign(C),$$

where the sum is over all pentagonal compositions of n.

2.8.35 For a positive integer k, define $\sigma(k)$ to be the sum of the positive integral divisors of k. Use logarithmic derivative in (2.7) to show that the number of partitions $P(n)$ of n is given by

$$P(n) = \frac{1}{n} \sum_{k=1}^{n} \sigma(k) P(n-k).$$

2.8.36 Give a combinatorial proof of the following identity:

$$1 + \sum_{k=1}^{n} \frac{q^k}{(1-q)(1-q^2)\cdots(1-q^k)} = \frac{1}{(1-q)(1-q^2)\cdots(1-q^n)}.$$

2.8.37

(a) Make use of Euler's theorem to show that the number of partitions of n into distinct even parts is equal to the number of partitions of n whose parts are congruent to 2 modulo 4.

(b) Derive the identity

$$\prod_{m \geq 1} (1 + q^{2m}) = \prod_{m \geq 1} \frac{1}{1 - q^{4m-2}}.$$

2.8.38 Let $A(n, m)$ denote the generating function of Ferrers diagrams that fit into the following angle:

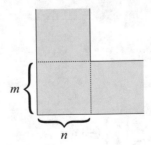

(a) Show that $A(n, m)$ satisfies the recursion

$$A(n, m) = A(n, m-1) + \frac{q^{m(n+1)}}{(1-q)(1-q^2)\cdots(1-q^n)(1-q)(1-q^2)\cdots(1-q^m)}.$$

(b) Calculate $A(1, m)$.

(c) In general, prove the formula

$$A(n, m) = \frac{1}{(1 - q)(1 - q^2) \cdots (1 - q^n)} \sum_{k=0}^{m} \frac{q^{k(n+1)}}{(1 - q)(1 - q^2) \cdots (1 - q^k)}.$$

(d) Derive the identity

$$\frac{1}{(1 - q)(1 - q^2) \cdots (1 - q^n)} \sum_{k=0}^{m} \frac{q^{k(n+1)}}{(1 - q)(1 - q^2) \cdots (1 - q^k)} =$$

$$\frac{1}{(1 - q)(1 - q^2) \cdots (1 - q^m)} \sum_{k=0}^{n} \frac{q^{k(m+1)}}{(1 - q)(1 - q^2) \cdots (1 - q^k)}.$$

(e) In part (d) let $m \to \infty$ to obtain

$$\frac{1}{(1 - q)(1 - q^2) \cdots (1 - q^n)} \sum_{k \geq 0} \frac{q^{k(n+1)}}{(1 - q)(1 - q^2) \cdots (1 - q^k)} = \prod_{k \geq 1} \frac{1}{1 - q^k}.$$

(For $n = 0$ this is the classification of Ferrers diagrams by the number of parts).

2.8.39 Put

$$F(x, q) = \sum_{m \geq 0} \frac{q^m x^m}{(1 - q)(1 - q^2) \cdots (1 - q^m)}.$$

Show that

$$\lim_{q \to 1} F((1 - q)x, q) = e^x,$$

$$\lim_{q \to 1} F((1 - q^{-1})x, q) = e^{-x}.$$

2.8.40 Prove the identity

$$\sum_{k=0}^{n} \begin{bmatrix} n \\ k \end{bmatrix} (-1)^k q^{\frac{k(k-1)}{2}} = 0$$

for $n > 0$.

2.8.41 In the discussion leading up to Theorem 2.5.1 we transposed two nonadjacent symbols σ_i and σ_{i+d} in a permutation by a sequence of $2d - 1$ adjacent transpositions. However, there may be other sequences of adjacent transpositions to achieve this. Show that any one of these necessarily involves an odd number of adjacent transpositions.

2.8.42 Show that if σ is the permutation on $A \cup B$ defined by the values of the permutation τ on A and the permutation ρ on B with $A \cap B = \emptyset$, then $\text{sign}(\sigma) = \text{sign}(\tau)\text{sign}(\rho)$.

2.8.43 Give a formal proof of Theorem 2.5.1.

2.8.44 Show that the Gaussian polynomials satisfy

(a) $\begin{bmatrix} n \\ k \end{bmatrix} = \frac{1-q^n}{1-q^{n-k}} \begin{bmatrix} n-1 \\ k \end{bmatrix}$

(b) $\begin{bmatrix} n \\ k \end{bmatrix} = \begin{bmatrix} n-1 \\ k \end{bmatrix} + q^{n-k} \begin{bmatrix} n-1 \\ k-1 \end{bmatrix}$

(c) Use part (b) to prove the q-binomial theorem by induction on n.
(d) Prove by induction on n (or combinatorially) that

$$\begin{bmatrix} n+m+1 \\ m+1 \end{bmatrix} = \sum_{j=0}^{n} q^j \begin{bmatrix} m+j \\ m \end{bmatrix}.$$

2.8.45 Use properties of Gaussian polynomials and induction on n to prove the the identity

$$\prod_{k=1}^{n}(1+q^k x^{-1})(1+q^{k-1}x) = \sum_{k=-n}^{n} \begin{bmatrix} 2n \\ n+k \end{bmatrix} q^{\frac{k(k-1)}{2}} x^k.$$

2.9   Sample Quiz for Chapter 2

1. Let E denote the collection of Ferrers diagrams produced by the following expression:

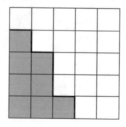

   (a) Construct all Ferrers diagrams with a total of 8 cells in E.
   (b) Give three Ferrers diagrams in E that are self-conjugate.
   (c) Construct the generating function of E.

2. Construct the generating function of partitions whose parts are divisible by 3.
3. Show that the Gaussian polynomials satisfy

$$\begin{bmatrix} n \\ k \end{bmatrix} = \begin{bmatrix} n-1 \\ k \end{bmatrix} + q^{n-k} \begin{bmatrix} n-1 \\ k-1 \end{bmatrix}$$

4. (a) Construct the word over the alphabet $\{1, 2\}$ which encodes the following partition in the 5×5 square:

   (b) Construct the generating function of all partitions that fit in this square.
5. Consider the following permutation σ:

$$\sigma = \begin{pmatrix} 1\ 2\ 3\ 4\ 5\ 6\ 7\ 8 \\ 3\ 1\ 2\ 5\ 4\ 7\ 6\ 8 \end{pmatrix}$$

   (a) Construct the inversion table of σ.
   (b) Calculate the number of inversions of σ.
   (c) Find the rank of σ.

Chapter 3
Planar Trees and the Lagrange Inversion Formula

3.1 Planar Trees

The trees we shall be concerned with here will be trees drawn on a plane. A few of them are shown in Figure 3.1.

Fig. 3.1 Examples of planar trees.

For such trees we have a very colorful terminology that is best introduced by examples (see Figure 3.2).

For a planar tree the number of branches *emanating upwards* from a given node will be called the *degree* of that node. For instance, the root of the tree in Figure 3.3 is a node of degree 4.

Given a node a the nodes which are immediately above a on the tree will be called the *children* of a or the *successors* of a as shown in Figure 3.4.

We see that for a planar tree, the degree of a node is nothing but the number of its children.

© Springer Nature Switzerland AG 2021
Ö. Eğecioğlu, A. M. Garsia, *Lessons in Enumerative Combinatorics*, Graduate
Texts in Mathematics 290, https://doi.org/10.1007/978-3-030-71250-1_3

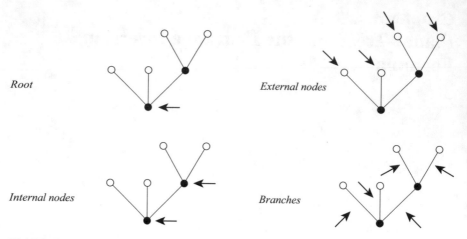

Root   *External nodes*

Internal nodes   *Branches*

Fig. 3.2 Some nomenclature: terminology of planar trees.

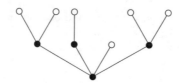

Fig. 3.3 The root of this tree has degree 4.

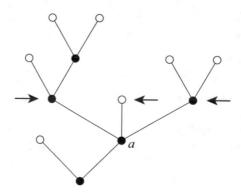

Fig. 3.4 The indicated nodes are the successors (children) of the node a.

3.1.1 Depth First Order

Drawing these trees on a plane enables us to *total order* the nodes of each of these trees in an unambiguous manner. This is done as follows.

First of all, given a planar tree T and its root a, let us call $a_1, a_2, \ldots, a_k$ the successors of a from "left" to "right" as depicted in Figure 3.5.

Secondly, let us denote by $T_1, T_2, \ldots, T_k$ the subtrees of T, respectively, stemming from $a_1, a_2, \ldots, a_k$. We shall sometimes refer to these subtrees as the

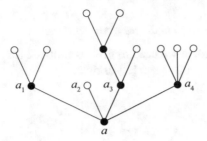

Fig. 3.5 Successors of the root a.

principal subtrees of T. So for the tree in Figure 3.5, we have the principal subtrees in Figure 3.6. This given, the order of T is defined recursively as follows

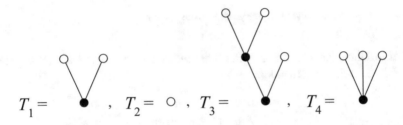

Fig. 3.6 Principal subtrees of the tree in Figure 3.5.

$$order\ T\ =\ root\ T,\ order\ T_1,\ order\ T_2, \ldots,\ order\ T_k\ .$$

We shall refer to this as the *Depth First Order* of T (abbreviated to DFO).

In other words, in the DFO of T its root comes first, then come all the nodes of T_1 in the DFO of T_1, then come all the nodes of T_2 in the DFO of T_2, etc. This is best illustrated by the examples in Figure 3.7. Trees may be stored in a computer by

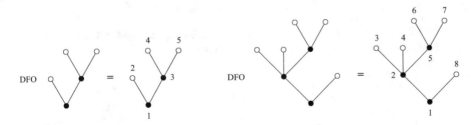

Fig. 3.7 Depth first order.

giving all father and child relationships. The special features that distinguish *planar* trees from other trees are the following:

1. *There is a root,*
2. *The children of any given node are totally ordered.*

So to store a planar tree we need to give its root and for each internal node, list its children in the given order from left to right.

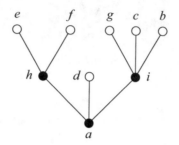

Fig. 3.8 Tree to be stored with arbitrary labels given to the nodes.

For instance, the tree in Figure 3.8 (here the nodes are labeled at random) can be stored in the following manner:

$$
\begin{array}{rc}
Root: & a \\
Parent & Children \\
a & h, d, i \\
h & e, f \\
i & g, c, b
\end{array}
$$

3.1.2   Tree Multiplication

We can define a useful form of multiplication for trees as follows. If $T_1, T_2, \ldots, T_k$ are given planar trees, we shall denote by

$$T_1 \times T_2 \times \cdots \times T_k$$

the tree T whose principal subtrees are $T_1, T_2, \ldots, T_k$, and we refer to T as the *product* of $T_1, T_2, \ldots, T_k$. This is shown in Figure 3.9.

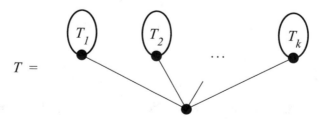

Fig. 3.9 The product of $T_1, T_2, \ldots, T_k$: $T = T_1 \times T_2 \times \cdots \times T_k$.

Thus, for instance, we have the products in Figure 3.10.

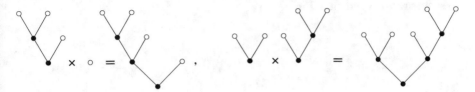

Fig. 3.10 Examples of tree multiplication.

3.1.3   The Word of a Tree

To each tree T we shall associate a word in the letters

$$x_0, x_1, x_2, x_3, \ldots.$$

This word, which is denoted by $w(T)$, is recursively defined in the following manner:

1. If T is the trivial tree "o," then

$$w(o) = x_0 .$$

2. If T is not trivial and $T_1, T_2, \ldots, T_k$ are the principal subtrees of T, then

$$w(T) = x_k w(T_1) w(T_2) \cdots w(T_k) . \tag{3.1}$$

An example of this is given in Figure 3.11.

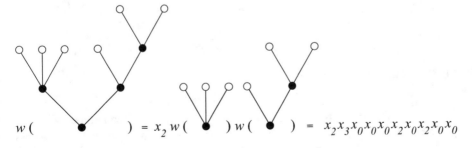

Fig. 3.11 Word of a tree.

Another way of doing this is to label each node of degree i by x_i, then read all these labels in the DFO, as in Figure 3.12. Thus this two-step procedure gives again the same word as shown in Figure 3.12.

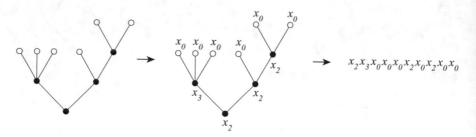

Fig. 3.12 Word of a tree using DFO.

Remark 3.1.1 We should note the following important fact. Namely

$$w(T_1 \times T_2 \times \cdots \times T_k) = x_k w(T_1) w(T_2) \cdots w(T_k) .$$

3.2 Planar Binary Trees

A planar binary tree, briefly a *binary tree* is simply a planar tree whose nodes have degrees 0 or 2 only. Examples of binary trees are shown in Figure 3.13.

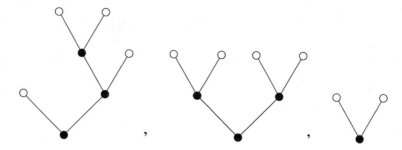

Fig. 3.13 Examples of binary trees.

Our first task will be to enumerate the binary trees with a given number of nodes.

Note that if $I(T)$ and $E(T)$ denote, respectively, the number of *internal* and *external* nodes of T, then we have the relation

$$E(T) = I(T) + 1 . \tag{3.2}$$

This is easily shown by induction of the number of internal nodes of T. Indeed, (3.2) is clearly true for the trivial tree o, since in that case

$$E(o) = 1, \quad I(o) = 0 .$$

Let us assume that (3.2) is true for all trees with fewer than n internal nodes. Let T be a tree with n internal nodes and let T_1 and T_2 be the principal subtrees of T. Since T_1 and T_2 will have fewer than n internal nodes, we shall have

$$E(T_1) = I(T_1) + 1 \quad \text{and} \quad E(T_2) = I(T_2) + 1.$$

This gives

$$E(T) = E(T_1 \times T_2) = E(T_1) + E(T_2) = I(T_1) + 1 + I(T_2) + 1 = I(T) + 1 .$$

Thus (3.2) must hold for all binary trees with n nodes as well, and the induction is complete.

This given, we shall enumerate binary trees by the number of internal nodes. More precisely, let c_n denote the number of binary trees with n internal nodes. Our goal is to obtain a formula which will enable us to calculate c_n for any value of n.

To this end, let us denote by C the formal sum of all binary trees as shown in Figure 3.14.

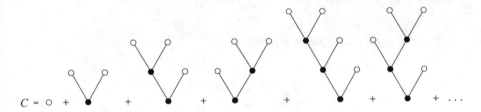

Fig. 3.14 Formal sum of all binary trees.

Let us also denote by C_n the formal sum of all binary trees with n internal nodes. A few of these are shown in Figure 3.15.

Clearly we have $c_n = |C_n|$ (cardinality of C_n). Note also that we have

$$C_{n+1} = \sum_{k=0}^{n} C_k \times C_{n-k} . \tag{3.3}$$

In particular we see that

$$C_3 = C_0 \times C_2 + C_1 \times C_1 + C_2 \times C_0$$

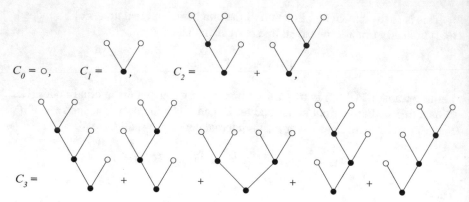

$$C_0 = o, \qquad C_1 = , \qquad C_2 = + , $$

$$C_3 = + + + + $$

Fig. 3.15 Formal sum of binary trees by internal nodes.

or in pictures, as shown in Figure 3.16.

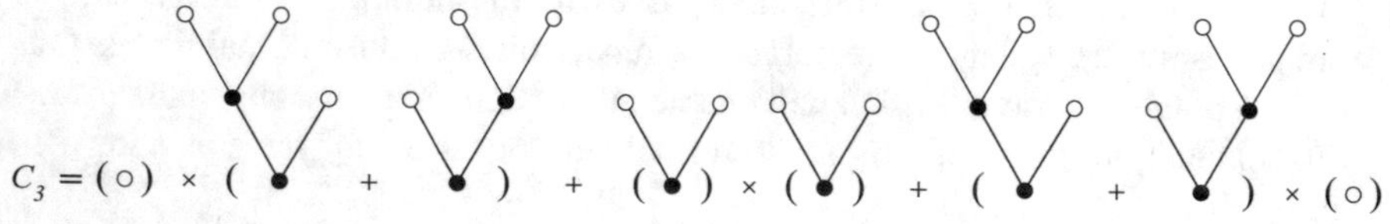

$$C_3 = (o) \times (+) + () \times () + (+) \times (o)$$

Fig. 3.16 The decomposition of C_3.

Thus (3.3) gives the recurrence relation

$$c_{n+1} = \sum_{k=0}^{n} c_k c_{n-k}, \qquad (c_0 = 1) .$$

This recurrence is nonlinear and as such it is difficult to solve directly. The alternate idea is to work with the identity

$$C = o + C \times C . \tag{3.4}$$

This simply expresses the fact that a binary tree T is either trivial or it splits uniquely into the product of a left subtree T_1 times a right subtree T_2.

Note that the generating function of c_n is given by the expression

$$C(x) = \sum_{n \geq 0} c_n x^n = \sum_{T \in C} x^{I(T)} . \tag{3.5}$$

In other words, to obtain $C(x)$ we just replace each tree T in Figure 3.14 by the corresponding monomial $x^{I(T)}$. Now since $T = T_1 \times T_2$ implies that

$$x^{I(T)} = x \cdot x^{I(T_1)} x^{I(T_2)} \, ,$$

the relation (3.4) yields the equation

$$C(x) = 1 + xC(x)C(x)$$

or better

$$x(C(x))^2 - C(x) + 1 = 0 \, .$$

Using the familiar formula for the quadratic equation gives

$$C(x) = \frac{1 + \sqrt{1 - 4x}}{2x} \quad \text{or} \quad C(x) = \frac{1 - \sqrt{1 - 4x}}{2x} \, . \tag{3.6}$$

Now, since $c_0 = 1$, we deduce from (3.5) that

$$C(0) = 1.$$

This excludes the first of the two possibilities in (3.6), and we get

$$C(x) = \frac{1 - \sqrt{1 - 4x}}{2x} \, . \tag{3.7}$$

Note that since

$$\frac{1 - \sqrt{1 - 4x}}{2x} = \frac{2}{1 + \sqrt{1 - 4x}} \, ,$$

$C(x)$ is indeed equal to 1 when $x = 0$ as desired.

We shall next use (3.7) to obtain a power series expansion for $C(x)$. To this end we recall Newton's expansion formula

$$(1 + x)^\alpha = 1 + \sum_{n \geq 1} \binom{\alpha}{n} x^n \, ,$$

where

$$\binom{\alpha}{n} = \frac{\alpha(\alpha - 1) \cdots (\alpha - n + 1)}{n!} \, .$$

For $\alpha = \frac{1}{2}$ we have

$$\binom{\frac{1}{2}}{n} = \tfrac{1}{2}(\tfrac{1}{2} - 1) \cdots (\tfrac{1}{2} - n + 1)\frac{1}{n!} \ .$$

To simplify this expression we note that

$$
\begin{aligned}
\tfrac{1}{2}(\tfrac{1}{2} - 1) \cdots (\tfrac{1}{2} - n + 1)\frac{1}{n!} &= (-1)^{n-1}\frac{(2-1)(4-1)\cdots(2(n-1)-1)}{2^n n!} \\
&= (-1)^{n-1}\frac{1 \cdot 3 \cdot 5 \cdots (2n-3)}{2^n n!} \\
&= (-1)^{n-1}\frac{1 \cdot 2 \cdot 3 \cdot 4 \cdot 5 \cdots (2n-2)}{2^{2n-1} n!(n-1)!} \\
&= (-1)^{n-1}\binom{2n-1}{n}\frac{2}{4^n(2n-1)} \ .
\end{aligned}
$$

This gives

$$(1+x)^{\frac{1}{2}} = 1 - 2\sum_{n \geq 1}(-1)^n\binom{2n-1}{n}\frac{x^n}{4^n(2n-1)} \ .$$

Replacing x by $-4x$ then yields

$$\sqrt{1 - 4x} = 1 - 2\sum_{n \geq 1}\binom{2n-1}{n}\frac{x^n}{2n-1} \ .$$

Substituting in (3.7) we finally obtain

$$C(x) = \sum_{n \geq 1}\binom{2n-1}{n}\frac{x^{n-1}}{2n-1} = \sum_{n \geq 0}\binom{2n+1}{n}\frac{x^n}{2n+1} \ ,$$

from which we deduce that

$$c_n = \frac{1}{2n+1}\binom{2n+1}{n} \ . \tag{3.8}$$

Using this formula we see that

$$c_0 = 1, \ c_1 = 1, \ c_2 = 2, \ c_3 = 5, \ c_4 = 14, \ c_5 = 42, \ldots$$

which agrees with our previous constructions.

3.3   Ternary Trees

The next simplest class of trees is that consisting of trees with nodes of degree 0 or 3 only. We shall call these *ternary trees* (see Figure 3.17).

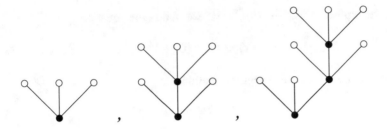

Fig. 3.17   Examples of ternary trees.

For convenience, let us again denote by C the formal sum of all ternary trees as shown in Figure 3.18.

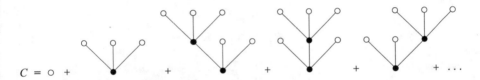

Fig. 3.18   Formal sum of ternary trees.

In this case it is not immediately clear what the relationship is between the numbers of *internal* and *external* nodes. Thus we shall use two variables in our generating function and we shall set

$$C(x) = \sum_{T \in C} t^{E(T)} x^{I(T)} . \qquad (3.9)$$

This gives

$$C(x) = t + t^3 x + 3t^5 x^2 + \cdots .$$

It is good to think of this series as a formal power series in x with coefficients that are polynomials in t. Let us organize the terms in (3.9) according to increasing powers of x and write

$$C(x) = t + \sum_{n \geq 1} c_n(t) x^n . \qquad (3.10)$$

Our goal is to give a formula for the polynomials $c_n(t)$.

To this end, we can follow the same method we used for binary trees up to a point. Indeed, we have for the same reason as before the formal series identity

$$C = o + C \times C \times C . \tag{3.11}$$

Making the replacement $T \to t^{E(T)} x^{I(T)}$ we obtain the equation

$$C(x) = t + x(C(x))^3. \tag{3.12}$$

Thus we are to solve the third degree equation

$$x(C(x))^3 - C(x) + t = 0 .$$

Rather than using the standard form for the solution of the cubic, which is not very familiar to most people, we shall show how to solve a general class of equations which occur in many other combinatorial situations. This is a very beautiful but not too well known result of Lagrange [20] which can be stated as follows:

Theorem 3.3.1 (Lagrange Inversion Formula) *Let $R(x)$ be the formal power series*

$$R(x) = R_0 + R_1 x + R_2 x^2 + \cdots \tag{3.13}$$

and let

$$f(x) = f_1 x + f_2 x^2 + f_3 x^3 + \cdots$$

be the formal power series solution of the equation

$$f(x) = x R(f(x)) . \tag{3.14}$$

Then

$$f_n = \frac{1}{n} (R(x))^n \Big|_{x^{n-1}} \tag{3.15}$$

(where here $\Big|_{x^n}$ denotes the operation of taking the coefficient of x^n in a series).

This result is best understood and appreciated if we use it to solve some special equations. Let us start with equation (3.12). To reduce this equation to the form (3.14), we set

$$f(x) = C(x) - t$$

and (3.12) becomes

$$f(x) = x(t + f(x))^3 \tag{3.16}$$

which is precisely of the form (3.14) with

$$R(x) = (t + x)^3 = t^3 + 3t^2x + 3tx^2 + x^3 .$$

Thus we have (3.13) with

$$R_0 = t^3, \quad R_1 = 3t^2, \quad R_2 = 3t, \quad R_3 = 1 \quad (R_n = 0 \text{ for } n \geq 4) .$$

Formula (3.15) then gives in this case

$$f_n = \frac{1}{n}(t + x)^{3n} \bigg|_{x^{n-1}} .$$

Using the binomial theorem we get

$$f_n = \frac{1}{n} \sum_{k=0}^{3n} \binom{3n}{k} x^k t^{3n-k} \bigg|_{x^{n-1}} = \frac{1}{n} \binom{3n}{n-1} t^{3n-(n-1)} .$$

This gives

$$C(x) = t + \sum_{n \geq 1} \frac{1}{n} \binom{3n}{n-1} t^{2n+1} x^n$$

and we find that the polynomial $c_n(t)$ is given by

$$c_n(t) = \frac{1}{n} \binom{3n}{n-1} t^{2n+1} .$$

Finally we derive the following two facts:

1. The number of ternary trees with n internal nodes is

$$\frac{1}{n} \binom{3n}{n-1} . \tag{3.17}$$

2. A ternary tree with n internal nodes has $2n + 1$ external nodes.

The reader may verify by experimentation that this is indeed the case.

We note that formula (3.17) for the number of ternary trees with n internal nodes can be written in the alternate form

$$\frac{1}{3n+1}\binom{3n+1}{n},$$

indicating a certain pattern going from binary trees to ternary trees in view of the formula in (3.8).

3.4   Lattice Path Representation for Planar Trees

The construction of the word $w(T)$ of a given tree we gave in Section 3.1 can be reversed and the tree T can be reconstructed from $w(T)$. Consequently the problem of enumerating planar binary trees may thus be transformed into the problem of counting *tree words*. However, this leads to a new question:

What words are planar tree words?

We can give this question a beautiful answer by means of a graphical representation of planar trees by lattice paths. We recall that a *lattice path* is a polygonal path in the x, y-plane whose vertices are the lattice points (i, j) with integer coordinates. Given a tree T we construct a lattice path $P(T)$ using the successive letters of $w(T)$ as a guide in the following manner.

We start at the origin and proceed in steps obtained by replacing each letter in $w(T)$ by a vector according to the following rule

$$x_i \;\to\; (1, i-1). \tag{3.18}$$

This is best illustrated by an example. Consider the tree T in Figure 3.19.

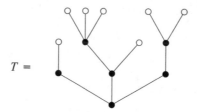

$T =$

Fig. 3.19 A planar tree T.

Thus

$$w(T) = x_3 x_1 x_0 x_2 x_3 x_0 x_0 x_0 x_0 x_1 x_2 x_0 x_0$$

and the replacements

$$x_0 \to (1, -1), \; x_1 \to (1, 0), \; x_2 \to (1, 1), \; x_3 \to (1, 2)$$

yield the path in Figure 3.20.

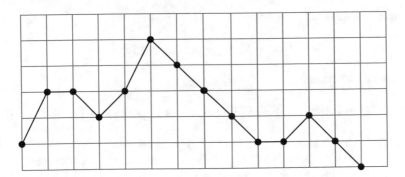

Fig. 3.20 The path $P(T)$ of the tree T in Figure 3.19.

It will be good to use the letter n to denote the total number of nodes in a tree. This given, we notice two properties of $P(T)$ neither of which is accidental:

(a) *the path ends at the point* $(n, -1)$, *(here* $n = 13$)   (3.19)

(b) *all but the last edge of* $P(T)$ *lies "above" the* $x - axis$.

Indeed we have the following remarkable theorem.

Theorem 3.4.1 *A word* $w = x_{k_1} x_{k_2} \cdots x_{k_n}$ *is a tree word if and only if the path* P *corresponding to it by the above construction has properties a) and b) of (3.19).*

Proof We shall only give a brief idea of why this theorem is true. The reader can convince himself much better by working out a few examples.

First of all we note that property (a) of (3.19) is an immediate consequence of the Euler formula for trees. Indeed, let $e(T)$ and $n(T)$ denote, respectively, the number of edges and nodes of T. Euler's identity (easily proved by induction) is

$$e(T) = n(T) - 1 .$$ (3.20)

On the other hand, let

$$w(T) = x_{k_1} x_{k_2} \cdots x_{k_n} .$$

Since we make our replacements according to the rule

$$x_{k_i} \rightarrow (1, k_i - 1)$$

we derive that the final vertex of $P(T)$ is the point

$$(n, k_1 - 1 + k_2 - 1 + \cdots + k_n - 1) .$$

However, we can easily see that

$$k_1 + k_2 + \cdots + k_n = e(T) \, .$$

Thus using (3.20) we get

$$k_1 - 1 + k_2 - 1 + \cdots + k_n - 1 = e(T) - n(T) = -1 \tag{3.21}$$

as asserted. This shows that condition (3.19) (a) is necessary.

The necessity of (3.19) (b) is slightly more sophisticated. Nevertheless, we can prove it again using induction. More precisely, we show that if (3.19) (b) holds for every one of the principal subtrees of a given tree T, then it must hold for T as well. We shall illustrate the idea in the case that

$$T = T_1 \times T_2 \times T_3.$$

Assume then that $P(T_1)$, $P(T_2)$ and $P(T_3)$ have properties (a) and (b) in (3.19), and let

$$w(T) = x_3 w(T_1) w(T_2) w(T_3) \, .$$

This given we can easily see that the path $P(T)$ will necessarily have the general behavior depicted in Figure 3.21.

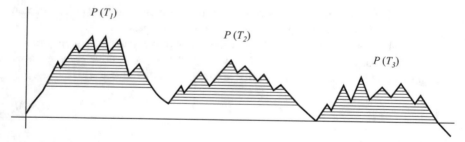

Fig. 3.21 The structure of $P(T)$.

Indeed note that this path must *factorize* into the juxtaposition of the initial step given by the vector $(1, 2)$ corresponding to the root of T, followed by the paths $P(T_1)$, $P(T_2)$, and $P(T_3)$ corresponding to the principal subtrees of T. Note now that the edge $(0, 0) \rightarrow (1, 2)$ brings the path $P(T)$ up to the level $y = 2$. After this step, the path $P(T_1)$ succeeds only in bringing $P(T)$ down to the level $y = 1$. Similarly, $P(T_2)$ succeeds only in bringing $P(T)$ down to level $y = 0$. Finally, it is $P(T_3)$ that brings $P(T)$ down to level $y = -1$. This forces $P(T)$ to lie above the x-axis all the way but the very end. This proves that $P(T)$ has property (3.19) (b). A look at the display in Figure 3.21 should convince the reader of the validity of property (3.19) (b) for all tree paths.

To complete our proof we must show that these conditions are also sufficient. That is, given a word w whose corresponding path $P(w)$ satisfies (3.19) (a) and (b) then there is a tree T whose word is w. Clearly this is so if w has only one letter. We can thus proceed by induction on the number of letters of w. Assume then that w has n letters and that the result is true for words with fewer than n letters.

For simplicity assume that the first letter of w is again x_3.

Our first step is to show that w can be factored in the form

$$w = x_3 w_1 w_2 w_3, \tag{3.22}$$

where each w_i is a tree word. This corresponds to factoring the path $P(w)$ into the initial step $(0,0) \rightarrow (1,2)$ followed by paths P_1, P_2, P_3 each satisfying the conditions in (3.19).

A look at Figure 3.21 should immediately reveal how this factorization may be achieved. Indeed, we simply let P_1, P_2, P_3 be the paths obtained by breaking up $P(w)$ just before it first descends to levels $-1, 0, 1$. This construction assures that the paths P_1, P_2, P_3 satisfy the conditions in (3.19). The induction hypothesis will then guarantee that the corresponding words w_1, w_2, w_3 will be tree words. This gives the desired factorization (3.22).

However, now we are done. Indeed, if T_1, T_2, T_3 are the trees corresponding to w_1, w_2, w_3, our desired tree T is simply the product

$$T = T_1 \times T_2 \times T_3 .$$

This completes our argument.   □

As we shall see, Theorem 3.4.1 not only will enable us to enumerate, purely combinatorially, all planar trees, but it will lead to a purely combinatorial proof of the Lagrange inversion formula itself. To see how this comes about, it is best to look again at the case of binary trees.

3.5 Combinatorial Enumeration of Binary Trees

For a binary tree T the path $P(T)$ has a very simple nature. Indeed, it consists only of edges with slopes 1 or -1. For example, when T is as in Figure 3.22, we have

$$w(T) = x_2 x_2 x_0 x_2 x_2 x_0 x_0 x_0 x_0 . \tag{3.23}$$

This gives the path in Figure 3.23.

Note that the word in (3.23) has $I(T) (= 4)$ x_2's and $E(T) (= 5)$ x_0's. Let $R(x_0^5 x_2^4)$ denote the set of all words which are rearrangements of the 9 letters $x_0^5 x_2^4$. Clearly, any word $w(T)$ which corresponds to a tree T with 5 external and

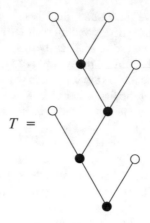

$T =$

Fig. 3.22 A binary tree T.

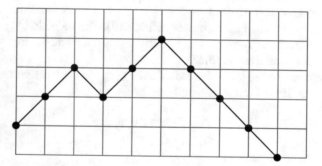

Fig. 3.23 The path $P(T)$ of the tree T in Figure 3.22.

4 internal nodes belongs to $R(x_0^5 x_2^4)$. Note that there are all together

$$\binom{9}{5}$$

words in $R(x_0^5 x_2^4)$. Indeed, to construct any one of them we need only choose the positions of the x_0's in the word. Note also that the path corresponding to any one of the words of $R(x_0^5 x_2^4)$ will necessarily have property $a)$ of (3.19). However, only a subset of $R(x_0^5 x_2^4)$ will also have property (3.19) (b).

We aim to find out the nature of this subset. For this purpose let us look again at an example. Let us pick at random a word with 4 x_2's and 5 x_0's. For instance,

$$w = x_2 x_0 x_0 x_2 x_2 x_2 x_0 x_0 x_0.$$

The path corresponding to this word is given in Figure 3.24.

Property (3.19) (b) does not hold for this path and we must conclude that w is not a tree word. However, let us extend this path by going back to the beginning of w and

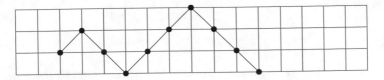

Fig. 3.24 The path $P(w)$ of the word $w = x_2 x_0 x_0 x_2 x_2 x_2 x_0 x_0 x_0$.

replacing letters by vectors as before. The resulting path is, of course, a juxtaposition of two identical copies of the previous path, as indicated in Figure 3.25.

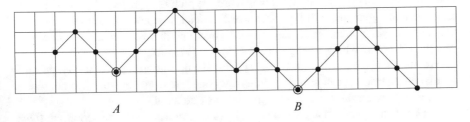

Fig. 3.25 The double path $PP(w)$ associated with $w = x_2 x_0 x_0 x_2 x_2 x_2 x_0 x_0 x_0$.

Let us call this the *double path* associated with w. Note the two pointed vertices in Figure 3.25. Call A the vertex on the left and B the one on the right. These two vertices can be located as follows:

(*a*) *B is the leftmost lowest point on the double path*,   (3.24)

(*b*) *A is the leftmost point on the double path that is exactly one level above B.*

Forget the rest of the double path and look at the portion P' that is between A and B. We can easily discern there a tree path! Indeed, this portion, by the definition of A and B must lie entirely above the level of A except for the last edge that must dip down to B. Note that P' is the path corresponding to the word

$$w' = x_2 x_2 x_2 x_0 x_0 x_0 x_2 x_0 x_0 .$$

Note further that this word can be obtained by breaking up the original word w after the third letter as indicated below

$$w = x_2 x_0 x_0 \mid x_2 x_2 x_2 x_0 x_0 x_0$$

and juxtaposing the resulting subwords in the reverse order. More precisely we have

$$w = w_1 w_2 \quad \text{and} \quad w' = w_2 w_1,$$

where

$$w_1 = x_2 x_0 x_0 \quad \text{and} \quad w_2 = x_2 x_2 x_2 x_0 x_0 x_0 .$$

A word w' obtained in this manner from w is called a *circular rearrangement* of w. Indeed, one way of obtaining such a word is to place the letters of w, say counterclockwise, on a circle, like beads on a necklace, then cut the necklace at some prescribed place. This imagery clearly indicates that there are exactly 9 circular rearrangements of w (including w itself). The crucial fact is that only one of these circular rearrangements corresponds to a tree word.

It is easy to convince ourselves after experimenting with a few examples that this holds true in general. That is, whatever word we pick in $R(x_0^5 x_2^4)$, only one of its circular rearrangements will be a tree word. Indeed, it is not difficult to see that in the corresponding double path there must be one (and there can be at most one) pair of points A and B which have properties (a) and (b) in (3.24).

We must thus conclude that $R(x_0^5 x_2^4)$ breaks up into groups each containing 9 words. Furthermore, each group contains only one tree word, and all the other words in the group are circular rearrangements of this tree word.

Thus we can count binary trees with 5 external nodes by simply counting these groups. However, these groups are easily counted. Indeed, suppose there are g of them all together. Since each of them contains 9 words, $9 \times g$ must be equal to the number of elements in $R(x_0^5 x_2^4)$. This gives

$$g = \frac{1}{9}\binom{9}{5} .$$

Clearly there is absolutely nothing special about the case we have studied. To enumerate binary trees with n internal and $n + 1$ external nodes we start of course with the class $R(x_0^{n+1} x_2^n)$ consisting of all words with $n + 1$ x_0's and n x_2's. We can easily discover that this class consists of groups of $2n + 1$ words each. The reason for this is that the only way two circular rearrangements of a word w can be identical is if w is of the form $w = u^p$ for some word u and an integer $p > 1$, and this is not possible if w has $n + 1$ x_0's and n x_2's.

In each of these groups of $2n + 1$ words, just like before, there is exactly one tree word, and each of the words in the group is a circular rearrangement of this tree word. Given a word w in $R(x_0^{n+1} x_2^n)$, we can construct the tree word w' in the group of w again by locating the points A and B on the double path of w according to the recipe in (3.24).

There are of course

$$\binom{2n + 1}{n}$$

words in $R(x_0^{n+1} x_2^n)$, and since each of the groups has $2n + 1$ words we must conclude that the number of tree words in $R(x_0^{n+1} x_2^n)$, or, which is the same, the number c_n of binary trees with n internal nodes, is given by the formula

$$c_n = \frac{1}{2n+1} \binom{2n+1}{n} .$$

This is in complete agreement with the formula we obtained in Section 3.2 by analytic methods.

The reader may notice that a very similar reasoning applies to the case of ternary trees as well. However, we shall not work with this further special case here since with only minor additions we will be in a position to enumerate much more general classes of planar trees.

3.6 A Combinatorial Proof of the Lagrange Inversion Formula

Let us denote by $\mathcal{T}(p_0, p_1, \ldots, p_N)$ the collection of all trees with p_0 nodes of degree 0, p_1 nodes of degree 1, p_2 nodes of degree 2, ..., p_N nodes of degree N. Our main goal here is to use the methods of the previous section to enumerate the class $\mathcal{T}(p_0, p_1, \ldots, p_N)$. As a by-product we shall obtain a purely combinatorial derivation of Theorem 3.3.1.

We shall use the same notation as in the previous sections. In particular, for a given $T \in T(p_0, p_1, \ldots, p_N)$ we let $w(T)$ and $P(T)$, respectively, denote the *word* of T and the *path* of T obtained by the constructions of Sections 3.1 and 3.4. It will also be useful to denote by $PP(w)$ the *double path* corresponding to the word w by the construction of Chapter 5.

Finally let us denote by $R(x_0^{p_0} x_1^{p_1} \cdots x_N^{p_N})$ the class of all words w which are rearrangements of the letters

$$x_0^{p_0} x_1^{p_1} \cdots x_N^{p_N} .$$

Clearly, for each $T \in \mathcal{T}(p_0, p_1, \ldots, p_N)$ the corresponding word $w(T)$ will be in $R(x_0^{p_0} x_1^{p_1} \cdots x_N^{p_N})$.

Here and in the following we shall use the letter n to denote the total number of nodes in a tree. That means that in $\mathcal{T}(p_0, p_1, \ldots, p_N)$ we shall have

$$n = p_0 + p_1 + \cdots + p_N . \tag{3.25}$$

This given, it is not difficult to see that Theorem 3.4.1 can be restated in a purely analytic manner as follows:

Theorem 3.6.1 *A word*

$$w = x_{k_1} x_{k_2} \cdots x_{k_n}$$

in $R(x_0{}^{p_0} x_1{}^{p_1} \cdots x_N{}^{p_N})$ *is the word of a tree in* $\mathcal{T}(p_0, p_1, \ldots, p_N)$ *if and only if*

$$(a) \quad k_1 + k_2 + \cdots + k_n = n - 1 , \tag{3.26}$$

$$(b) \quad k_1 + k_2 + \cdots + k_i > i - 1 \qquad (i = 1, 2, \ldots, n - 1) . \tag{3.27}$$

Proceeding as we did in Section 3.5 , we deduce that the collection

$$R(x_0{}^{p_0} x_1{}^{p_1} \cdots x_N{}^{p_N})$$

breaks up into groups each consisting of exactly n words. Furthermore, each group contains one and only one tree word, and all the other words in the group are circular rearrangements of this tree word.

The reader should verify, looking at some examples, that, even in this more general situation, if we pick a word w in $R(x_0{}^{p_0} x_1{}^{p_1} \cdots x_N{}^{p_N})$, the tree word in the group of w can again be constructed by locating on its double path $PP(w)$ the points A and B whose definition is given in (3.24).

This given, we can easily derive the following basic result:

Theorem 3.6.2 *The number of planar trees with n nodes of which p_0 are of degree 0, p_1 are of degree 1, ..., p_N are of degree N is*

$$\frac{1}{n} \frac{n!}{p_0! p_1! \cdots p_N!} . \tag{3.28}$$

Our next task is to put together a generating function for planar trees.

For simplicity, it will be good to place a bound on the degrees of the nodes of the trees we work with. Thus let C here be the collection of all planar trees *all of whose nodes are of degree less than or equal to some fixed integer N.*

Let us denote by $W(C)$ the formal sum of all words of trees in C. In symbols

$$W(C) = \sum_{T \in C} w(T) . \tag{3.29}$$

We have the following remarkable result:

Theorem 3.6.3 *The formal sum $W(C)$ in (3.29) satisfies the following algebraic equation:*

$$W(C) = x_0 + x_1 W(C) + x_2 (W(C))^2 + \cdots + x_N (W(C))^N . \tag{3.30}$$

Proof We simply recall that if a tree T is the product of k subtrees $T_1, T_2, \ldots, T_k$, then the word of T factorizes into the product

$$w(T) = x_k w(T_1) w(T_2) \cdots w(T_k) .$$

Moreover, if $T \in C$ then each factor $w(T_i)$ is a summand of $W(C)$. Conversely, every word obtained in this manner (for $k \leq N$) is necessarily one of the words coming from a tree T in C. This given, formula (3.30) simply expresses the fact that each tree, consisting of more than one node, factors uniquely into a product of $k \geq 1$ principal subtrees. Of course, the first term on the right hand side of (3.30) must be included to take care of the tree with no internal nodes.   □

This theorem will enable us to construct, purely combinatorially, series solutions to algebraic equations.

Let us associate to each tree T a monomial $m(T)$ which keeps track simultaneously of how many nodes T has all together, and of those how many are of degree 0, how many of degree 1, etc. More precisely, for a tree with n nodes and p_i nodes of degree i we set

$$m(T) = x^n R_0{}^{p_0} R_1{}^{p_1} \cdots R_N{}^{p_N} . \qquad (3.31)$$

Clearly, $m(T)$ can be obtained from $W(T)$ by simply replacing the variable x_i by $x R_i$ and rearranging the resulting monomial so that it is of the form given in (3.31). This given, let us set

$$f(x) = \sum_{T \in C} m(T) . \qquad (3.32)$$

It is good to view this expression as a formal series in x with coefficients that are polynomials in the variables $R_0, R_1, \ldots, R_N$. This is very similar to what we did in Section 3.3 for ternary trees.

From Theorem 3.6.3 we can immediately derive the following corollary:

Theorem 3.6.4 *The formal series $f(x)$ satisfies the equation*

$$f(x) = x R(f(x)), \qquad (3.33)$$

where

$$R(s) = R_0 + R_1 s + R_2 s^2 + \cdots + R_N s^N . \qquad (3.34)$$

Proof The equality in (3.30) is clearly preserved if we replace x_i by $x R_i$. However, when we do this, $W(C)$ simply becomes $f(x)$. Thus the formal series $f(x)$ satisfies the equation

$$f(x) = x R_0 + x R_1 f(x) + x R_2 (f(x))^2 + \cdots + x R_N (f(x))^N$$

and this is clearly the same as equation (3.33) (provided of course $R(x)$ is given by (3.34)). □

We are now in a position to give a combinatorial proof of the Lagrange inversion formula. To start, let C_n denote the subset of C consisting of all trees with n nodes. Let us set

$$M(C_n) = \sum_{T \in C_n} m(T) .$$

We might say that this polynomial "enumerates" C_n by (keeping track of) the number of nodes of various degrees. Of course, since each $T \in C_n$ has n nodes, each term in $M(C_n)$ will have x^n as a factor. Extracting this common factor we can write

$$M(C_n) = x^n Q(C_n) . \tag{3.35}$$

Now, we claim that we have

$$Q(C_n) = \frac{1}{n} \sum_{*} \frac{n!}{p_0! p_1! \cdots p_N!} R_0{}^{p_0} R_1{}^{p_1} \cdots R_N{}^{p_N}, \tag{3.36}$$

where the "$*$" is to indicate that the sum in (3.36) is *restricted* to the terms for which we have

(a)　$p_0 + p_1 + \cdots + p_N = n$,

(b)　$p_1 + 2p_2 + \cdots + N p_N = n - 1$.

To see why (3.36) holds, we note that a tree with n nodes, of which p_i are of degree i will contribute to $Q(C)$ a monomial

$$R_0{}^{p_0} R_1{}^{p_1} \cdots R_N{}^{p_N}$$

whose exponents will satisfy the conditions (a) and (b) given above. Indeed, condition (a) expresses the fact that the tree has n nodes and condition (b), as we observed before, is a simple consequence of Euler's formula.

Finally, Theorem 3.6.2 tells us that the number of trees contributing the monomial $R_0{}^{p_0} R_1{}^{p_1} \cdots R_N{}^{p_N}$ is equal to

$$\frac{1}{n} \frac{n!}{p_0! p_1! \cdots p_N!} .$$

Thus (3.36) must hold as asserted.

We claim next that we have the following identity

$$\sum_* \frac{n!}{p_0! p_1! \cdots p_N!} R_0{}^{p_0} R_1{}^{p_1} \cdots R_N{}^{p_N} = \left(R_0 + R_1 s + R_2 s^2 + \cdots + R_N s^N \right)^n \Big|_{s^{n-1}}.$$
(3.37)

To see this note that by the multinomial theorem we have

$$(R_0 + R_1 s + \cdots + R_N s^N)^n$$

$$= \sum_{p_0 + p_1 + \cdots + p_N = n} \frac{n!}{p_0! p_1! \cdots p_N!} (R_0)^{p_0} (s R_1)^{p_1} \cdots (s^N R_N)^{p_N} \qquad (3.38)$$

$$= \sum_{p_0 + p_1 + \cdots + p_N = n} \frac{n!}{p_0! p_1! \cdots p_N!} R_0{}^{p_0} R_1{}^{p_1} \cdots R_N{}^{p_N} s^{p_1 + 2p_2 + \cdots + N p_N}.$$

In this last expression the terms in which s appears raised to the power $n - 1$ are exactly those for which both conditions in (3.37) are satisfied. Thus the coefficient of s^{n-1} in this expression will yield precisely the left hand side of (3.37).

Combining all these observations, we obtain the following version of the Lagrange inversion formula.

Theorem 3.6.5 *The formal power series solution* $f(x)$ *of the equation*

$$f(x) = x \left[R_0 + R_1 f(x) + R_2 (f(x))^2 + \cdots + R_N (f(x))^N \right]$$

is given by the formula

$$f(x) = \sum_{n \geq 1} \frac{x^n}{n} (R_0 + R_1 s + R_2 s^2 + \cdots + R_N s^N)^n \Big|_{s^{n-1}}. \qquad (3.39)$$

Proof From the very definition of $f(x)$ we have

$$f(x) = \sum_{n \geq 1} M(C_n) .$$

But formulas (3.35), (3.36), and (3.37) combined give

$$M(C_N) = \frac{x^n}{n} \left(R_0 + R_1 s + \cdots + R_N s^N \right)^n \Big|_{s^{n-1}}.$$

This gives (3.39).   □

This result can be considerably extended to yield not only $f(x)$ but indeed any power of $f(x)$. We shall not give proofs but only state the result and give a brief idea of the combinatorial arguments involved.

To get this further extension, we work rather than with tree words, with words of the form

$$W = W(T_1)W(T_2) \cdots W(T_k) \qquad (3.40)$$

with $T_1, T_2, \ldots, T_k$ elements of C. We might say that these are words corresponding to *forests* with k trees.

We have the following characterization of these words:

Theorem 3.6.6 *A word*

$$W = x_{k_1} x_{k_2} \cdots x_{k_n}$$

is a k-forest word (i.e., it is of the form (3.40)) if and only if

(a) $k_1 + k_2 + \cdots + k_n = n - k$,

(b) $k_1 + k_2 + \cdots + k_i > i - k \qquad (i = 1, 2, \ldots, n - 1)$.

This is entirely analogous to Theorem 3.6.1, and it can be proved in exactly the same way.

Now, just as before we consider the collection $R(x_0{}^{p_0} x_1{}^{p_1} \cdots x_N{}^{p_N})$ consisting of all words which are rearrangements of the letters

$$x_0{}^{p_0} x_1{}^{p_1} \cdots x_N{}^{p_N} .$$

However, now we shall work with sets of exponents which satisfy the conditions

$$p_0 + p_1 + \cdots + p_N = n , \qquad (3.41)$$

$$p_1 + 2p_2 + \cdots + Np_N = n - k .$$

This given we discover that $R(x_0{}^{p_0} x_1{}^{p_1} \cdots x_N{}^{p_N})$ breaks up into groups again. But now the number of k-forest words in each group is *exactly* k, and each word in the group is a circular rearrangement of any one of the k-forest words in the group.

This fact may again be established by the "double path" argument. Indeed, given $W \in R(p_0, p_1, \ldots, p_N)$, we find in its double path two points A and B defined by the following conditions:

(a) *B is the leftmost lowest point in the double path,* $\qquad (3.42)$

(b) *A is the leftmost point on the double path that is k levels above B.*

Using only the portion of the double path of W that is between A and B, we can easily locate the k distinct k-forest words in the group of W.

The consequence of these observations can be stated in the following generalization of Theorem 3.6.2.

Theorem 3.6.7 *The number of k-forests of planar trees with n nodes, of which p_0 are of degree 0, p_1 are of degree 1, ..., p_N are of degree N is*

$$\frac{k}{n}\frac{n!}{p_0!p_1!\cdots p_N!}.$$ (3.43)

This combinatorial fact can be then translated into the following analytical result.

Theorem 3.6.8 *Let $f(x)$ be the formal series solution of the equation*

$$f(x) = xR(f(x))$$

with

$$R(s) = R_0 + R_1s + \cdots + R_Ns^N.$$

Then for any integer k we have

$$(f(x))^k = \sum_{n\geq k} x^n \frac{k}{n}\left(R_0 + R_1s + \cdots + R_Ns^N\right)^n \Bigg|_{s^{n-k}}.$$ (3.44)

3.7 Miscellaneous Applications and Examples

The developments of this chapter have numerous applications. We shall limit ourselves to discussing only a few that may be presented without getting too much out of context.

3.7.1 Incomplete Binary Trees

These trees are binary-tree-like structures, which may be described as follows. Each node may, like in a binary tree, have either no children at all or two children, but in addition the possibility is included that a node may have *only a left child* or *only a right child*.

A few examples shown in Figure 3.26 should suffice to illustrate this definition.

It so happens that the number d_m of incomplete binary trees with m nodes is the same as the number of binary trees with m *internal* nodes. Thus from formula (3.8) we get

$$d_m = \frac{1}{2m+1}\binom{2m+1}{m}.$$ (3.45)

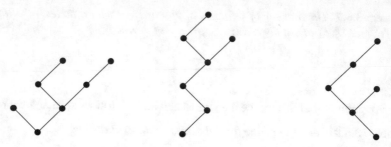

Fig. 3.26 Examples of incomplete binary trees.

The reason for this fact is very simple. Each incomplete binary tree with m nodes can be *completed* to a binary tree with m internal nodes by adding the *missing* external nodes. Thus the trees in Figure 3.26 are completed to the binary trees in Figure 3.27.

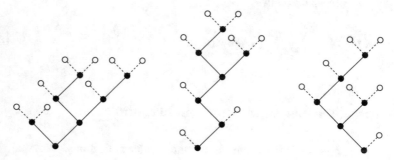

Fig. 3.27 Completion of the incomplete binary trees in Figure 3.26.

In other words, an incomplete binary tree is what one obtains from a binary tree upon removing all the external nodes together with the branches that are attached to them.

3.7.2 0-1-2 Trees

Theorems 3.6.4 and 3.6.5 may be used in a more flexible manner than their derivation may suggest. Indeed, we can now enumerate different classes of trees by simply specializing the variables R_n.

More precisely, rather than considering binary trees, ternary trees, or other similarly restricted classes separately, we may simply work with the general planar tree and obtain information about a desired class be setting equal to zero the R_n's corresponding to unwanted degrees. As a result, the unwanted trees will have zero weight and will not yield a contribution to the resulting series $f(x)$ in (3.32).

To be specific, suppose that we want to enumerate the class C_{012} consisting of trees whose nodes have degrees 0, 1, or 2 only. We can do this by setting $R_n = 0$ for $n \geq 3$. Moreover, if we want to keep track only of the total number of nodes (forgetting how many nodes we have of each particular degree) we simply set $R_0 = R_1 = R_2 = 1$. Doing this reduces (3.31) to the weight

$$m(T) = x^{n(T)} \qquad (n(T) = \# \text{ nodes of } T) . \qquad (3.46)$$

Theorem 3.6.4 then yields that the formal series

$$f(x) = \sum_{T \in C_{012}} x^{n(T)}$$

satisfies the equation

$$f(x) = x \left(1 + f(x) + f^2(x) \right) .$$

From Theorem 3.6.5 we then deduce that

$$f(x) = \sum_{n \geq 1} \frac{x^n}{n} \left(1 + s + s^2 \right)^n \bigg|_{s^{n-1}} .$$

An easy calculation yields that

$$\left(1 + s + s^2 \right)^n \bigg|_{s^{n-1}} = \sum_{k \leq (n-1)/2} \binom{n}{k} \binom{n-k}{k+1} .$$

Thus we must conclude that the number M_n of 0-1-2 trees with n nodes is given by the formula

$$M_n = \frac{1}{n} \sum_{k \leq (n-1)/2} \binom{n}{k} \binom{n-k}{k+1} . \qquad (3.47)$$

3.7.3 Enumeration of Binary Trees by External Nodes

Suppose we want to enumerate the class C of binary trees by keeping track also of the number of external nodes. In other words, we would like to obtain the series

$$\phi(x, t) = \sum_{T \in C} x^{n(T)} t^{E(T)} \qquad (E(T) = \# \text{ external nodes of } T). \qquad (3.48)$$

We see that this corresponds to setting

$$R_0 = t, \quad R_1 = 0, \quad R_2 = 1, \quad R_n = 0 \qquad (n \geq 3)$$

in (3.31) and (3.32).

Theorem 3.6.4 then yields that $\phi(x, t)$ satisfies the equation

$$\phi(x, t) = xt + x(\phi(x, t))^2 . \tag{3.49}$$

From Theorem 3.6.5 we then deduce that

$$\phi(x, t) = \sum_{n \geq 1} \frac{x^n}{n} \left(t + s^2 \right)^n \bigg|_{s^{n-1}} . \tag{3.50}$$

Note that the term

$$\frac{1}{n} \left(t + s^2 \right)^n \bigg|_{s^{n-1}} \tag{3.51}$$

is different from zero only when $n - 1$ is even. This immediately implies that binary trees always have an odd number of nodes (confirming a fact that the reader may already have noticed).

Setting $n = 2k - 1$ with $k \geq 1$ in (3.51), a simple calculation yields that this term is equal to

$$\frac{t^k}{2k - 1} \binom{2k - 1}{k} .$$

Using this result we may rewrite (3.50) in the form

$$\phi(x, t) = \sum_{k \geq 1} x^{2k-1} t^k \frac{1}{2k - 1} \binom{2k - 1}{k} . \tag{3.52}$$

We thereby conclude that the number of binary trees with k external nodes is equal to

$$\frac{1}{2k - 1} \binom{2k - 1}{k} .$$

This is in complete agreement with formula (3.8) (since a binary tree with m internal nodes has $m + 1$ external nodes, this formula should agree with (3.8) when $m + 1 = k$).

It will be good to point out that if we set $x = 1$ in (3.48), (3.49), and (3.52), we deduce that the enumerator of binary trees by external nodes, namely the series

$$g(t) = \sum_{T \in C} t^{E(T)} = \sum_{n \geq 1} t^k \frac{1}{2k+1} \binom{2k+1}{k}$$

satisfies the equation

$$g(t) = t + g^2(t) . \tag{3.53}$$

3.7.4 The Number of Planar Trees

Clearly we need not insist on the requirement that each of our trees should have nodes of degree $\leq N$ for some fixed N. We placed this requirement to simplify our exposition. If we drop it, the only change occurs in $R(s)$ (formula (3.34)) which has to be replaced by the formal series

$$R(s) = \sum_{k \geq 0} R_k s^k .$$

This given, we may find that Theorem 3.6.5 has some rather surprising consequences.

Indeed, suppose that we wish to enumerate the class $\mathcal{P}$ of all planar trees and we want to keep track only of the total number of nodes. This corresponds once more to choosing the weight given in (3.46), which in the present context is obtained by making the choice

$$1 = R_0 = R_1 = R_2 = \cdots = R_n = \cdots .$$

This done, from Theorem 3.6.4 we get that the formal series

$$p(x) = \sum_{T \in \mathcal{P}} m(T) = \sum_{n \geq 1} p_n x^n$$

satisfies the equation

$$p(x) = x + xp(x) + xp^2(x) + \cdots + xp^n(x) + \cdots$$
$$= x \left(1 + p(x) + p^2(x) + \cdots \right)$$
$$= \frac{x}{1 - p(x)} .$$

This may also be rewritten in the form

$$p(x) - p^2(x) = x$$

or better

$$p(x) = x + p^2(x) \,.$$

Comparing with equation (3.53) we deduce that p and g must be the same series! This forces us to conclude that

The number p_m of planar trees with a total of m nodes is equal to

the number of binary trees with m external nodes.

Thus we have again

$$p_m = \frac{1}{2m-1}\binom{2m-1}{m} \,. \tag{3.54}$$

It is tempting at this point to ask whether we can put together a direct bijection between these two classes of trees. Now it so happens that there is a rather simple construction which transforms an arbitrarily given planar tree with m nodes into an incomplete binary tree with $m - 1$ nodes. This construction, combined with the observations in Section 3.7.1, yields a direct verification of the statement of equality of these two classes of trees.

The basis for this construction is the replacement of *fan-outs* by *racks*. We recall that the fan-out of a node consists of the collection of edges which come out of it. Thus the fan-out of a node of degree 4 is the configuration on the left of Figure 3.28, and in our construction this configuration is replaced by the 4-rack on the right in Figure 3.28.

Fig. 3.28 Replacing a fan-out of a node of degree 4 by a 4-rack.

In general, each n-fan-out is replaced by the corresponding n-rack as shown in Figure 3.29.

The construction is best communicated by illustrating it on an example. For instance, making these replacements on the planar tree in Figure 3.30 yields the object in Figure 3.31, and this is transformed into the desired incomplete binary tree upon removing the bottom edge and tilting the resulting figure 45 degrees, obtaining the incomplete binary tree in Figure 3.32.

The reader should not have any difficulty verifying that this construction is reversible and thus defines a bijection between our two classes of trees.

Fig. 3.29 Replacing an n-fan-out by an n-rack.

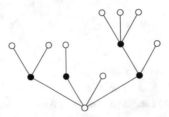

Fig. 3.30 A planar tree.

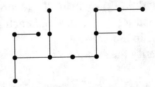

Fig. 3.31 The object obtained from the planar tree in Figure 3.30 after replacing the fan-outs by racks.

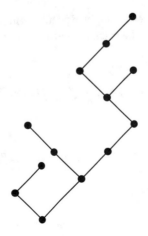

Fig. 3.32 The incomplete binary tree corresponding to the planar tree in Figure 3.30.

3.8   Exercises for Chapter 3

3.8.1 For the trees given below:

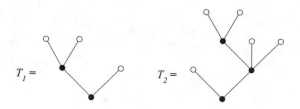

(a) Construct the depth first order of T_1 and T_2.
(b) Draw the trees $T_1 \times T_2$ and $T_2 \times T_1$.
(c) Comparing the trees you have obtained in part (b), conclude that tree multiplication is not commutative.

3.8.2 For the trees T_1 and T_2 of problem 3.8.1:

(a) Find the words $w(T_1)$ and $w(T_2)$ by using the recursion (3.1).
(b) Find the words $w(T_1)$ and $w(T_2)$ by labeling each node of degree i by x_i and then reading these labels in the DFO.
(c) Find $w(T_1 \times T_2)$ and verify the relation

$$w(T_1 \times T_2) = x_2 w(T_1) w(T_2).$$

3.8.3 Construct the lattice paths corresponding to the trees T_1 and T_2 of problem 3.8.1.

3.8.4 Let $I(T)$ and $E(T)$, respectively, denote the number of internal and external nodes of a planar tree T. For the trees T_1 and T_2 of problem 3.8.1:

(a) compute the numbers

$$I(T_1), \ E(T_1), \ I(T_2), \ E(T_2), \ I(T_1 \times T_2), \ E(T_1 \times T_2),$$

(b) verify the relations

$$I(T_1 \times T_2) = 1 + I(T_1) + I(T_2), \quad E(T_1 \times T_2) = E(T_1) + E(T_2).$$

3.8.5 Argue that given planar trees $T_1, T_2, \ldots, T_k$,

$$I(T_1 \times T_2 \times \cdots \times T_k) = 1 + I(T_1) + I(T_2) + \cdots + I(T_k),$$
$$E(T_1 \times T_2 \times \cdots \times T_k) = E(T_1) + E(T_2) + \cdots + E(T_k).$$

3.8.6 If S and T are planar trees, when is $T \times S = S \times T$?

3.8.7

(a) Construct three planar trees T_1, T_2, and T_3 for which $T_1 \times (T_2 \times T_3) \neq (T_1 \times T_2) \times T_3$.
(b) Describe T_1, T_2, and T_3 if $T_1 \times (T_2 \times T_3) = (T_1 \times T_2) \times T_3$.

3.8.8 Consider the formal sums of trees S_1 and S_2.

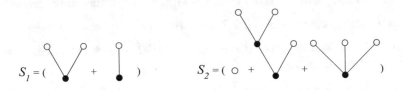

(a) Calculate the product $S_{12} = S_1 \times S_2$.
(b) Let $S_{12}(t, x)$, $S_1(t, x)$, $S_2(t, x)$ denote the polynomials obtained by replacing each tree T in S_{12}, S_1, S_2, respectively, by the monomial $x^{E(T)} t^{I(T)}$. Verify the relation

$$S_{12}(t, x) = t\, S_1(t, x) S_2(t, x).$$

3.8.9 Calculate the number of binary trees with

(a) 4 internal nodes,
(b) 6 external nodes,
(c) a total number of 13 nodes.

3.8.10 Verify that

$$\frac{1}{2m + 1}\binom{2m + 1}{m} = \frac{1}{m + 1}\binom{2m}{m}.$$

3.8.11 Make a list of all ternary trees with 3 internal nodes.

3.8.12 Find the total number of nodes of a 0-5 tree with 7 internal nodes.

3.8.13 Calculate the number of planar trees which have a total number of 15 nodes of which

(a) 9 are of degree 0, 4 are of degree 2 and 2 are of degree 3,
(b) 10 are of degree 0, 3 are of degree 2 and 2 are of degree 4.

3.8.14 Compute the number of ternary trees with

(a) 4 internal nodes,
(b) 7 external nodes,
(c) a total number of 16 nodes.

3.8.15 Calculate the number of planar trees which have 7 external nodes, 2 nodes of degree 2 and 2 nodes of degree 3.

3.8.16 What is the number of planar trees with 7 nodes of degree 0, 4 nodes of degree 2 and 1 node of degree 3?

3.8.17

(a) What is the number of external nodes of a 0-4 tree with 5 internal nodes?
(b) Calculate the number of 0-4 trees with 5 internal nodes.

3.8.18 How many external nodes does a 0-7 tree with 4 internal nodes have?

3.8.19 Calculate the number of 0-2-3 trees with a total number of 6 nodes.

3.8.20 What is the number of external nodes of a 0-2-3 tree with 6 nodes of degree 2 and 4 nodes of degree 3?

3.8.21 Calculate the total number of nodes of a 0-6 tree with 9 internal nodes.

3.8.22 Find the number of planar trees which have the same distribution of degrees as the following tree:

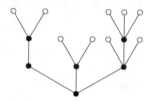

3.8.23 What is the number of planar trees with a total number of 7 nodes?

3.8.24 Making use of the formula (3.47) of Section 3.7, calculate the number of 0-1-2 trees with a total of 6 nodes.

3.8.25 Let T be a 0-2-4 tree. Suppose that T has 11 external nodes and no more than one node of degree 2. How many nodes of degree 4 does T have?

3.8.26 Let C denote the formal sum of all 0-2-4 trees

(a) Show that C satisfies the relation

$$C = o + C \times C + C \times C \times C \times C .$$

(b) Replace each tree T in C by its monomial $m(T) = x^{n(T)}$ where $n(T)$ is the total number of nodes of T. Set

$$f(x) = \sum_{T \in C} m(T)$$

and show that

$$f(x) = x + xf(x)^2 + xf(x)^4 .$$

Using the Lagrange Inversion formula deduce that

$$f(x)\Big|_{x^n} = \frac{1}{n}\left(1 + s^2 + s^4\right)^n \Big|_{s^{n-1}} .$$

(c) Expand $(1 + s^2 + s^4)^n$ by the multinomial theorem and show that a 0-2-4 tree must have an odd number of nodes.

(d) Finally show that the number of 0-2-4 trees with a total of $2m + 1$ nodes is given by

$$\frac{1}{2m+1} \sum_{k \le \frac{m}{2}} \binom{2m+1}{k}\binom{2m+1-k}{m-2k} .$$

3.8.27 Construct the lattice paths corresponding to the following trees:

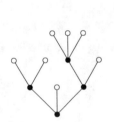 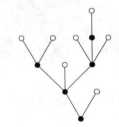 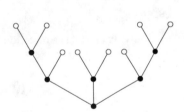

3.8.28 Which of the following words are planar tree words?

(a) $x_2 x_2 x_0 x_0 x_2 x_0 x_2 x_0 x_0$,
(b) $x_3 x_0 x_1 x_1 x_3 x_0 x_0 x_0 x_1$,
(c) $x_1 x_1 x_1 x_2 x_0 x_1 x_1 x_1 x_0$,
(d) $x_1 x_2 x_4 x_0 x_0 x_1 x_0 x_0 x_0$,
(e) $x_2 x_0 x_2 x_2 x_0 x_0 x_0 x_2 x_0$.

3.8.29 Find the unique tree word which is a circular rearrangement of

(a) $x_0 x_0 x_2 x_0 x_1 x_0 x_2 x_3 x_0$,
(b) $x_2 x_0 x_0 x_4 x_2 x_0 x_0 x_0 x_0$,
(c) $x_2 x_2 x_0 x_0 x_0 x_0 x_0 x_2 x_2$.

3.8.30

(a) Construct the tree T which corresponds to the word

$$w(T) = x_3 x_2 x_2 x_0 x_0 x_0 x_3 x_0 x_0 x_0 x_2 x_0 x_2 x_0 x_0.$$

(b) Draw the lattice path $P(T)$ of T.

3.8.31

(a) Find the unique tree word $w(T)$ which is a circular rearrangement of

$$x_3 x_0 x_0 x_0 x_3 x_0 x_0 x_0 x_2 x_2 x_0.$$

(b) Construct the planar tree T which corresponds to $w(T)$.

3.8.32

(a) Construct the trees T_1, T_2, T_3 which correspond to the circular rearrangements
of the words

$$x_0 x_2 x_2 x_0 x_0, \quad x_0 x_0 x_3 x_0, \quad x_3 x_0 x_2 x_0 x_0 x_0,$$

respectively.

(b) What is the word $w(T)$ of the tree $T = T_1 \times T_2 \times T_3$?
(c) Verify that $w(T_3 \times T_2 \times T_1) = x_3 w(T_3) w(T_2) w(T_1)$.

3.8.33

(a) Show that there is a bijection between the ways of parenthesizing the expression
$a_1 \cdot a_2 \cdots a_n$, e.g.,

$$(((a_1 \cdot a_2) \cdot a_3) \cdot (a_4 \cdot a_5)) \cdot a_6$$

and binary trees with n external nodes (Chapter 1, Section 1.4).
(b) Construct the parenthesization of $a_1 \cdot a_2 \cdots a_6$ that corresponds to following
binary trees

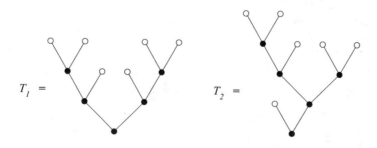

3.8.34 Use Theorem 1.4.4 of Chapter 1 to construct the binary trees that correspond to the following Young tableaux

3	5	6
1	2	4

3	5	6	8
1	2	4	7

2	5	8	9	10
1	3	4	6	7

3.8.35 Suppose T is a binary tree with principal subtrees T_1 and T_2. We have that if T is trivial, then $w(T) = x_0$, otherwise

$$w(T) = x_2 w(T_1) w(T_2),$$

and that this word corresponds to reading the labels of T in DFO where internal nodes are labeled with x_2 and external nodes are labeled x_0. In case of binary trees, DFO is also called *preorder*. Consider the following words $POST(T)$ (for *postorder*) and $IN(T)$ (for *inorder*) defined by setting $POST(T) = IN(T) = x_0$ if T is the trivial tree, and

$$POST(T) = POST(T_1)POST(T_2)x_2$$

$$IN(T) = IN(T_1)x_2 IN(T_2),$$

otherwise.

(a) Construct $POST(T)$ and $IN(T)$ if

    (a) $w(T) = x_2 x_2 x_0 x_2 x_2 x_0 x_0 x_0 x_0$,
    (b) $w(T) = x_2 x_2 x_2 x_0 x_0 x_0 x_2 x_0 x_2 x_0 x_0$,
    (c) $w(T) = x_2 x_2 x_0 x_2 x_0 x_0 x_2 x_0 x_0$.

(b) Construct the lattice path corresponding to T if

    (a) $POST(T) = x_0 x_0 x_0 x_2 x_0 x_2 x_0 x_2 x_2$,
    (b) $IN(T) = x_0 x_2 x_0 x_2 x_0 x_2 x_0 x_2 x_0$.

3.8.36 A parenthesized arithmetic expression involving the binary operators

$$+, \ -, \ *, \ /, \ \uparrow$$

can be described conveniently by a binary tree (here $-$ is subtraction and $\uparrow$ is exponentiation). For example,

$$((a_1 - a_2) * a_3) + ((a_4 + a_5)/a_6) \tag{3.55}$$

can be represented as

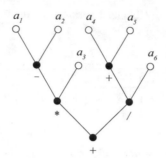

Traversing T in postorder produces the string

$$a_1 a_2 - a_3 * a_4 a_5 + a_6 / +$$

from which the sequence of operations indicated by (3.55) can be unambiguously recovered. This is referred to as the *postfix* form of the *infix* expression in (3.55).

(a) Find the postfix forms of the following expressions

    (a) $(a_1 \uparrow (a_2 + a_3))/(a_5 - a_6)$,
    (b) $((a_1/(a_2 * a_3)) - a_4) * (a_5/a_6)$.

(b) Convert the following expressions given in postfix notation to infix form

    (a) $a_1 a_2 + a_3 * a_4 a_5 a_6 * \uparrow /$,
    (b) $a_1 a_2 a_3 a_4 a_5 a_6 / / / / /$.

3.8.37 Construct a tree structure to represent arithmetic expressions when the unary operator—(negative) is allowed.

3.8.38 Using the Lagrange Inversion formula, construct the formal power series solutions of the following equations:

(a) $f(x) = x(1 + f(x))^2$,
(b) $g(x) = x/(1 + g(x))^3$,
(c) $h(x)^4 - 2h(x) + x = 0$.

3.8.39 Let T_{03} denote the set of all ternary trees and put

$$f(t) = \sum_{T \in T_{03}} t^{I(T)},$$

where $I(T)$ is the number of internal nodes of T.

(a) Explain why $f(t)$ satisfies the algebraic equation

$$f(t) = 1 + t f(t)^3.$$

(b) Find $f(t)\Big|_{t^6}$.

3.8.40 Let T_{05} be the collection of all 0-5 trees and put

$$g(x) = \sum_{T \in T_{05}} x^{E(T)},$$

where $E(T)$ is the number of external nodes of T.

(a) Find an algebraic equation satisfied by $g(x)$.
(b) Calculate the number of 0-5 trees which have 13 external nodes.

3.8.41 Construct the series

$$f(x) = \sum_{n \geq 1} f_n x^n$$

which solves the algebraic equation

$$f(x) = x(1 + 2f(x))^2.$$

3.8.42 Let $f(x)$ be the formal power series solution of the equation

$$f(x) = x(1 - f(x)^3).$$

Using Theorem 3.6.8, construct the series $g(x) = f(x)^2$.

3.8.43 Use the Lagrange Inversion formula to find the formal power series solution of the equation

$$f(x) = xe^{f(x)}.$$

3.8.44 Construct the binary trees corresponding to the following planar trees:

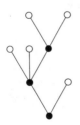  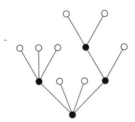

3.8.45

(a) Construct the planar trees which correspond to the following binary trees T_1 and T_2:

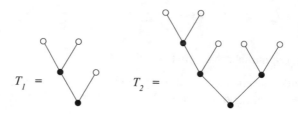

(b) Construct the planar trees that correspond to the binary trees $T_1 \times T_2$ and $T_2 \times T_1$.

3.8.46 Let H be the formal sum of all 0-1 trees,

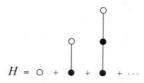

(a) Clearly there is only one 0-1 tree on n nodes for each $n \geq 1$. To verify this put

$$h(x) = \sum_{T \in H} x^{n(T)},$$

where $n(T)$ is the total number of nodes of T, and show that

$$h(x) = x(1 + h(x)).$$

(b) Using the Lagrange Inversion formula or otherwise, solve the above equation for $h(x)$ to get

$$h(x) = \frac{x}{1 - x}.$$

(c) Show that this implies there is only one 0-1 tree on n vertices for any n.

3.8.47 Let T_{04} be the collection of all 0-4 trees and put

$$g(t) = \sum_{T \in T_{04}} t^{I(T)}.$$

(a) Show that $g(t)$ satisfies

$$g(t) = 1 + tg(t)^4.$$

(b) Set $h(t) = g(t) - 1$. Show that $h(t)$ satisfies the equation

$$h(t) = t(1 + h(t))^4.$$

(c) Use the Lagrange Inversion formula to compute the number of 0-4 trees with n internal nodes.

3.8.48 Write down the equation satisfied by the generating function

$$f(x) = \sum_{T \in T_{024}} x^{E(T)},$$

where T_{024} denotes the family of 0-2-4 trees and $E(T)$ is the number of external nodes of T.

3.8.49 Let D be a set of nonnegative integers containing zero and denote by d_n the number of planar trees on n vertices with degrees from the set D. If we set

$$f_D(x) = \sum_{n \geq 1} d_n x^n,$$

show that $f_D(x)$ satisfies

$$f_D(x) = x \sum_{d \in D} f_D(x)^d.$$

3.8.50 Use Theorem 3.6.5 to show that if $f(x)$ is a power series of the form $f(x) = x R(x)$ with $R(0) \neq 0$ and $g(x) = f^{-1}(x)$ (i.e., $g(x)$ is a power series such that $f(g(x)) = x$), then

$$g(x) \bigg|_{x^n} = \frac{1}{n} \frac{1}{R(s)^n} \bigg|_{s^{n-1}}.$$

3.8.51 Use problem 3.8.40 to find the inverses of the following power series:

(a) $f(x) = xe^{-x}$,
(b) $f(x) = x(x + 1)^3$,
(c) $f(x) = x/(2 - x^3)$.

3.8.52 Let PE denote the collection of all planar trees all of whose nodes have *even* degrees.

(a) Let

$$f(x) = \sum_{T \in PE} x^{n(T)} \, .$$

Use Theorem 3.6.3 and the remarks following it to obtain an algebraic equation satisfied by $f(x)$.

(b) Let T_{03} denote the class of ternary trees and put

$$g(x) = \sum_{T \in T_{03}} x^{E(T)}.$$

Show that $g(x)$ satisfies the equation

$$g(x) = x + g(x)^3.$$

(c) Compare the results of (a) and (b) above. What conclusions can you draw?

3.8.53 Let $s \geq 1$ be an integer. Show that the number of planar trees with a total of n nodes where each node has degree a multiple of s is equal to the number of $0 - (s + 1)$ trees with n external nodes.

3.8.54 Suppose T is a planar tree with n nodes. An *increasing labeling* of T is an assignment of $\{1, 2, \ldots, n\}$ to the nodes of T such that the labels increase as we go up the tree starting at the root. For example,

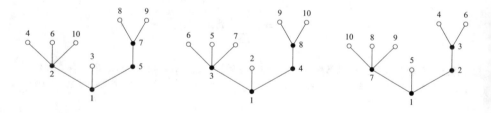

are all increasing labelings of the same underlying tree.

(a) Show that the number of increasing labelings of T is given by

$$\frac{n!}{\prod_{x \in T} |T_x|},$$

where $|T_x|$ denotes the total number of nodes in the subtree T_x of T rooted at vertex x.

(b) Consider the increasing labelings of T where we impose the additional condition that the labels of the sons of any node must increase from left to right. Use the bijection between planar trees and incomplete binary trees to find a formula for such increasing labelings of T.

3.8.55 Show that there is a natural one-to-one correspondence between permutations of $[n]$ and increasing labeled incomplete binary trees with n nodes.

3.8.56 Consider the class of *incomplete ternary trees* that are obtained from ternary trees by removing some of the external nodes.

(a) Calculate the number of incomplete ternary trees with n internal nodes.
(b) Show that the number of increasing labeled incomplete ternary trees on a total of n nodes is given by

$$1 \cdot 3 \cdot 5 \cdots (2n - 1).$$

3.8.57 For $m \geq 2$, the class of incomplete m-ary trees are defined similar to incomplete binary trees and incomplete ternary trees. Show that in general, the number of increasing labeled incomplete m-ary trees on n nodes is given by

$$\prod_{k=1}^{n-1} (k(m - 1) + 1).$$

3.9   Sample Quiz for Chapter 3

1. Let

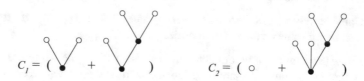

   (a) Calculate the product

$$C_{12} = C_1 \times C_2.$$

   (b) Calculate the polynomial

$$W(C_{12}) = \sum_{T \in C12} x^{n(T)} t^{E(T)},$$

   where $n(T)$ and $E(T)$ denote the total number of nodes and the number of
   external nodes of T, respectively.

2. Let

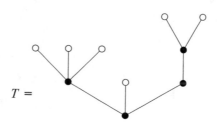

   (a) Construct the word $w(T)$ associated with T.
   (b) Construct the path $P(T)$ corresponding to T.

3. (a) Find the unique tree word w which is a circular rearrangement of the
      following word

$$x_0 x_2 x_0 x_0 x_2 x_2 x_0 x_3 x_0 x_0.$$

   (b) Construct the tree which corresponds to w.
4. Construct the series $f(x)$ which is a solution of the equation

$$f(x) = x(1 - f(x)^2).$$

5. Calculate the number of planar trees with item 6 nodes of degree 0, 3 nodes of degree 2 and a single node of degree 3.
6. How many ternary trees are there with 6 internal nodes? Justify your answer.

Chapter 4
Cayley Trees

4.1 Introduction

In this chapter we shall work with geometric figures obtained by joining a certain number of distinguished points of space by edges in such a manner that

$$(a) \quad \textit{The resulting figure is connected,} \qquad (4.1)$$

$$(b) \quad \textit{No circuits are formed.}$$

Examples of such structures are given in Figure 4.1.

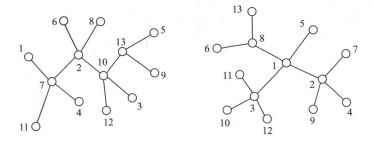

Fig. 4.1 Examples of Cayley trees.

We can see that these are tree-like structures, and indeed in most treatments they are simply called *trees* or *labeled trees*. Our everyday experience requires the presence of a root on a tree as well as an up-and-down orientation. However, these structures have neither of these features. For this reason it may be more appropriate to call them *spatial arborescences* to distinguish them from other structures that are more tree-like, such as planar rooted trees, for instance. But this terminology

© Springer Nature Switzerland AG 2021
Ö. Eğecioğlu, A. M. Garsia, *Lessons in Enumerative Combinatorics*, Graduate
Texts in Mathematics 290, https://doi.org/10.1007/978-3-030-71250-1_4

is rather arcane itself. As a compromise we shall refer to them as *Cayley trees*, in honor of Arthur Cayley who was the first to enumerate them [3].

It will be good to make some notational conventions. To begin with, n will here and after denote the number of vertices of a Cayley tree. The vertices themselves will usually be represented by the symbols

$$x_1, \; x_2, \ldots, x_n \; .$$

Since Cayley trees have no up or down orientation, rather than speaking of *parent* or *children* of a given node, it is more appropriate to speak of *neighbors*; where (as we may guess) the neighbors of a node are the nodes at the endpoints of the edges emanating from it.

The number of neighbors of a node x_i in a Cayley tree T is called its *valence* and denoted by $v_i(T)$.

Usually *degree* and *valence* are two terms that are used interchangeably for the number of neighbors of a node. However, for notational convenience, in this chapter:

When we speak of a Cayley tree, we will refer to $v_i - 1$, i.e., the number of neighbors of x_i minus one as the node's degree and denote it by the symbol $d_i(T)$.

This is similar to the notion of degree that we used for planar trees in Chapter 3. Thus for the tree T in Figure 4.2, we have

$$d_1(T) = 3, \; d_2(T) = 0, \; d_3(T) = 1, \; d_4(T) = 2, \ldots$$

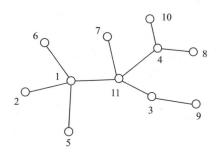

Fig. 4.2 A Cayley tree with $n = 11$ vertices.

Note that the valence of a node in a Cayley tree is simply one more than its degree.

For computational reasons it will be good to require Cayley trees to have at least two vertices. Thus the simplest Cayley tree is the structure consisting of two vertices x_1 and x_2 joined by an edge (see Figure 4.3).

$$1 \; \text{O}\!\!-\!\!-\!\!-\!\!-\!\!\text{O} \; 2$$

Fig. 4.3 The simplest Cayley tree has $n = 2$ vertices.

For $n = 3$ we have 3 possibilities as shown in Figure 4.4.

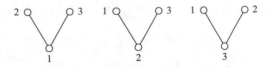

Fig. 4.4 Cayley trees with $n = 3$ vertices.

The reader may draw all the Cayley trees with 4 vertices. There are 16 of them.
It will be convenient in the following to denote the collection of all Cayley trees
with n nodes by the symbol C_n. We have the first few values as

$$|C_2| = 1, \ |C_3| = 3, \ |C_4| = 16 .$$

In this chapter we shall derive a number of formulas enumerating several classes of
Cayley trees. In particular we shall deduce the following remarkable result.

Theorem 4.1.1 (Cayley) *The number of Cayley trees with n nodes is*

$$|C_n| = n^{n-2} . \tag{4.2}$$

4.2 The Monomial and the Word of a Cayley Tree

Given a Cayley tree T, the expression

$$m(T) = x_1^{d_1(T)} x_2^{d_2(T)} \cdots x_n^{d_n(T)} \tag{4.3}$$

will be referred to as the *monomial* of T . Thus, for instance, for the tree in
Figure 4.5,

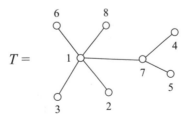

Fig. 4.5 A Cayley tree T with $n = 8$ vertices.

we have

$$m(T) = x_1^4 x_7^2 .$$

Note that in this example the algebraic degree of $m(T)$ is two less than the number of vertices in T. This is no accident, it holds in general.

To see why, observe that when T is a Cayley tree, the expression

$$e(T) = 1 + d_1(T) + d_2(T) + \cdots + d_n(T) \tag{4.4}$$

gives precisely the number of edges of T. In fact, if we imagine x_1 to be the root of T, and orient T away from x_1, then $d_i(T)$ (for each $i > 1$) gives the number of edges leading away from x_i. For the root x_1 itself the number of edges leading away is $1 + d_1(T)$. Thus (4.4) must hold as asserted. This given, by Euler's formula

$$d_1(T) + d_2(T) + \cdots + d_n(T) = n - 2 . \tag{4.5}$$

As a matter of fact we can associate to each tree T an $(n-2)$-letter word

$$w(T) = x_{i_1} x_{i_2} \cdots x_{i_{n-2}} \tag{4.6}$$

from which T itself can be unambiguously reconstructed. The letters of $w(T)$ are determined one by one by disassembling T edge by edge into the successive trees

$$S_0(T)(= T), \ S_1(T), \ S_2(T), \ldots, \ S_{n-2}(T)$$

defined by the following rule:

To obtain $S_i(T)$, we remove from $S_{i-1}(T)$ the highest numbered external node together with the edge that leads to it.

Let us now suppose that we are at the kth step of this procedure. Note that (for $k \le n - 2$) the edge we remove must lead to some other node of $S_{k-1}(T)$, and let this node be x_{i_k}. This given, we record this fact by placing x_{i_k} as the kth letter of $w(T)$.

This is best explained by some examples. Consider the tree T in Figure 4.6.

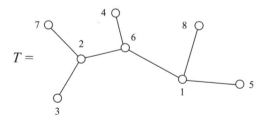

Fig. 4.6 A Cayley tree T on $n = 8$ vertices.

The successive trees and the words we obtain starting with T are given in Figure 4.7.

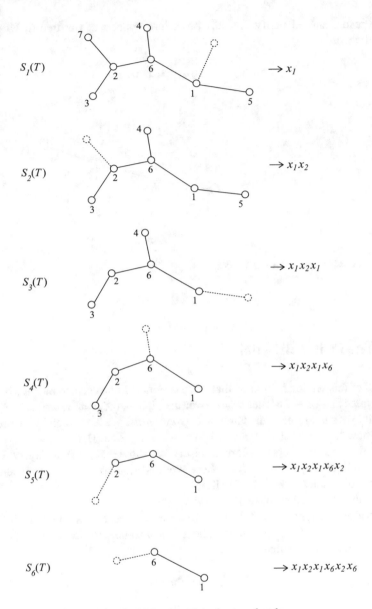

Fig. 4.7 Successively removing the highest numbered external node.

The resulting word is

$$w(T) = x_1 x_2 x_1 x_6 x_2 x_6 .$$

The reader should verify that this procedure applied to the tree in Figure 4.8 yields the word

$$w(T) = x_9 x_9 x_2 x_2 x_9 x_4 x_4 x_1 \ .$$

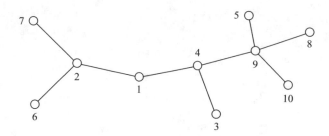

Fig. 4.8 A Cayley tree with $n = 10$ nodes.

4.3 The Prüfer Bijection

The goal of this section is to show that there is a one-to-one correspondence between Cayley trees and $(n-2)$-letter words over the alphabet $X_n = \{x_1, x_2, \ldots, x_n\}$. For simplicity we shall refer to the latter as *Cayley words*. We have seen in the previous section how to construct such a word from a Cayley tree. Here, we show that this construction can be reversed. More precisely, we show that given a Cayley word w there is a unique Cayley tree $T(w)$ whose word is equal to w. This correspondence is referred to as the *Prüfer bijection* [23].

To do this, we need to make some conventions.

First of all, an appropriate way to store a Cayley tree T in a computer is to give for each node of T the list of its neighbors. For instance, for the tree in Figure 4.6 we would store the information

$$
\begin{array}{ll}
\text{Node} & \text{neighbor list} \\
x_1: & x_5, x_6, x_8 \\
x_2: & x_3, x_6, x_7 \\
x_3: & x_2 \\
x_4: & x_6 \\
x_5: & x_1 \\
x_6: & x_1, x_2, x_4
\end{array}
\qquad (4.7)
$$

$$x_7 : \quad x_2$$

$$x_8 : \quad x_1$$

We shall refer to this as the *neighbor list* of the tree. Clearly, from such a list the original tree may be readily drawn.

Taking this into account, our program is to give an algorithm which, given an arbitrary $(n-2)$-letter word over X, will produce the neighbor list of the tree $T(w)$.

To describe our algorithm we need some notation.

Recall that the valence of a node is the number of edges coming out of it. Let then $v_{i,j}$ denote the valence of the node x_j in $S_i(T)$.

The table whose ith row consists of the integers

$$v_{i,1}, \ v_{i,2}, \ \ldots, \ v_{i,n}$$

will be referred to as the *valence table* of T.

For instance, for the tree T given in Figure 4.6, our disassembling procedure yields the following valence table:

Row	Tree	x_1	x_2	x_3	x_4	x_5	x_6	x_7	x_8	Successive words
0	$T:$	3	3	1	1	1	3	1	1	
1	$S_1(T):$	2	3	1	1	1	3	1	0	x_1
2	$S_2(T):$	2	2	1	1	1	3	0	0	$x_1 x_2$
3	$S_3(T):$	1	2	1	1	0	3	0	0	$x_1 x_2 x_1$
4	$S_4(T):$	1	2	1	0	0	2	0	0	$x_1 x_2 x_1 x_6$
5	$S_5(T):$	1	1	0	0	0	2	0	0	$x_1 x_2 x_1 x_6 x_2$
6	$S_6(T):$	1	0	0	0	0	1	0	0	$x_1 x_2 x_1 x_6 x_2 x_6$

$$(4.8)$$

Perhaps we should mention that valence zero simply means that the corresponding node has been removed all together. We should also note that row 0 gives the valences of each of the nodes in T itself.

Note that from this table we can easily obtain the neighbor list of T. For instance, by observing which valences decrease, we can infer from row 4 that in going from $S_3(T)$ to $S_4(T)$ the edge (x_4, x_6) was removed. This implies in particular that x_4 and x_6 are neighbors. Reading the table in this manner row by row produces the following pairs of neighbors:

$$(x_8, x_1), \ (x_7, x_2), \ (x_5, x_1), \ (x_4, x_6), \ (x_3, x_2), \ (x_2, x_6) . \qquad (4.9)$$

Moreover, the last row of the table yields that the last two remaining nodes are x_1 and x_6. Thus our last neighbor pair is

$$(x_1, x_6) \,.$$

$$(4.10)$$

Since T is disassembled by removing all its edges one by one, the rows of the table must yield every neighbor pair. However, knowing all the neighbor pairs is equivalent to knowing the neighbor list. The reader may verify that the pairs listed in (4.9) and (4.10) do indeed yield the neighbor list in (4.7).

Now the remarkable fact is that the valence table may be constructed directly from the word of T.

To show this, note first that every time we remove a node that is attached to x_2, for instance, we place the letter x_2 in $w(T)$. However, when we remove x_2 itself, we place its last remaining neighbor in $w(T)$. Thus the frequency of occurrence of x_2 in $w(T)$ is one less than its valence in T, and that is precisely the degree of x_2 in T. We can see that same holds true for each of the nodes.

Thus, the 0th row of our table may be constructed by adding 1 to the frequency of each letter in $w(T)$.

It so happens that the successive initial segments of $w(T)$ yield us the successive rows of the valence table. For instance, looking again at the table in (4.8), we see that the first letter of $w(T)$ being x_1 yields that the first edge removed was attached to x_1; thus the valence of x_1 in $S_1(T)$ must be one less than it was in T.

Note further that at any stage the external nodes are those of valence 1. From the first row of the table in (4.8) we deduce that the highest labeled external node is x_8, thus x_8 was the node that was removed at the first step of disassembly. Thus the valence of x_8 in the first row must be 0. This completely determines row 1.

We can easily see by the same reasoning that the row i of the table may be reconstructed from row $i - 1$ and the ith letter of $w(T)$. For instance, from row 3 and the knowledge that the 4th letter of $w(T)$ is x_6 we construct row 4 by the following reasoning: the node removed was attached to x_6 thus

the valence of x_6 in row 4 must be one less than in row 3.

Moreover in row 3 the highest node of valence one is x_4 and so x_4 is the highest external node in $S_3(T)$ and it must be the one that is removed to obtain $S_4(T)$. Therefore

the valence of x_4 in row 4 must be one less than in row 3.

All other nodes are unaffected in this step, thus

all the other entries in row 4 are left unchanged.

Carrying out this reasoning row by row produces the table in (4.8) directly from the word

$$x_1 x_2 x_1 x_6 x_2 x_6$$

without ever having to look at the tree in Figure 4.6!

It is not difficult to see that this procedure may be applied to any $(n-2)$-letter word over the alphabet X. The result is the valence table of the desired Cayley tree $T(w)$. Since as we have seen, the valence table yields the tree itself, we have thus obtained a one-to-one correspondence between Cayley words and Cayley trees.

To implement the construction of $T(w)$ from w in a computer, our procedure can be streamlined as follows. Rather than constructing the valence table of $T(w)$ and then from it deduce the neighbor list, these two operations can be carried out simultaneously. Indeed, since each row produces a neighbor pair, and each row together with w determines the next row, we need only use two arrays, one say W, of size $n-2$ to store w and one, say S, of size n to store the row that is being processed. Each step can then be carried out by an updating procedure whose input is the array S and a letter of w and whose output is the next row and the neighbor pair. The reader should have no difficulty in writing a program along these lines.

Remark 4.3.1 It should be noted that to carry out this construction by hand we can proceed as follows. Given the word

$$w = x_{i_1} x_{i_2} \cdots x_{i_{n-2}}$$

we write in two successive rows A and B the strings of letters $x_{i_1}, x_{i_2}, \ldots, x_{i_{n-2}}$ and $x_n, x_{n-1}, \ldots, x_1$ in that order, respectively. This done, we connect x_{i_1} to the first letter in row B which does not occur in row A. We then remove the connected pair of letters. This process is then repeated. At each step:

We connect the first letter of A to the first letter of B which does not occur in A. Then remove the connected pair of letters.

We proceed in this manner until there are no letters left in row A. We then connect and remove the two letters left in row B.

It is not difficult to see that the pairs we remove are precisely the neighbor pairs of the corresponding Cayley tree. This is best understood by working on an example. For instance, working with the word

$$w = x_1 x_2 x_1 x_6 x_2 x_6$$

we obtain the succession of rows and removed pairs given in Figure 4.9.

We see by comparing with (4.9) and Figure 4.9 that the pairs we obtain are precisely the same as those given by the previous construction.

Of course this will be so for any Cayley word. Indeed, we can see by comparing corresponding steps in the two procedures that the process by which we remove the pair at the ith stage here is only a disguised version of the process by which we construct the $(i+1)$st row of the valence table.

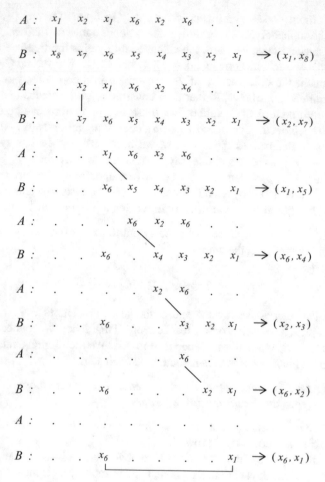

Fig. 4.9 Alternate construction of the neighbor list from a Cayley word.

4.4 Enumeration of Cayley Trees and Cayley Forests

The algorithm we have put together for constructing the word of a Cayley tree and a Cayley tree from a word, give us a bijection between the Cayley trees of C_n and the $(n-2)$-letter words over the alphabet $X_n = \{x_1, x_2, \ldots, x_n\}$. Thus it must be that the sum

$$\sum_{T \in C_n} w(T)$$

is simply what we obtain upon multiplying the formal sum

$$x_1 + x_2 + \cdots + x_n$$

by itself $n - 2$ times.

Thus we must conclude that the following identity holds.

Theorem 4.4.1

$$M(C_n) = \sum_{T \in C_n} m(T) = (x_1 + x_2 + \cdots + x_n)^{n-2} . \tag{4.11}$$

We see that Cayley's theorem follows at once from this upon setting $x_1 = x_2 = \cdots = x_n = 1$. However, our bijection yields several further results.

Theorem 4.4.2 *The number of Cayley trees with vertices $x_1, x_2, \ldots, x_n$ of degrees $d_1, d_2, \ldots, d_n$ is equal to*

$$\frac{(n-2)!}{d_1! d_2! \cdots d_n!} . \tag{4.12}$$

Proof Use the multinomial theorem and formula (4.11). □

Of course, we can also observe that the expression in (4.12) gives precisely the number of $(n - 2)$-letter words over X with the letter x_i occurring d_i times.

Theorem 4.4.3 *The number of Cayley trees on $x_1, x_2, \ldots, x_n$ with x_1 a node of degree k is equal to*

$$\binom{n-2}{k}(n-1)^{n-k-2} . \tag{4.13}$$

Proof This is simply the number of $(n - 2)$-letter words over X with the letter x_1 occurring exactly k times. Indeed, $\binom{n-2}{k}$ counts the number of ways of placing x_1 in such a word and $(n - 1)^{n-k-2}$ counts the number of ways of filling the remaining positions with the letters $x_2, x_3, \ldots, x_n$. □

Finally, we obtain the following result of Cayley:

Theorem 4.4.4 *The number $F(n, p)$ of Cayley forests on $x_1, x_2, \ldots, x_n$ which have a total of p trees and such that the nodes $x_1, x_2, \ldots, x_p$ belong to different trees is given by the formula*

$$F(n, p) = p\, n^{n-p-1} . \tag{4.14}$$

Proof We shall consider two separate collections of objects. The first, which we denote by $\mathcal{T}_n(x_{i_1}, x_{i_2}, \ldots, x_{i_p})$, is the collection of all Cayley trees on $x_0, x_1, \ldots, x_n$ in which the vertex x_0 is of valence p and is connected to the vertices

$$x_{i_1}, x_{i_2}, \ldots, x_{i_p} . \tag{4.15}$$

The second, which we denote by $\mathcal{F}_n(x_{i_1}, x_{i_2}, \ldots, x_{i_p})$, is the collection of all Cayley forests on $x_1, x_2, \ldots, x_n$ with a total of p trees and with the vertices in (4.15) belonging to different trees.

We note that these two collections have the same number of elements. Indeed, a tree in $\mathcal{T}_n(x_{i_1}, x_{i_2}, \ldots, x_{i_p})$ yields a forest in $\mathcal{F}_n(x_{i_1}, x_{i_2}, \ldots, x_{i_p})$ by the removal of x_0 and all the edges connected to it.

We shall briefly refer to this operation as the x_0-*pruning* of a tree.

Secondly, note that all the collections $\mathcal{F}_n(x_{i_1}, x_{i_2}, \ldots, x_{i_p})$ have the same number of elements. Indeed the elements of $\mathcal{F}_n(x_{i_1}, x_{i_2}, \ldots, x_{i_p})$ can be transformed into elements of $\mathcal{F}_n(x_1, x_2, \ldots, x_p)$ by any relabeling that sends the p-tuple $x_{i_1}, x_{i_2}, \ldots, x_{i_p}$ into the p-tuple $x_1, x_2, \ldots, x_p$. This gives that

$$\#\mathcal{F}_n(x_{i_1}, x_{i_2}, \ldots, x_{i_p}) = F(n, p) \ .$$

Finally, we observe that the collection $\mathbf{C}_{p-1}$ of all Cayley trees on $x_0, x_1, \ldots, x_n$ with x_0 of valence p, by Theorem 4.4.3, has

$$\binom{n+1-2}{p-1}(n+1-1)^{n+1-2-(p-1)} = \binom{n-1}{p-1}n^{n-p}$$

elements.

Note now that by x_0-pruning every tree in $\mathbf{C}_{p-1}$ we obtain all of the forests belonging to the collections $\mathcal{F}_n(x_{i_1}, x_{i_2}, \ldots, x_{i_p})$ for all possible choices of $x_{i_1}, x_{i_2}, \ldots, x_{i_p}$. Loosely speaking, we may say that $\mathbf{C}_{p-1}$ breaks up into the disjoint union of the collections $\mathcal{F}_n(x_{i_1}, x_{i_2}, \ldots, x_{i_p})$. Since each of these collections has $F(n, p)$ elements and there are $\binom{n}{p}$ possible ways of choosing the p-tuples $x_{i_1}, x_{i_2}, \ldots, x_{i_p}$ out of $x_1, x_2, \ldots, x_n$, we must conclude that

$$\binom{n}{p}F(n, p) = \binom{n-1}{p-1}n^{n-p} \ .$$

Making the appropriate cancellations we obtain (4.14) as asserted.   □

4.5 Functional Digraphs, the Joyal Encoding

Let $[n] = \{1, 2, \ldots, n\}$ and $[m] = \{1, 2, \ldots, m\}$. It is easy to see that the number of maps of $[n]$ into $[m]$ is equal to

$$m^n \ .$$

This is so since to define a map f of $[n]$ into $[m]$ we have m choices for each of the function values

$$f(1), \ f(2), \ldots, \ f(n) .$$

For this reason the collection of all maps from $[n]$ to $[m]$ is denoted by the symbol

$$[m]^{[n]} .$$

Note now that the number of *self*-maps of $[n]$ into $[n]$ is thus equal to

$$n^n .$$

Comparing with Cayley's theorem we see that

$$n^n = n^2 |C_n| . \tag{4.16}$$

This suggests that there should be a correspondence between Cayley trees and the collection of maps $[n]^{[n]}$.

A very beautiful and natural correspondence was put together by André Joyal [18]. To present it we need some further conventions.

There is a very convenient way of representing a map f of $[n]$ into $[n]$. We simply draw a directed graph with the elements of $[n]$ as vertices with arrows going from an element to is image under f. More precisely, for each $i = 1, 2, \ldots, n$ we draw an arrow from i to $f(i)$.

For instance, when $n = 11$, the function $f \in [n]^{[n]}$ given by the table in (4.17)

i	1	2	3	4	5	6	7	8	9	10	11
$f(i)$	7	11	10	2	5	5	11	2	5	7	10

(4.17)

is represented by the directed graph in Figure 4.10.

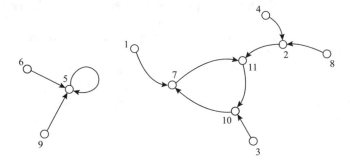

Fig. 4.10 The digraph of the function in (4.17).

The directed graphs corresponding in this manner to self-maps of a given set of elements are usually referred to as *functional digraphs* (digraph is an abbreviation for *directed graph*).

The Joyal idea is based on an ingenious interpretation of the right hand side of (4.16).

Note that the number of *rooted* Cayley trees on n nodes is

$$n|C_n| .$$

The reason for this is that we have n possible choices for the root. The Joyal idea is simply to interpret the right hand side of (4.16) as the number of *bi-rooted* Cayley trees on n nodes.

We may call the two roots the *Head* and *Tail* bones. The resulting configuration may then be referred to as a *vertebrate*. Figure 4.11 illustrates a vertebrate on 16 nodes. The head and tail bones are indicated by H and T, respectively.

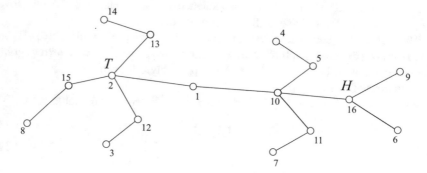

Fig. 4.11   A vertebrate on 16 nodes.

The proof of (4.16) consists in producing an algorithm which establishes a bijection between vertebrates on n nodes and the maps $[n]^{[n]}$.

The Joyal algorithm is best communicated by an example. We shall illustrate it on the vertebrate depicted in Figure 4.11.

The first step consists in removing the *vertebral column* of the given vertebrate, and orient its edges from tail to head. In this case we get the object in Figure 4.12. This resulting object is interpreted as a permutation of the integers appearing as labels. Here the resulting permutation, written in two-line notation, is

$$\begin{pmatrix} 1 & 2 & 10 & 16 \\ 2 & 1 & 10 & 16 \end{pmatrix} .$$

This permutation is then replaced by its cycle diagram as shown in Figure 4.13.

This done, the remaining portions of the vertebrate are appended as they were in the original vertebrate. Furthermore, each edge is given an orientation pointing *towards* the cycle diagram. The resulting object in our case is shown in Figure 4.14.

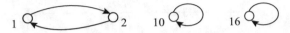

Fig. 4.12 Removing the vertebral column.

Fig. 4.13 The cycle diagram of the vertebral column interpreted as a permutation.

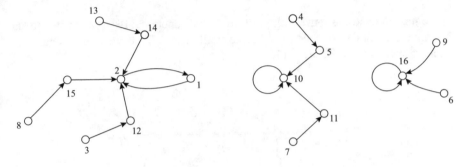

Fig. 4.14 Appending the other pieces of the vertebrate to the cycle diagram.

We can clearly see that this is a functional digraph.

The algorithm can be easily reversed. We illustrate this by constructing the vertebrate corresponding to the functional digraph depicted in Figure 4.10. To do this, we first remove the *cycle diagram* from the functional digraph, in the present case obtaining the diagram in Figure 4.15.

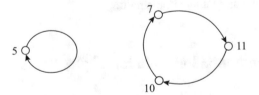

Fig. 4.15 Removing the cycle from the functional digraph of Figure 4.10.

We then construct the permutation corresponding to this diagram. Using two-line notation, we get in this case

$$\begin{pmatrix} 5 & 7 & 10 & 11 \\ 5 & 11 & 7 & 10 \end{pmatrix}.$$

The bottom line of the permutation then yields the vertebral column of the desired vertebrate oriented from tail bone "T" to head bone "H." This gives the object in Figure 4.16.

Fig. 4.16 The vertebral column.

We next append the remaining portions of the functional digraph on the nodes they were previously attached to and remove all arrows. Thus we finally obtain the resulting vertebrate in Figure 4.17.

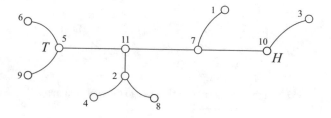

Fig. 4.17 The vertebrate corresponding the function in Figure 4.10.

It is not difficult to see that the algorithm we have presented yields a bijection between the vertebrates on n nodes and the function mapping $[n]$ to $[n]$.

This establishes the identity in (4.16).

Canceling the factor n^2 from both sides of (4.16) yields Cayley's formula by a completely different and rather simple path.

4.6   A Determinantal Formula for Cayley Trees

In this section we shall present a remarkable algebraic formula producing in a global manner all Cayley trees at once. This formula yields several results by appropriate specializations of the variables appearing in it. As we shall later see in Section 8.6.1 of Chapter 8, one of these specializations gives a useful result of graph theory often referred to as the *matrix-tree* theorem.

The reader should have no difficulty following our presentation up to the statement of the determinantal formula. The proof of the result itself is not difficult but since it may require a bit more stamina than the rest of the section, it may be skipped at first reading.

Our arguments here are based on yet a third encoding for Cayley trees, one which may be viewed as the most natural and the easiest to implement, it stems from the following considerations.

In a real life tree the sap flows from the roots to the outermost branches. We can use this idea to give an orientation to each branch of a rooted Cayley tree. More precisely, on each branch we draw an arrowhead giving the direction of the sap flow in a real tree. We might refer to this as the *sap* orientation.

For instance, consider the tree in Figure 4.18. The sap orientation of T rooted at 5, would be as shown in Figure 4.19.

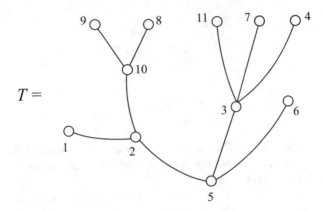

Fig. 4.18 A Cayley tree T on $n = 11$ vertices.

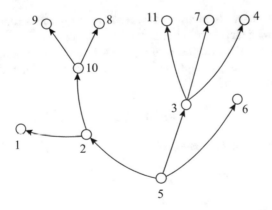

Fig. 4.19 The sap orientation of the tree T in Figure 4.18.

Our encoding is actually based on the opposite orientation, which we shall refer to as the *antisap* orientation. For instance, the antisap orientation of the tree T in Figure 4.18 is obtained by reversing all the arrows in Figure 4.19. This orientation of T is as shown in Figure 4.20.

From this orientation we construct what we shall call the *antisap* word of T, briefly denoted by $asw(T)$ as follows. For each edge (i, j) of T we place the letter $a_{i,j}$ in $asw(T)$ or the letter $a_{j,i}$ according as the antisap orientation of (i, j) is

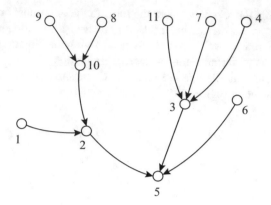

Fig. 4.20 The antisap orientation of the tree T in Figure 4.18.

from i to j or from j to i. When there is no possibility of confusion, we will use the symbols a_{ij} and a_{ji} without the comma in between the subscripts so as not to clutter up the notation unnecessarily.

The order in which these letters are inserted in $asw(T)$ itself is immaterial, but to be specific we shall write them in order of increasing first index. This is best communicated by an example. For instance, from Figure 4.20 we get

$$asw(T) = a_{12} \, a_{25} \, a_{35} \, a_{43} \, a_{65} \, a_{73} \, a_{8,10} \, a_{9,10} \, a_{10,2} \, a_{11,3} \, .$$

In Figure 4.21 we have listed all 16 Cayley trees on 4 nodes rooted at node with label 4 together with their corresponding antisap words.

Note that in each of the diagrams in Figure 4.21 there is one and only one arrow coming out of every node except the root. Thus for each of these words we must chose one and only one of

$$a_{12}, \, a_{13}, \, a_{14}$$

for the first letter, one and only one of

$$a_{21}, \, a_{23}, \, a_{24}$$

for the second letter, one and only one of

$$a_{31}, \, a_{32}, \, a_{34}$$

for the third letter. Indeed, the list of antisap words in Figure 4.21 was constructed by carrying out all the multiplications in the expression

$$(a_{12} + a_{13} + a_{14})(a_{21} + a_{23} + a_{24})(a_{31} + a_{32} + a_{34}) \, , \qquad (4.18)$$

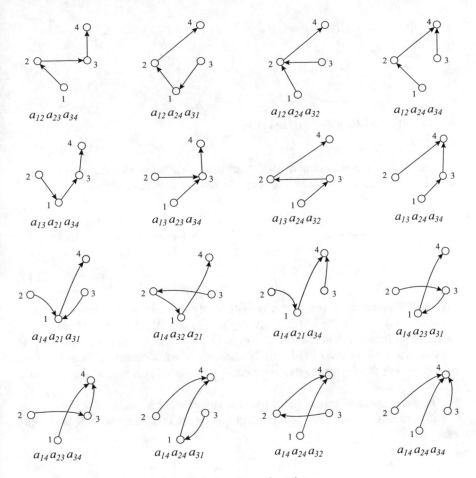

Fig. 4.21 Cayley trees on 4 nodes with their antisap orientation.

drawing a digraph from each of the resulting words, and discarding those that do not correspond to trees. Indeed, each of these words may be associated with a digraph in an obvious manner. For instance, the two words obtained after multiplying out (4.18)

$$a_{12}a_{23}a_{31} \quad \text{and} \quad a_{12}a_{21}a_{34}$$

suggest the digraphs in Figure 4.22.

A quick check reveals that each of the discarded words corresponds to a digraph with a cyclical component.

We may wonder at this point whether or not there might be an expression similar to that in (4.18) which yields only tree words. Indeed, we have a remarkable formula which is very closely related to the expression in (4.18). For convenience let us

Fig. 4.22 The digraphs of $a_{12}a_{23}a_{31}$ and $a_{12}a_{21}a_{34}$.

denote by $C_n(i)$ the collection of all Cayley trees on n nodes rooted at the node with label i. For the case of 4 nodes, our formula may be stated as follows

$$\sum_{T \in C_4(4)} asw(T) = \det \begin{bmatrix} a_{12} + a_{13} + a_{14} & -a_{12} & -a_{13} \\ -a_{21} & a_{21} + a_{23} + a_{24} & -a_{23} \\ -a_{31} & -a_{32} & a_{31} + a_{32} + a_{34} \end{bmatrix}$$

$$(4.19)$$

With a little patience the reader can verify directly that this determinant is precisely equal to the sum of all the words appearing in Figure 4.21, no more no less.

It should be noted that the expression in (4.18) is the product corresponding to the diagonal of the matrix in (4.19). Remarkably, the additional products obtained by calculating the determinants do precisely the job of eliminating the unwanted words.

The same result of course holds true for any number of nodes. For instance, to obtain the sum of all antisap words of Cayley trees on 3 nodes rooted at 3, we simply evaluate

$$\det \begin{bmatrix} a_{12} + a_{13} & -a_{12} \\ -a_{21} & a_{21} + a_{23} \end{bmatrix}.$$

This is

$$(a_{12} + a_{13})(a_{21} + a_{23}) - a_{12}a_{21} = a_{12}a_{23} + a_{13}a_{21} + a_{13}a_{23} ,$$

and as we see, the oriented trees in question are the ones in Figure 4.23 with their antisap words

$$a_{13}a_{21}, \ a_{12}a_{23}, \ a_{13}a_{23} .$$

We aim to state and prove the corresponding result for an arbitrary number of nodes. It is convenient to let the number of nodes be $n + 1$. In this case we use $n(n + 1)$ letters

$$a_{ij} \ \text{for} \ i, j = 1, 2, \ldots, n + 1, \ (i \neq j) .$$

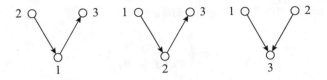

Fig. 4.23 Cayley trees on 3 nodes with antisap words $a_{13}a_{21}, a_{12}a_{23}, a_{13}a_{23}$.

To simplify our presentation it is good to make the convention that

$$a_{ii} = 0 \quad \text{for} \quad i = 1, 2, \ldots, n+1, \qquad (4.20)$$

and set

$$R_i = a_{i1} + a_{i2} + \cdots + a_{i,n+1} \qquad (i = 1, 2, \ldots, n+1). \qquad (4.21)$$

Also, let us make the convention that the simpler notation

$$C_{n+1}$$

denotes Cayley trees on $n+1$ nodes rooted at $n+1$, i.e., what we called $C_{n+1}(n+1)$. This given, the general result may be stated as follows.

Theorem 4.6.1 *The sum of all antisap words of Cayley trees on $n + 1$ nodes rooted at $n + 1$ is given by the formula*

$$\sum_{T \in C_{n+1}} asw(T) = \det \begin{bmatrix} R_1 & -a_{12} & -a_{13} & \ldots & -a_{1n} \\ -a_{21} & R_2 & -a_{23} & \ldots & -a_{2n} \\ -a_{31} & -a_{32} & R_3 & \ldots & -a_{3n} \\ \vdots & \vdots & \vdots & & \vdots \\ -a_{n1} & -a_{n2} & -a_{n3} & \ldots & R_n \end{bmatrix}. \qquad (4.22)$$

Before we can proceed with a proof of this theorem we need a few preliminary observations.

Recall first that the determinant of an $n \times n$ matrix

$$B = [b_{i,j}]$$

is given by the formula

$$\det B = \sum_{\sigma} (-1)^{i(\sigma)} b_{1\sigma_1} b_{2\sigma_2} \cdots b_{n\sigma_n}, \qquad (4.23)$$

where the sum is carried out over all permutations

$$\sigma = \begin{pmatrix} 1 & 2 & \dots & n \\ \sigma_1 & \sigma_2 & \dots & \sigma_n \end{pmatrix}$$

and $i(\sigma)$ denotes the number of inversions of σ.

To calculate the determinant of the matrix in (4.22) we need to set in (4.23)

$$b_{i,j} = \begin{cases} R_i & \text{if } i = j, \\ -a_{ij} & \text{if } i \neq j. \end{cases} \qquad (4.24)$$

Using the convention in (4.20) we may rewrite this in the form

$$b_{i,j} = R_i \delta_{i,j} - a_{ij},$$

where

$$\delta_{i,j} = \begin{cases} 1 & \text{if } i = j, \\ 0 & \text{if } i \neq j. \end{cases} \qquad (4.25)$$

For convenience let M denote the matrix in (4.22).

With these conventions we obtain that

$$\det M = \sum_{\sigma} (-1)^{i(\sigma)} \prod_{i=1}^{n} \left(R_i \delta_{i\sigma_i} - a_{i\sigma_i} \right) . \qquad (4.26)$$

The product in this expression may be expanded in the form

$$\prod_{i=1}^{n} \left(R_i \delta_{i\sigma_i} - a_{i\sigma_i} \right) = \sum_{S \subseteq [n]} \prod_{i \in \overline{S}} (R_i \delta_{i\sigma_i}) \prod_{i \in S} (-a_{i\sigma_i}),$$

where $\overline{S}$ denotes the complement of S in $[n]$.

Substituting in (4.26) and changing order of summation yields

$$\det M = \sum_{S \subseteq [n]} \sum_{\sigma} (-1)^{i(\sigma)} \prod_{i \in \overline{S}} (R_i \delta_{i\sigma_i}) \prod_{i \in S} (-a_{i\sigma_i}) . \qquad (4.27)$$

Note now that the product

$$\prod_{i \in \overline{S}} R_i \delta_{i\sigma_i}$$

is different from zero if and only if each of the factors δ_{i,σ_i} is different from zero. In view of (4.25), this holds if and only if for the permutation σ we have

$$\sigma_i = i \quad \text{for each } i \in \overline{S}. \tag{4.28}$$

Moreover, since each a_{ii} is equal to zero, the product

$$\prod_{i \in S}(-a_{i\sigma_i})$$

is different from zero if and only if

$$\sigma_i \neq i \quad \text{for each } i \in S. \tag{4.29}$$

We shall refer to a permutation σ satisfying the conditions in (4.28) and (4.29) as a *derangement* of the set S. The set of all such permutations will be denoted by $D(S)$. This given, we may rewrite (4.27) in the form

$$\det M = \sum_{S \subseteq [n]} \sum_{\sigma \in D(S)} (-1)^{i(\sigma)+|S|} \prod_{i \in \overline{S}} R_i \prod_{i \in S} a_{i\sigma_i}. \tag{4.30}$$

Some further simplifications are needed. First of all recall that if a permutation σ acts on a set S, Theorem 2.5.1 of Chapter 2 implies that

$$\text{sign}(\sigma) = (-1)^{i(\sigma)} = (-1)^{|S|+k(\sigma)}, \tag{4.31}$$

where $k(\sigma)$ is the number of cycles in the diagram of σ. For a given σ acting on a set S we may also set

$$a(\sigma) = \prod_{i \in S} a_{i\sigma_i} \tag{4.32}$$

and refer to it as the *word* of σ. Clearly from $a(\sigma)$ we may recover σ itself. Indeed, we may construct the cycle diagram of σ from $a(\sigma)$ precisely in the same manner we construct a rooted Cayley tree T with its antisap orientation from $asw(T)$.

Finally we must expand the product

$$\prod_{i \in \overline{S}} R_i$$

and give it a combinatorial interpretation. This is best accomplished if we work on a special case. Suppose that $n = 4$ and $\overline{S} = \{1, 3, 4\}$. Then in view of (4.21) we have

$$\prod_{i \in \overline{S}} R_i = R_1 R_3 R_4 = (a_{12}+a_{13}+a_{14}+a_{15})(a_{31}+a_{32}+a_{34}+a_{35})(a_{41}+a_{42}+a_{43}+a_{45}) \,.$$

(4.33)

Now each word obtained by carrying out the multiplications in this expression corresponds to a functional digraph. For instance, the words

$$a_{13}a_{35}a_{42}, \ a_{13}a_{34}a_{41}$$

correspond to the digraphs in Figure 4.24.

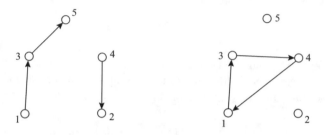

Fig. 4.24 Digraphs of the words $a_{13}a_{35}a_{42}, a_{13}a_{34}a_{41}$.

Two features are common to all the digraphs corresponding to the terms coming out of (4.33). They all represent maps of the set $\{1, 3, 4\}$ into $\{1, 2, 3, 4, 5\}$. Furthermore, because of the absence of the letters

$$a_{11}, \ a_{33}, \ a_{44} \,,$$

none of these maps have fixed points, or which is the same, none of the corresponding digraphs have loops.

It will be convenient, given a subset $S \subseteq [n]$, to denote by $M(\overline{S})$ the collection of all fixed point free maps of $\overline{S}$ into $[n + 1]$. To each $f \in M(\overline{S})$ we shall associate a word $w(f)$ by setting

$$w(f) = \prod_{i \in \overline{S}} a_{i, f(i)} \,.$$

Clearly, the same word $w(f)$ can be obtained by drawing the functional digraph of f, replacing each edge $i \to j$ (oriented from i to j) by the letter a_{ij} and inserting these letters in $w(f)$ in order of increasing first index.

This given, we see that the expression in (4.33) may be rewritten in the form

$$R_1 R_3 R_4 = \sum_{f \in M(\{1,3,4\})} w(f).$$

In general we may write

$$\prod_{i \in \bar{S}} R_i = \sum_{f \in M(\bar{S})} w(f).$$

Substituting (4.31), (4.32) and this expression in (4.30) we finally obtain that

$$\det M = \sum_{S \subseteq [n]} \sum_{\sigma \in D(S)} \sum_{f \in M(\bar{S})} (-1)^{k(\sigma)} w(f) a(\sigma). \tag{4.34}$$

We are now finally in a position to establish Theorem 4.6.1.

Let f be a map of [5] into [6] and let Δ denote its corresponding digraph. Note that if f is cycle-free then Δ must be a tree. Moreover, Δ has 5 edges, namely

$$1 \to f(1), \ 2 \to f(2), \ 3 \to f(3), \ 4 \to f(4), \ 5 \to f(5).$$

Now, a tree with 5 edges has (by Euler's formula) 6 nodes. Thus 6 must be one of the nodes of Δ. This means that we must have $f(i) = 6$ for some i. Since each of the nodes $1, 2, \ldots, 5$ has an edge coming out of it, if we take the successive images by f of any node $j = 1, 2, \ldots, 5$ we must eventually end up at 6. To put it bluntly Δ must necessarily be a Cayley tree rooted at 6 with its antisap orientation.

An example of a cycle-free map f of [5] into [6] could be

i	1	2	3	4	5
$f(i)$	4	3	4	6	3

with the corresponding digraph as shown in Figure 4.25.

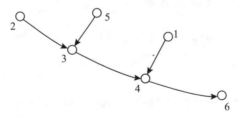

Fig. 4.25 The digraph of the cycle-free map f.

and we can easily recognize our Cayley tree here.

Conversely, every Cayley tree on 6 nodes rooted at 6 with its antisap orientation may be interpreted as the digraph of a cycle-free map of [5] into [6].

Clearly the same reasoning applies for any number of nodes. The conclusion is that Cayley trees on $n+1$ nodes rooted at $n+1$ and functional digraphs of cycle-free maps of $[n]$ into $[n+1]$ are one and the same thing. Thus letting $CFM(n)$ denote the collection of all cycle-free maps of $[n]$ into $[n+1]$ we must have

$$\sum_{T \in C_{n+1}} as\, w(T) = \sum_{f \in CFM(n)} w(f) \,.$$

This given, in view of (4.34), our theorem will follow if we prove that

$$\sum_{S \subseteq [n]} \sum_{\sigma \in D(S)} \sum_{f \in M(\overline{S})} (-1)^{k(\sigma)} w(f) a(\sigma) = \sum_{f \in CFM(n)} w(f) \,. \tag{4.35}$$

Proof Let us represent each term $w(f)a(\sigma)$ by a *two-color* digraph obtained by superimposing the digraph of f (say in Blue) with the cycle diagram of σ (say in Red).

For instance, when $n = 15$, $S = \{3, 4, 8, 9, 10, 11, 12\}$ we may have the digraph in Figure 4.26 where the colors are indicated by B and R.

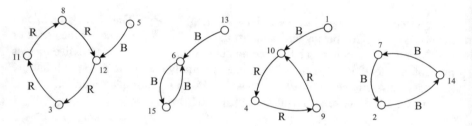

Fig. 4.26 Two-color digraph of f and σ (B and R, respectively).

The reader is urged to color the edges labeled B in blue and those labeled R in red for a better view of the argument.

From the blue digraph we then get

$$w(f) = a_{1,10}\, a_{2,14}\, a_{5,12}\, a_{6,15}\, a_{7,2}\, a_{13,6}\, a_{14,7}\, a_{15,6} \,, \tag{4.36}$$

and the red one gives

$$a(\sigma) = a_{3,11}\, a_{4,9}\, a_{8,12}\, a_{9,10}\, a_{10,4}\, a_{11,8}\, a_{12,3} \,.$$

Since $k(\sigma) = 2$, the combined digraph contributes the term

$$(-1)^2 w(f)a(\sigma) = w(f)a(\sigma) \tag{4.37}$$

$$= a_{1,10}\, a_{2,14}\, a_{5,12}\, a_{6,15}\, a_{7,2}\, a_{13,6}\, a_{14,7}\, a_{15,6}\, a_{3,11}\, a_{4,9}\, a_{8,12}\, a_{9,10}\, a_{10,4}\, a_{11,8}\, a_{12,3}$$

to the left hand side of (4.35).

Observe now that the digraph obtained by changing the color of the cycle $(2, 14, 7)$ from blue to red, namely the digraph in Figure 4.27, is also one of the two-color digraphs coming out of (4.35).

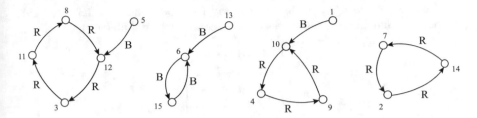

Fig. 4.27 Changing the color of the cycle $(2, 14, 7)$ from B to R.

More precisely, this is the composite digraph corresponding to the pair f', σ' with associated words

$$w(f') = a_{1,10}\, a_{5,12}\, a_{6,15}\, a_{13,6}\, a_{15,6} ,$$

$$a(\sigma') = a_{2,14}\, a_{3,11}\, a_{4,9}\, a_{7,2}\, a_{8,12}\, a_{9,10}\, a_{10,4}\, a_{11,8}\, a_{12,3}\, a_{14,7} .$$

Note that because of the factor $(-1)^{k(\sigma)}$ the contribution of the digraph in Figure 4.27 to the left hand side of (4.35) is now

$$(-1)^3 w(f') a(\sigma') = -w(f) a(\sigma) .$$

Comparing with (4.37) we see that the pair of digraphs in Figure 4.26 and Figure 4.27 contribute nothing to the sum on the left hand side of (4.35) since their corresponding terms cancel each other out.

We discover that each two-color digraph with cycles is a member of such a pair. Indeed, the color change that led to Figure 4.27 from Figure 4.26 is a particular case of a general rule which may be formulated as follows:

> Given a two-color digraph with cycles, look for the cycle which contains the node with smallest label. If this cycle is red change it to blue and if it is blue change it to red.

It is easy to see that this produces a pairing of two-color digraphs whose corresponding terms, as was the case for the example considered, cancel each other.

From this pairing we deduce that if we carry out the sum on the left hand side of (4.35) the only terms that survive are those which correspond to digraphs with no cycles. Forcedly such digraphs have to be all blue. Moreover they must be digraphs of cycle-free maps of $[n]$ into $[n + 1]$.

In other words, what remains of the sum on the left hand side of (4.35), after all these cancellations is precisely the sum on the right hand side.

This completes the proof of Theorem 4.6.1.   □

Remark 4.6.1 In the sequel we shall derive a number of applications of formula (4.22). However, here we cannot escape the temptation to mention a rather curious fact which comes out of (4.22). This is seen as follows.

If we replace each a_{ij} by 1 in (4.22), say for $n = 4$ then the left hand side becomes the number of Cayley trees on 5 nodes while the right hand side becomes

$$\det \begin{bmatrix} 4 & -1 & -1 & -1 \\ -1 & 4 & -1 & -1 \\ -1 & -1 & 4 & -1 \\ -1 & -1 & -1 & 4 \end{bmatrix}.$$

In other words Theorem 4.6.1 implies that the value of this determinant is 5^3.
More generally, the same specialization of the a_{ij}'s yields

$$\det \begin{bmatrix} n & -1 & -1 & \dots & -1 \\ -1 & n & -1 & \dots & -1 \\ -1 & -1 & n & \dots & -1 \\ \vdots & \vdots & \vdots & & \vdots \\ -1 & -1 & -1 & \dots & n \end{bmatrix} = (n+1)^{n-1}. \tag{4.38}$$

The reader is urged to find a direct proof of this determinantal identity and thereby derive from Theorem 4.6.1 yet a third proof of Cayley's theorem.

Remark 4.6.2 Note that the Cayley trees on 3 nodes rooted at 2 with antisap orientation are as shown in Figure 4.28.

Fig. 4.28 Cayley trees in $C_3(2)$ with the corresponding words $a_{12}a_{31}, a_{12}a_{32}, a_{13}a_{32}$.

The sum of the corresponding antisap words is

$$\sum_{T \in C_3(2)} asw(T) = a_{12}a_{31} + a_{12}a_{32} + a_{13}a_{32}. \tag{4.39}$$

If we calculate the cofactor of any element of the second row of the matrix

$$\begin{bmatrix} a_{12} + a_{13} & -a_{12} & -a_{13} \\ -a_{21} & a_{21} + a_{23} & -a_{23} \\ -a_{31} & -a_{32} & a_{31} + a_{32} \end{bmatrix}, \tag{4.40}$$

the result will be precisely the sum in (4.39). We recall that the cofactor of the (i, j)th element of a matrix B is defined to be the product of $(-1)^{i+j}$ and the determinant of the matrix obtained by erasing the ith row and the jth column of B.

For instance, the cofactor of the $(2, 1)$th element of the matrix in (4.40) is

$$(-1)^{1+2} \det \begin{bmatrix} -a_{12} & -a_{13} \\ -a_{32} & a_{31} + a_{32} \end{bmatrix} = -(-a_{12}a_{31} - a_{12}a_{32} - a_{13}a_{32}) \, ,$$

which is precisely the expression in (4.39).

This is no accident. The general result may be formulated as follows:

Theorem 4.6.2 *The cofactor of any element of the ith row of the $(n + 1) \times (n + 1)$ matrix*

$$\begin{bmatrix} R_1 & -a_{12} & -a_{13} & \cdots & -a_{1,n+1} \\ -a_{21} & R_2 & -a_{23} & \cdots & -a_{2,n+1} \\ -a_{31} & -a_{32} & R_3 & \cdots & -a_{3,n+1} \\ \vdots & \vdots & \vdots & & \vdots \\ -a_{n+1,1} & -a_{n+1,2} & -a_{n+1,3} & \cdots & R_{n+1} \end{bmatrix} \quad (4.41)$$

is equal to

$$\sum_{T \in C_{n+1}(i)} asw(T) \, ,$$

i.e., the sum of all antisap words of Cayley trees on $n + 1$ nodes rooted at i.

This result of course contains Theorem 4.6.1 as a special case in which $i = n+1$. The assertion concerning cofactors of diagonal elements of the matrix in (4.41) is very easy to derive from Theorem 4.6.1. For the off-diagonal elements the proof demands additional considerations not pertinent to our treatment here and will not be given.

The reader is urged to discover by experimentation the meaning of the determinant of the matrix obtained by removing 2 rows and 2 columns from (4.41) or better yet, any number k of rows and k columns.

4.7 Extensions and Applications

By our determinantal formula of Theorem 4.6.1 we have

$$\sum_{T \in C_{n+1}} asw(T) = \det \begin{bmatrix} R_1 & -a_{12} & -a_{13} & \cdots & -a_{1n} \\ -a_{21} & R_2 & -a_{23} & \cdots & -a_{2n} \\ -a_{31} & -a_{32} & R_3 & \cdots & -a_{3n} \\ \vdots & \vdots & \vdots & & \vdots \\ -a_{n1} & -a_{n2} & -a_{n3} & \cdots & R_n \end{bmatrix} \, . \quad (4.42)$$

Here C_{n+1} denotes all Cayley trees on $n+1$ nodes rooted at $n+1$. In our encoding, we use letters a_{ij} indexed by $i, j = 1, 2, \ldots, n+1$ with

$$a_{ii} = 0 \text{ for } i = 1, 2, \ldots, n+1,$$

and

$$R_i = a_{i1} + a_{i2} + \cdots + a_{i,n+1} \text{ for } i = 1, 2, \ldots, n+1.$$

We recall that $asw(T)$ is obtained by putting together for every edge (i, j) in T the letter a_{ij} or the letter a_{ji} in $asw(T)$ according as the orientation of (i, j) is from i to j or from j to i towards the root $n+1$ in T.

We can use the formula (4.42) by specializing the variables in the $asw(T)$ in various ways. Suppose, for example, that we are interested in keeping track of the number of *fall* edges, that is edges that are oriented from i to j with $i > j$ in T. Let $F(T)$ denote the number of fall edges of $T \in C_{n+1}$ and define the monomial of T as

$$m(T) = x^{F(T)}.$$

For example, for the tree T in Figure 4.30 we have two fall edges $5 \to 2$ and $6 \to 4$, and so

$$m(T) = x^2.$$

This monomial can be obtained from $asw(T)$ by setting $a_{ij} = 1$ if $i < j$ and $a_{ij} = x$ if $i > j$. We may as well do this substitution before we evaluate the determinant in (4.42). The diagonal term R_i with this substitution becomes

$$(i - 1)x + n - i + 1$$

for $i = 1, 2, \ldots, n$. Therefore the generating function for Cayley trees in C_{n+1} by the number of fall edges in their antisap orientation is given by

$$\sum_{T \in C_{n+1}} x^{F(T)} = \det \begin{bmatrix} n & -1 & -1 & \cdots & -1 \\ -x & x+n-1 & -1 & \cdots & -1 \\ -x & -x & 2x+n-2 & \cdots & -1 \\ \vdots & \vdots & \vdots & & \vdots \\ -x & -x & -x & \cdots & (n-1)x+1 \end{bmatrix}.$$

This determinant can be evaluated with some algebraic effort, and it turns out that

$$\sum_{T \in C_{n+1}} x^{F(T)} = (x+n)(2x+n-1)(3x+n-2)\cdots((n-1)x+2). \qquad (4.43)$$

In this section, we consider another bijection for the number of Cayley trees. The bijection not only gives another proof of Cayley's theorem, but has a number of natural weight-preserving properties that immediately provide proofs of identities similar to (4.43) without resorting to algebraic manipulations for determinant evaluation.

To this end, let us denote by $\mathcal{F}_{n+1}$ the set of functions f from $\{2, 3, \ldots, n\}$ into $\{1, 2 \ldots, n+1\}$. Evidently the number of such functions is

$$|\mathcal{F}_{n+1}| = (n+1)^{n-1}.$$

As before C_{n+1} denotes the set of Cayley trees on $n+1$ nodes with their antisap orientation.

Our bijection *ER* [5] is between $\mathcal{F}_{n+1}$ and C_{n+1}. It is most easily described by referring to an explicit example. Suppose $n+1 = 21$ and f is given by

i	2	3	4	5	6	7	8	9	10	11	12	13	14	15	16	17	18	19	20
$f(i)$	5	4	5	3	21	7	12	1	4	4	20	19	19	6	1	16	6	7	12

We again consider the digraph of f, i.e., we draw f as a directed graph with vertex set [21] by putting a directed edge from i to j if $f(i) = j$. For example, the digraph for f given above is pictured on the upper part of Figure 4.29.

A moment's thought will convince one that in general, the digraph corresponding to an $f \in \mathcal{F}_{n+1}$ will consist of two trees rooted at 1 and $n+1$, respectively, with all edges directed toward their roots; plus a number of directed cycles of length ≥ 1 where for each vertex v on any given cycle there is possibly a tree attached to v with v as the root and all edges directed toward v. Note that there are trees rooted at 1 and $n+1$ due to the fact that these are not in the domain of f, and consequently there are no directed edges out of 1 or $n+1$. Note also that cycles of length 1 or loops simply correspond to fixed points of f.

To describe the bijection, we imagine the digraph of f is drawn as in Figure 4.29 so that

(a) *The trees rooted at 1 and $n+1$ are drawn on the extreme left and extreme right, respectively, with their edges directed upwards,*

(b) *The cycles are drawn so that their vertices form a directed path on the line between 1 and $n+1$ with one backedge above the line. The tree attached to any vertex on a cycle is drawn below the line between 1 and $n+1$ with edges directed upwards,*

(c) *Each cycle is arranged so that its smallest element is at the right and the cycles themselves are drawn ordered from left to right by increasing smallest element.*

Once the directed graph for f is drawn as above, let us refer the rightmost element in the ith cycle reading from left to right as r_i, and the leftmost element in the ith cycle as l_i. Thus for the f given above,

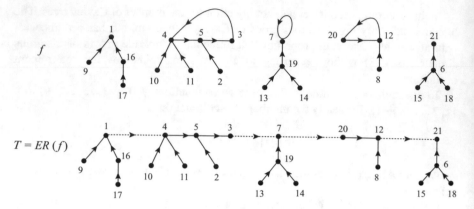

Fig. 4.29 The bijection: the digraph of f above, and the corresponding Cayley tree $ER(f)$ below.

$$l_1 = 4, r_1 = 3, \quad l_2 = r_2 = 7, \quad l_3 = 20, r_3 = 12 .$$

Once the digraph of an f is drawn in this way, it is easy to describe $ER(f)$:

If the directed graph of f has k cycles where $k > 0$, simply eliminate the backedges

$$r_i \rightarrow l_i, \quad \text{for} \quad i = 1, 2, \ldots, k$$

and add the edges

$$1 \rightarrow l_1, \; r_l \rightarrow l_2, \; r_2 \rightarrow l_3, \ldots, \; r_k \rightarrow n + 1 .$$

For example, in Figure 4.29, we eliminate the backedges

$$3 \rightarrow 4, \; 7 \rightarrow 7, \; 12 \rightarrow 20 ,$$

from the digraph of f given on the upper part of the figure and add the edges

$$1 \rightarrow 4, \; 3 \rightarrow 7, \; 7 \rightarrow 20, \quad \text{and} \quad 12 \rightarrow 21$$

which are dotted at the bottom of the figure for emphasis. If there are no cycles in the directed graph of f, i.e., if $k = 0$, then we simply add the edge $1 \rightarrow n + 1$ to the digraph of f.

Note that it is immediate that ER is a bijection between $\mathcal{F}_{n+1}$ and $\mathcal{C}_{n+1}$ since given any Cayley tree $T \in \mathcal{C}_{n+1}$, we can easily recover the directed graph of the $f \in \mathcal{F}_{n+1}$ such that $ER(f) = T$.

The key point here is that by our conventions for ordering the cycles of f when we draw its digraph, it is easy to recover the sequence of nodes $r_1, r_2, \ldots, r_k$; r_1 is the smallest element on the path between 1 and $n + 1$ in T, r_2 is the smallest element on the path between r_1 and $n + 1$, etc., and clearly, knowing $r_1, r_2, \ldots, r_k$ allows us to recover f from T.

Of course the bijection ER proves Cayley's theorem that

$$|C_{n+1}| = (n + 1)^{n-1} .$$

But in fact it proves much more. To state the general result we need some notation.

The sense of direction of the antisap orientation allows us to define a weight on every edge. If edge $e = (i, j)$ is oriented from i to j in the antisap orientation, we denote this by $i \to j$ and assign it the weight

$$w(i \to j) = \begin{cases} xq^i t^j & \text{if } i > j \ \text{(a fall)}, \\ yp^i s^j & \text{if } i < j \ \text{(a rise)}. \end{cases}$$

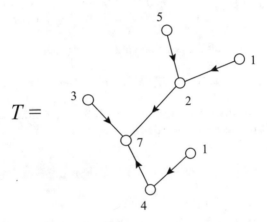

Fig. 4.30 A Cayley tree T with $n + 1 = 7$ nodes rooted at 7 with $asw(T) = a_{12}a_{27}a_{37}a_{47}a_{52}a_{64}$.

So, for example, for the tree T pictured in Figure 4.30, we have

$$w(5 \to 2) = xq^5 t^2 \quad \text{and} \quad w(1 \to 2) = yps^2 .$$

We define the weight of the whole tree as the product of the weights of its edges:

$$w(T) = \prod_{e \in E(T)} w(e) .$$

For the tree given in Figure 4.30 this gives the weight

$$w(T) = (yps^2)(xq^5 t^2)(yp^2 s^7)(yp^3 s^7)(yp^3 s^7)(xq^6 t^4) .$$

We have the following generalization of Cayley's theorem:

Theorem 4.7.1

$$\sum_{T \in C_{n+1}} w(T) = y p s^{n+1} \prod_{i=2}^{n} \left[x q^i (t + t^2 + \cdots + t^{i-1}) + y p^i (s^i + s^{i+1} + \cdots + s^{n+1}) \right].$$

$$(4.44)$$

Proof Given $f \in \mathcal{F}_{n+1}$, we define the weight of f, which we denote by $W(f)$, as follows:

$$W(f) = \prod_{i=2}^{n} W(f, i),$$

where

$$W(f, i) = \begin{cases} x q^i t^j & \text{if } f(i) = j \text{ and } i > j \quad \text{(a fall)}, \\ y p^i s^j & \text{if } f(i) = j \text{ and } i \leq j \quad \text{(a rise)}. \end{cases}$$

For fixed i, the sum over all the possible values of $W(f, i)$ is simply

$$x q^i t + x q^i t^2 + \cdots + x q^i t^{i-1} + y p^i s^i + y p^i s^{i+1} + \cdots + y p^i s^{n+1}. \qquad (4.45)$$

Since the choices for the values of f at $i = 2, 3, \ldots, n$ are made independently according to (4.45), it easily follows that

$$\sum_{f \in \mathcal{F}_{n+1}} W(f) = \prod_{i=2}^{n} \left[x q^i (t + t^2 + \cdots + t^{i-1}) + y p^i (s^i + s^{i+1} + \cdots + s^{n+1}) \right].$$

$$(4.46)$$

To prove (4.44), it is enough to show that when going from f to the corresponding Cayley tree $T = ER(f)$, the weight changes as follows:

$$w(T) = y p s^{n+1} W(f). \qquad (4.47)$$

When the digraph of f has no cycles, the only edge added is $1 \rightarrow n + 1$, which contributes the factor $y p s^{n+1}$ to $W(f)$, so (4.47) holds in this case.

When the digraph of f has $k > 0$ cycles, the only difference between the contribution of the weights of f and T are due to the difference between the weights

$$r_1 \rightarrow l_1, \ r_2 \rightarrow l_2, \ldots, \ r_k \rightarrow l_k$$

which are deleted from the functional digraph and

$$1 \rightarrow l_1, \ r_1 \rightarrow l_2, \ldots, \ r_{k-1} \rightarrow l_k, \ r_k \rightarrow n + 1$$

subsequently added to obtain T.

Let us consider the case of our example function f and $T = ER(f)$. Looking at the backedges that are eliminated in Figure 4.29 from left to right, we see that we lose the factors

$$yp^3s^4, \ yp^7s^7, \ yp^{12}s^{20} \tag{4.48}$$

from $W(f)$. Note that our cycles have their smallest element on the right and therefore these lost weights are all of the type yp^is^j. This also holds for $i = j$ because of the way we defined the weight of $W(f, i)$ by making $f(i) = i$ a rise.

Therefore

$$W(f) = (yp^3s^4)(yp^7s^7)(yp^{12}s^{20})\Theta(f), \tag{4.49}$$

where $\Theta(f)$ is the product of the weights of the remaining edges in the digraph of f.

The edges added to obtain T from left to right, which are shown as dotted lines in Figure 4.29, contribute the weights

$$yps^4, \ yp^3s^7, \ yp^7s^{20}, \ yp^{12}s^{21} \ . \tag{4.50}$$

Therefore for $w(T)$, in addition to $\Theta(f)$, we need the product of the weights in (4.49), i.e.,

$$w(T) = (yps^4)(yp^3s^7)(yp^7s^{20})(yp^{12}s^{21})\Theta(f) \ . \tag{4.51}$$

Putting (4.49) and (4.51) together, we have

$$w(T) = \frac{(yps^4)(yp^3s^7)(yp^7s^{20})(yp^{12}s^{21})}{(yp^3s^4)(yp^7s^7)(yp^{12}s^{20})} W(f) = yps^{21} W(f) \ . \tag{4.52}$$

We note that for this example $n + 1 = 21$.

It is not a coincidence of course that the canceling the corresponding factors from the numerator and the denominator in (4.52) leaves nothing but yps^{n+1}. This holds for the general case as well. □

We can check that if we specialize the general result in (4.44) by setting $p = q = r = s = y = 1$ and thus leaving only x in the weight of the tree, we immediately obtain the expression in (4.43) for the generating function for Cayley trees by the number of fall edges.

With minor modifications, the bijection ER can be tailored to work for Cayley trees rooted at an arbitrary vertex labeled i, not just $n + 1$. Furthermore essentially the same idea with cycles and breaking backedges can be used to derive bijective proofs for number of spanning trees of various other graph families as well.

It should be mentioned that these bijections can be viewed as a kind of merging of the Joyal encoding and the so-called *fundamental transform* of Foata on permutations [9].

4.8 Exercises for Chapter 4

4.8.1 By constructing their neighbor lists, show that the following two figures

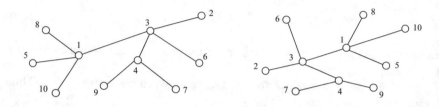

represent the same Cayley tree.

4.8.2 Construct the monomials of the following trees:

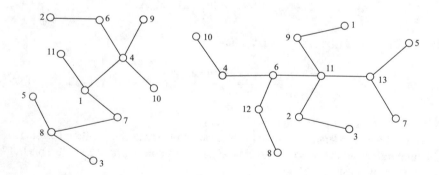

4.8.3 Construct the word of the following Cayley tree

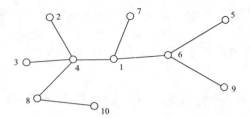

4.8.4 Put together the Cayley tree T which corresponds to the word

$$x_1^4 x_3 x_5^2 .$$

4.8.5 Construct the words corresponding to the trees T_1, T_2, T_3.

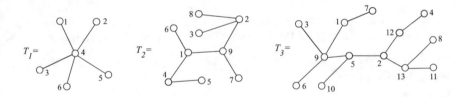

4.8.6 Put together the Cayley trees corresponding to the following words:

(a) $x_2 x_1 x_2 x_4 x_3 x_2$,

(b) $x_5 x_4 x_3 x_2 x_1 x_5 x_3$,

(c) $x_2 x_9 x_3 x_3 x_3 x_6 x_2 x_1$.

4.8.7 Indicate how a Cayley tree can be stored in a computer by specifying the neighbors of nodes of positive degree only.

4.8.8 Given a Cayley tree $T \in C_n$, put

$$v(T) = x_1^{v_1(T)} x_2^{v_2(T)} \cdots x_n^{v_n(T)},$$

where $v_i(T)$ is the valence of the node i. Show that

$$\sum_{T \in C_n} v(T) = x_1 x_2 \cdots x_n (x_1 + x_2 + \cdots + x_n)^{n-2} .$$

4.8.9 What is the number of Cayley trees on 7 nodes where the node labeled 5 has degree 2 ?

4.8.10 Calculate the number of Cayley trees on $x_1, x_2, \ldots, x_9$ where x_1 and x_2 both have degree 3.

4.8.11 What is the number of Cayley trees on $x_1, x_2, \ldots, x_9$ with the following degrees:

$$d_1 = d_2 = d_3 = d_4 = d_5 = 0, \; d_6 = 1, \; d_7 = d_8 = d_9 = 2 \,.$$

Give an example of a tree where the degrees of the nodes have the distribution given above.

4.8.12 What is the number of Cayley trees on $x_1, x_2, \ldots, x_{12}$ with the following degrees:

$$d_1 = d_2 = d_3 = 0, \; d_4 = 5, \; d_5 = 1, \; d_6 = d_7 = 2 \,.$$

Give an example of a tree where the degrees of the nodes have the distribution given above.

4.8.13 Calculate the number of Cayley forests on $x_1, x_2, \ldots, x_9$ with 4 trees such that x_1, x_2, x_3, x_4 belong to different trees. Give an example of such a forest.

4.8.14 What is the number of forests of rooted Cayley trees on $x_1, x_2, \ldots, x_9$ having exactly 4 trees?

4.8.15

(a) Construct the functional digraphs corresponding to the following functions in $[11]^{[11]}$:

i	1	2	3	4	5	6	7	8	9	10	11
$f(i)$	6	10	2	2	3	7	11	4	7	3	9

i	1	2	3	4	5	6	7	8	9	10	11
$g(i)$	7	11	10	2	5	5	5	2	5	7	1

(b) Put together the vertebrates on 11 nodes corresponding to the functions f and g in part (a).

4.8.16 Construct the functions in $[11]^{[11]}$ corresponding to the vertebrates C_1 and C_2.

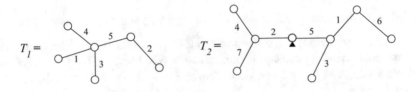

4.8.17 For $n \geq 3$, let E_n denote the collection of *edge-labeled* trees and RE_n the collection of *rooted* edge-labeled trees on n nodes. For instance, T_1 and T_2

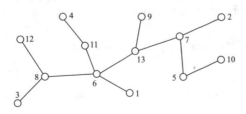

are elements of E_6 and RE_8, respectively.

(a) List the trees in E_4 and RE_3.

(b) Construct a Cayley tree corresponding to a tree $T \in RE_n$ as follows: First slide the labels away from the root of T to the nearest vertex. Next, label the root node n.

   Show that this defines a bijection between RE_n and Cayley trees on n nodes.

(c) Use part (b) to conclude that

$$|E_n| = n^{n-3}.$$

(d) Construct the rooted edge-labeled tree which corresponds to the following Cayley tree:

4.8.18

(a) Find the antisap words $asw(T_1)$ and $asw(T_2)$ of the following trees in $C_8(8)$:

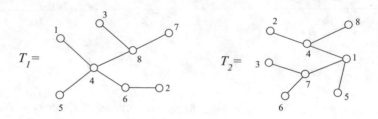

(b) What are the corresponding words if we regard T_1 and T_2 as elements of $C_8(1)$?

4.8.19 Let $C_n(i)$ denote the collection of all Cayley trees on n nodes rooted at i, with the corresponding antisap orientation as in 6.3. For $T \in C_n(i)$, replace each variable a_{rs} in $asw(T)$ by 1 or x according as $r < s$ or $r > s$. Denote the resulting monomial by $x^{F(T)}$.

(a) Verify that for the trees $C_4(1)$ depicted in Figure 4.21, the corresponding monomials $x^{F(T)}$ are as follows:

$$1 \quad x \quad x \quad 1$$
$$x \quad 1 \quad x \quad 1$$
$$x^2 \quad x^2 \quad x \quad x$$
$$1 \quad x \quad x \quad 1$$

(b) Conclude that

$$\sum_{T \in C_4(4)} x^{F(T)} = 6 + 8x + 2x^2 = (x+3)(2x+2) .$$

(c) Specialize the variables a_{rs} in (4.22) as indicated above and evaluate the resulting determinant to show that

$$\sum_{T \in C_{n+1}(n+1)} x^{F(T)} = (x+n)(2x+n-1)(3x+n-2)\cdots((n-1)x+2).$$

(d) Draw all the trees in $C_4(1)$ with their corresponding antisap orientation. Show that

$$\sum_{T \in C_4(1)} x^{F(T)} = 6x + 8x^2 + 6x^3 = x(2x+2)(3x+1) .$$

(e) Use Theorem 4.6.2 to prove that in general

$$\sum_{T \in C_{n+1}(1)} x^{F(T)} = x(2x + n - 1)(3x + n - 2) \cdots (nx + 1) .$$

4.8.20

(a) Show that there are $n!$ trees T in $C_{n+1}(n + 1)$ with $F(T) = 0$ and $(n - 1)!$ trees with $F(T) = n - 1$.
(b) Similarly, prove that there are $(n - 1)!$ trees T in $C_{n+1}(1)$ with $F(T) = 1$ and $n!$ trees with $F(T) = n$. Explain why there are no trees in $C_{n+1}(1)$ with $F(T) = 0$.

4.8.21 Replace each variable a_{rs} in $asw(T)$ by 1 or q^r according to if $r < s$ or $r > s$. Denote the resulting monomial by $q^{\text{Lmaj}(T)}$.

(a) Show that for the trees in $C_4(4)$ given in Figure 4.21 the monomials $q^{\text{Lmaj}(T)}$ are

$$1 \ q^3 \ q^3 \ 1$$
$$q^2 \ 1 \ q^3 \ 1$$
$$q^5 \ q^5 \ q^3 \ q^3$$
$$1 \ q^3 \ q^3 \ 1$$

(b) Conclude that

$$\sum_{T \in C_4(4)} q^{\text{Lmaj}(T)} = 6 + 2q^2 + 6q^3 + 2q^5 = (q^2 + 3)(2q^3 + 2) .$$

(c) Use Theorem 4.6.1 or Theorem 4.7.1 to show that in general the following formula holds:

$$\sum_{T \in C_{n+1}(n+1)} q^{\text{Lmaj}(T)} = (q^2 + n)(2q^3 + 2) \cdots ((n - 1)q^n + 2).$$

(d) Using Theorem 4.6.2, state and prove a similar formula for the sum

$$\sum_{T \in C_{n+1}(1)} q^{\text{Lmaj}(T)} .$$

4.8.22

(a) Show that there are $n!$ trees T in $C_{n+1}(n + 1)$ with $\text{Lmaj}(T) = 0$ and $(n - 1)!$ trees with $\text{Lmaj}(T) = \frac{1}{2}(n - 1)(n + 2)$.
(b) Calculate the number of trees T in $C_{n+1}(1)$ having the minimum and the maximum possible values of $\text{Lmaj}(T)$ for this family of trees.

4.8.23 Replace each variable a_{rs} in $asw(T)$ by 1 or $q^r t^s$ according to if $r < s$ or $r > s$. Denote the monomial obtained from $asw(T)$ in this manner by $w(T)$.

(a) Verify that for the collection $C_4(4)$ (see Figure 4.21) the corresponding monomials $w(T)$ are

$$
\begin{array}{cccc}
1 & q^3 t & q^3 t^2 & 1 \\
q^2 t & 1 & q^3 t^2 & 1 \\
q^5 t^2 & q^5 t^3 & q^3 t & q^3 t \\
1 & q^3 t & q^3 t^2 & 1
\end{array}
$$

(b) Show that

$$
\sum_{T \in C_4(4)} w(T) = (q^2 t + 3)(q^3(t^3 + t) + 2).
$$

(c) Use Theorem 4.6.1 or Theorem 4.7.1 to prove

$$
\sum_{T \in C_{n+1}(n+1)} w(T) = (q^2 t + n)(q^3(t + t^2) + n - 1) \cdots (q^n(t + t^2 + \cdots + t^{n-1}) + 2).
$$

(d) State and use Theorem 4.6.2 to prove a similar formula for the expression

$$
\sum_{T \in C_{n+1}(1)} w(T).
$$

4.8.24 Replace a_{rs} in $asw(T)$ by 1 or t^s according to if $r < s$ or $r > s$. Use Theorem 4.7.1 to show that the sum of the monomials of the trees in $C_{n+1}(n+1)$ is given by

$$
(t + n)(t^2 + t + n - 1)(t^3 + t^2 + t + n - 2) \cdots (t^{n-1} + \cdots + t^2 + t + 2).
$$

4.8.25 Fix a $j \in \{2, 3, \ldots, n\}$ and define $f : \{2, 3, \ldots, n\} \to [n + 1]$ by $f(i) = j$, $i = 2, 3, \ldots, n$. Describe the Cayley tree $T = ER(f)$.

4.8.26 Define $g : \{2, 3, \ldots, n\} \to [n + 1]$ by $g(i) = n + 1, i = 2, 3, \ldots, n$. Describe the Cayley tree $T = ER(g)$.

4.8.27 Define $h : \{2, 3, \ldots, n\} \to [n + 1]$ by $h(i) = i, i = 2, 3, \ldots, n$. Describe the Cayley tree $T = ER(h)$.

4.8.28 Show that the all edges of $T = ER(f) \in C_{n+1}$ are rise edges if and only if $f : \{2, 3, \ldots, n\} \to [n + 1]$ is such that $f(i) \geq i$ for $i = 2, 3, \ldots, n$.

4.8.29 Give a characterization of the Cayley trees $T = ER(f) \in C_{n+1}$ corresponding to functions $f : \{2, 3, \ldots, n\} \to [n+1]$ with $f(i) > i$ for $i = 2, 3, \ldots, n$.

4.8.30 Suppose $n + 1 = 21$ and $f : \{2, 3, \ldots, 20\} \to [21]$ is given by

i	2	3	4	5	6	7	8	9	10	11	12	13	14	15	16	17	18	19	20
$f(i)$	4	1	19	13	17	8	1	21	9	5	16	11	12	17	20	21	8	2	14

Construct the Cayley tree $ER(f)$.

4.8.31 Construct the function $f : \{2, 3, \ldots, 10\} \to [11]$ which corresponds to the tree T in Figure 4.18 under the ER bijection.

4.8.32 Construct the digraph of the function that corresponds to the following Cayley tree under the ER bijection.

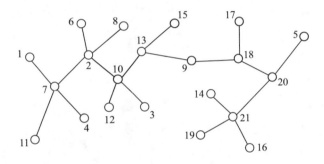

4.8.33 Show that the expected number of fall edges $F(T)$ over $T \in C_{n+1}$ (Cayley trees on $n + 1$ nodes rooted at $n + 1$) is

$$\frac{(n-1)(n)}{2(n+1)}.$$

4.8.34 For $T \in C_{n+1}$ define

$$\mathrm{Lmaj}(T) = \sum_{i \to j \in T} i \chi (i > j)$$

$$\mathrm{Rmaj}(T) = \sum_{i \to j \in T} j \chi (i > j).$$

Prove that these quantities have the following expected values:

$$E(\mathrm{Lmaj}) = \frac{1}{3}n(n-1), \quad E(\mathrm{Rmaj}) = \frac{1}{6}n(n-1).$$

4.8.35 For $T \in C_{n+1}$ define

$$\mathrm{Lmin}(T) = \sum_{i \to j \in T} i \chi (i < j)$$

$$\mathrm{Rmin}(T) = \sum_{i \to j \in T} j \chi (i < j) .$$

Prove that these quantities have the following expected values:

$$E(\mathrm{Lmin}) = \frac{1}{6} n(n + 5) , \quad E(\mathrm{Rmin}) = \frac{1}{3} n(n + 1) .$$

4.9 Sample Quiz for Chapter 4

1. Construct the word w which corresponds to the following Cayley tree:

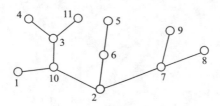

2. Put together the Cayley tree T which corresponds to the word $x_1 x_1 x_8 x_8 x_6 x_2 x_9$.
3. Calculate the number of Cayley trees on 5 nodes in which the nodes labeled 2 and 5 both have degree 1 .
4. Calculate the number of Cayley trees on 11 nodes with degrees

$$d_1 = \cdots = d_7 = 0, \quad d_8 = 4, \quad d_9 = 1, \quad d_{10} = d_{11} = 2 \, .$$

5. Calculate the number of Cayley forests on $x_1, x_2, \ldots, x_{11}$ having exactly four trees.
6. Construct the vertebrate that corresponds to the following functional digraph:

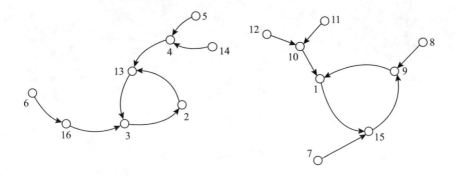

Chapter 5
The Cayley–Hamilton Theorem

5.1 The Cayley–Hamilton Theorem

We recall that this famed theorem of linear algebra is stated as follows:

Theorem 5.1.1 *Let $A = [a_{i,j}]$ be an $n \times n$ matrix and let*

$$\det \begin{bmatrix} x - a_{11} & -a_{12} & \cdots & -a_{1n} \\ -a_{21} & x - a_{22} & \cdots & -a_{2n} \\ \vdots & \vdots & & \vdots \\ -a_{n1} & -a_{n2} & \cdots & x - a_{nn} \end{bmatrix} = x^n + c_1 x^{n-1} + c_2 x^{n-2} + \cdots + c_n . \quad (5.1)$$

Then

$$A^n + c_1 A^{n-1} + c_2 A^{n-2} + \cdots + c_n I = 0 . \quad (5.2)$$

(Here I denotes the identity matrix and 0 stands for the matrix whose entries are all equal to zero).

The cases $n = 2, 3$ of this result were proved by Cayley [4], $n = 4$ case by Hamilton [17], and the general case by Frobenius [12].

Denoting by $P_A(x)$ the polynomial on the right hand side of (5.1), we recall that the equation

$$P_A(x) = 0$$

is usually referred to as the *characteristic equation* of the matrix A and the polynomial $P_A(x)$ is called the *characteristic polynomial* of A. Thus the Cayley–Hamilton theorem simply says that each matrix satisfies its own characteristic

© Springer Nature Switzerland AG 2021
Ö. Eğecioğlu, A. M. Garsia, *Lessons in Enumerative Combinatorics*, Graduate Texts in Mathematics 290, https://doi.org/10.1007/978-3-030-71250-1_5

equation. For the benefit of the reader who is not familiar with this result, it is good to illustrate it by an example.

For instance, if

$$A = \begin{bmatrix} 1 & -1 \\ 2 & 3 \end{bmatrix}$$

then

$$P_A(x) = \det \begin{bmatrix} x-1 & 1 \\ -2 & x-3 \end{bmatrix} = (x-1)(x-3) + 2 = x^2 - 4x + 5.$$

Thus in this case (5.2) reduces to

$$A^2 - 4A + 5I = 0.$$

Now we see that since

$$A^2 = \begin{bmatrix} 1 & -1 \\ 2 & 3 \end{bmatrix} \begin{bmatrix} 1 & -1 \\ 2 & 3 \end{bmatrix} = \begin{bmatrix} -1 & -4 \\ 8 & 7 \end{bmatrix}$$

equation (5.2) simply states that

$$\begin{bmatrix} -1 & -4 \\ 8 & 7 \end{bmatrix} - 4\begin{bmatrix} 1 & -1 \\ 2 & 3 \end{bmatrix} + 5\begin{bmatrix} 1 & 0 \\ 0 & 1 \end{bmatrix} = \begin{bmatrix} 0 & 0 \\ 0 & 0 \end{bmatrix}.$$

This is clearly so since the left hand side is

$$\begin{bmatrix} -1-4+5 & -4+4+0 \\ 8-8+0 & 7-12+5 \end{bmatrix}.$$

This most remarkable and useful result is traditionally presented in algebra courses and is not usually associated with digraphs or permutation diagrams.

However, upon comparing the determinants occurring in (4.22) and (5.1) we may rightly wonder if Theorem 5.1.1 can be given a proof similar to that we have given Theorem 4.6.1 of Chapter 4.

Now precisely such a proof was recently discovered by Howard Straubing [27]. In fact, Straubing shows that (5.2) may be interpreted as a statement concerning certain pairs of two-colored digraphs. Since these arguments are very much in the spirit of the developments in the previous chapter, and we cannot escape the temptation to reproduce them here.

For completeness, we restate some of the definitions necessary for our treatment. Recall first that the determinant of an $n \times n$ matrix

$$B = [b_{i,j}]$$

is given by the formula

$$\det B = \sum_{\sigma} (-1)^{i(\sigma)} b_{1\sigma_1} b_{2\sigma_2} \cdots b_{n\sigma_n}, \tag{5.3}$$

where the sum is carried out over all permutations σ of $[n]$ and $i(\sigma)$ denotes the number of inversions of σ.

Our point of departure is again formula (5.3). Here we set

$$b_{i,j} = x\delta_{i,j} - a_{i,j}$$

with $\delta_{i,j}$ again defined by

$$\delta_{i,j} = \begin{cases} 1 & \text{if } i = j, \\ 0 & \text{if } i \neq j. \end{cases}$$

This given, from formula (5.2) we obtain the expansion

$$P_A(x) = \det[x\delta_{i,j} - a_{i,j}] = \sum_{\sigma} (-1)^{i(\sigma)} \prod_{i=1}^{n} (x\delta_{i,\sigma_i} - a_{i,\sigma_i}) . \tag{5.4}$$

Now, as before we write

$$\prod_{i=1}^{n} (x\delta_{i,\sigma_i} - a_{i,\sigma_i}) = \sum_{S \subseteq [n]} \prod_{i \in \overline{S}} x\delta_{i,\sigma_i} \prod_{i \in S} (-a_{i,\sigma_i}) .$$

Substituting in (5.4) and interchanging orders of summation give

$$P_A(x) = \sum_{S \subseteq [n]} \sum_{\sigma} (-1)^{i(\sigma)} \prod_{i \in \overline{S}} x\delta_{i,\sigma_i} \prod_{i \in S} (-a_{i,\sigma_i}) . \tag{5.5}$$

Note now that the product

$$\prod_{i \in \overline{S}} x\delta_{i,\sigma_i}$$

is different from zero if and only if

$$\sigma_i = i \quad \text{for each} \quad i \in \overline{S} . \tag{5.6}$$

Moreover, when it is different from zero this product is equal to

$$x^{n-|S|} .$$

Note further that condition (5.6) simply says that σ leaves unchanged the elements of the complement of S. This implies in particular that σ permutes the elements of S among themselves. Clearly such a σ may be viewed as a permutation of the set S. Denote then by $P(S)$ the collection of all permutations of S and define as before for each $\sigma \in P(S)$ a word

$$a(\sigma) = \prod_{i \in S} a_{i,\sigma_i} .$$

These considerations allow us to write the inner sum in (5.5) in much simpler form. Namely, for each $S \subseteq [n]$ we get

$$\sum_{\sigma} (-1)^{i(\sigma)} \prod_{i \in \overline{S}} x \delta_{i,\sigma i} \prod_{i \in S} (-a_{i,\sigma i}) = \sum_{\sigma \in P(S)} (-1)^{i(\sigma)+|S|} x^{n-|S|} a(\sigma) .$$

Substituting in (5.5) and using again formula

$$(-1)^{i(\sigma)} = (-1)^{|S|+k(\sigma)},$$

where $k(\sigma)$ is the number of cycles in the cycle decomposition of σ, we obtain that

$$P_A(x) = \sum_{S \subseteq [n]} \sum_{\sigma \in P(S)} (-1)^{k(\sigma)} x^{n-|S|} a(\sigma) . \tag{5.7}$$

We now group terms yielding the same power of x by breaking up the first sum according to the cardinality of S. More precisely, we rewrite (5.7) in the form

$$P_A(x) = \sum_{p=0}^{n} \sum_{\substack{S \subseteq [n] \\ |S|=p}} \sum_{\sigma \in P(S)} (-1)^{k(\sigma)} x^{n-p} a(\sigma) . \tag{5.8}$$

Thus we finally obtain

$$P_A(x) = \det[x \delta_{i,j} - a_{i,j}] = \sum_{p=0}^{n} x^{n-p} \sum_{\substack{S \subseteq [n] \\ |S|=p}} \sum_{\sigma \in P(S)} (-1)^{k(\sigma)} a(\sigma) .$$

Comparing with (5.2) we see that the coefficients $c_1, c_2, \ldots, c_n$ are given by the formula

$$c_p = \sum_{\substack{S \subseteq [n] \\ |S|=p}} \sum_{\sigma \in P(S)} (-1)^{k(\sigma)} a(\sigma) \qquad (p = 1, 2, \ldots, n).$$

Thus, equation (5.2) becomes

$$\sum_{p=0}^{n} A^{n-p} \sum_{\substack{S \subseteq [n] \\ |S|=p}} \sum_{\sigma \in P(S)} (-1)^{k(\sigma)} a(\sigma) = 0 . \tag{5.9}$$

Here we use the convention that A^0 is to represent the identity matrix I.

Now we recall that the (i, j)th entry of the qth power of the $n \times n$ matrix $A = [a_{i,j}]$ is given by the expression

$$\sum_{i_1=1}^{n} \sum_{i_2=1}^{n} \cdots \sum_{i_{q-1}=1}^{n} a_{i,i_1} a_{i_1,i_2} \cdots a_{i_{q-1},j} . \tag{5.10}$$

This expression may be viewed as the generating function of the collection $\Pi_{i,j}(q)$ of directed paths which go from i to j in q steps.

More precisely, if we interpret the elements $a_{i,j}$ as letters of an alphabet, and we associate to the path

$$\pi = (i \to i_1 \to i_2 \to \cdots \to i_{q-1} \to j) ,$$

the word

$$w(\pi) = a_{i,i_1} a_{i_1,i_2} \cdots a_{i_{q-1},j} , \tag{5.11}$$

then the sum in (5.10) may be rewritten in the form

$$\sum_{\pi \in \Pi_{i,j}(q)} w(\pi) .$$

This given, the (i, j)th entry of the matrix on the left hand side of (5.9) becomes

$$\sum_{p=0}^{n} \sum_{\pi \in \Pi_{i,j}(n-p)} w(\pi) \sum_{\substack{S \subseteq [n] \\ |S|=p}} \sum_{\sigma \in P(S)} (-1)^{k(\sigma)} a(\sigma) . \tag{5.12}$$

Thus to prove the Cayley–Hamilton theorem we need to show that for each i and j we do have

$$\sum_{p=0}^{n} \sum_{\pi \in \Pi_{i,j}(n-p)} \sum_{\substack{S \subseteq [n] \\ |S|=p}} \sum_{\sigma \in P(S)} (-1)^{k(\sigma)} w(\pi) a(\sigma) = 0 .$$ (5.13)

As we did in the proof of Theorem 4.6.1 of Chapter 4, we may represent each term $w(\pi)a(\sigma)$ in this multiple sum as a two-colored digraph obtained by superimposing the directed path π (say in Blue) with the cycle diagram of σ (say in Red).

In Figure 5.1 we have depicted one such two-colored digraph for the case $n = 11$, $p = 4$, $i = 6$, $j = 8$, and $S = \{1, 4, 9, 10\}$, using the letters B and R for the colors.

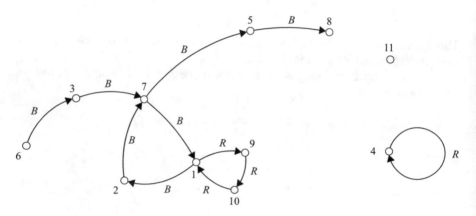

Fig. 5.1 A two-colored digraph.

Another example for the case $n = 10$, $p = 5$, $i = 3$, $j = 4$, and $S = \{1, 5, 6, 8, 10\}$ could be as in Figure 5.2.

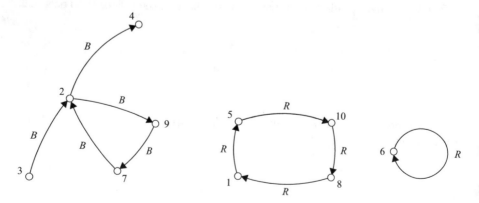

Fig. 5.2 Another two-colored digraph.

The reader is urged to color the edges labeled B in blue and those labeled R in red for a better view of the argument. Note that in the first example the blue digraph touches a red cycle while in the second example there is no contact between blue and red digraphs, but the blue digraph itself has a cycle.

What is true in general is that at least one of these two circumstances must always occur. More precisely we have the following basic property:

If the blue digraph has no cycles then it must touch a red cycle.   (5.14)

The reason for this is very simple. For a given value of n, let π, σ be one of our pairs and let σ act on a set S with p elements. The way our pairs are constructed (see formula (5.12)) the path π must necessarily have exactly $n - p$ edges. Now, if π does not have a cycle, it will visit exactly $n - p + 1$ vertices. On the other hand, the cycle diagram of σ has exactly p vertices. Since there are only n vertices all together we must conclude that π and the cycle diagram of σ must share at least one vertex. This proves (5.14).

We are now in a position to prove our identity (5.13) very much in the same way we proved Theorem 4.6.1 of Chapter 4. Using (5.14), we can construct a transformation ϕ which pairs off any given two-colored digraph (π, σ) appearing in (5.13) with another such digraph

$$(\pi', \sigma') = \phi(\pi, \sigma)$$

in such a manner that

$$(a) \quad w(\pi')a(\sigma') = w(\pi)a(\sigma) , \qquad\qquad (5.15)$$
$$(b) \quad (-1)^{k(\sigma')} = -(-1)^{k(\sigma)} .$$

The transformation ϕ can be defined as follows. Let (π, σ) be a two-colored digraph of (5.13). Note that, in view of (5.14), as we follow the blue path from i to j, one of the following two alternatives must hold true:

A : We go through a complete blue cycle before (if ever) touching a red cycle.

B : We touch a red cycle before (if ever) completing a blue cycle.

This given, the image (π', σ') of (π, σ) by ϕ is constructed by the following simple rule:

In case A :   Change that blue cycle of π to red,   (5.16)

In case B :   Change that red cycle of σ to blue.

Note that for the two-colored digraph in Figure 5.1 we are in case B and for the digraph in (5.12) we are in case A. Thus applying this rule, the image by ϕ of the digraph in Figure 5.1 is the digraph in Figure 5.3, and the image of the digraph in Figure 5.2 is the one in Figure 5.4.

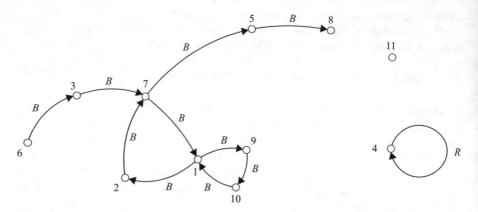

Fig. 5.3 Image of the digraph in Figure 5.1.

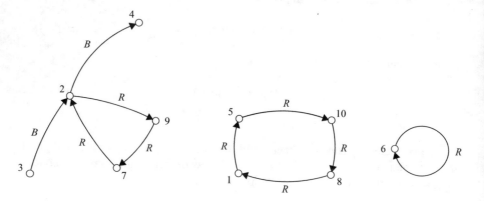

Fig. 5.4 Image of the digraph in Figure 5.2.

Note that our transformation ϕ has the following basic properties:

    (a) *It does not remove or add any edges to a two-colored digraph,*

    (b) *It always changes by one the number of red cycles,*   (5.17)

    (c) *If we apply it twice we get back where we started.*

We can easily see that properties (5.17) (a) and (b) imply properties (5.15) (a) and (b), respectively. Note further that (5.17) (c) implies that the collection of two-colored digraphs of (5.13) breaks up into a union of disjoint pairs

$$(\pi, \sigma) \longleftrightarrow (\pi', \sigma')$$

the members of which are interchanged by ϕ.

Putting all this together, we may conclude that the sum in (5.13) is in point of fact a sum of pairs of terms having opposite signs and must therefore be equal to zero as asserted.

This completes the combinatorial proof of the Cayley–Hamilton theorem.

5.2 Miscellaneous Applications and Examples

We shall limit ourselves here to illustrating some uses of the Cayley–Hamilton theorem.

Basically our aim is to show that the characteristic equation (5.2) may be used to simplify the calculation of expressions involving the given matrix.

To give an idea of how this comes about we shall start by showing that the inverse of a matrix A may always be expressed as a polynomial in A. Let us work first with an example. For instance let

$$A = \begin{bmatrix} -1 & 0 & 3 \\ 0 & 2 & 0 \\ 3 & 0 & -1 \end{bmatrix}. \tag{5.18}$$

We then get

$$P_A(x) = \det \begin{bmatrix} x+1 & 0 & -3 \\ 0 & x-2 & 0 \\ -3 & 0 & x+1 \end{bmatrix} = x^3 - 12x + 16.$$

Thus (5.2) in this case reduces to

$$A^3 - 12A + 16I = 0. \tag{5.19}$$

Note that from (5.18) we get that

$$P_A(0) = \det \begin{bmatrix} 1 & 0 & -3 \\ 0 & -2 & 0 \\ -3 & 0 & 1 \end{bmatrix} = (-1)^3 \det A = -16.$$

Thus $\det A \neq 0$ and A has an inverse. However, we can easily see from (5.19) that

$$A(12I - A^2) = 16I$$

from which we derive that

$$A^{-1} = \frac{1}{16}\left(12I - A^2\right) = \frac{1}{16}\begin{bmatrix} 2 & 0 & 6 \\ 0 & 8 & 0 \\ 6 & 0 & 2 \end{bmatrix} .$$

The reader may verify that the expression on the right hand side of this equation does indeed give the inverse of our matrix.

The same idea can of course be used to express the inverse of any $n \times n$ matrix A with nonzero determinant as a polynomial in A of degree at most $n - 1$. Indeed, we see that the coefficient of the identity matrix in (5.2) is given by the formula

$$c_n = (-1)^n \det A .$$

Thus this coefficient is different from zero whenever $\det A \neq 0$. So, if that is the case, we may rewrite (5.2) in the form

$$A^{-1} = -\frac{1}{c_n}\left(A^{n-1} + c_1 A^{n-2} + \cdots + c_{n-1}I\right)$$

and this is our desired formula for the inverse.

The calculation of the inverse or any other function of a matrix may be simplified even further by some additional manipulations stemming from the Cayley–Hamilton theorem. To do this we need some preliminary observations.

First of all note that any relation between polynomials of a variable x remains valid if x is replaced by any square matrix. For instance, the relation

$$(x - 2)(x + 4) = x^2 + 2x - 8 ,$$

implies that for any matrix A we have

$$(A - 2I)(A + 4I) = A^2 + 2A - 8I .$$

More generally, if $P(x)$, $Q(x)$, and $R(x)$ are polynomials and

$$P(x)Q(x) = R(x),$$

then for any matrix A we have

$$P(A)Q(A) = R(A) .$$

The same of course holds true for any other polynomial identity. This general principle follows from the fact that in establishing identities involving polynomials in x the only relation we are allowed to use is

$$x^n \cdot x^m = x^{n+m} .$$

This given such identities will necessarily have to hold as well if x is replaced by any algebraic quantity A which satisfies

$$A^n \cdot A^m = A^{n+m} . \tag{5.20}$$

However, we know (and it is not difficult to show) that (5.20) holds for any square matrix.

Starting from this observation we can reduce the calculation of any given polynomial of an $n \times n$ matrix A to that of a polynomial of degree $n - 1$ at most and even less in some cases.

Let us observe that the matrix in (5.18) also satisfies the simpler equation

$$A^2 + 2A - 8I = 0 \tag{5.21}$$

(the reader may verify this directly, we shall later see how it comes about).

For convenience set

$$m(x) = x^2 + 2x - 8 . \tag{5.22}$$

Note that from (5.21) we deduce that

$$A^2 = -2A + 8I . \tag{5.23}$$

Thus

$$\begin{aligned} A^3 &= -2A^2 + 8A \\ &= -2(-2A + 8I) + 8A \\ &= 12A - 16I , \end{aligned} \tag{5.24}$$

which in turn gives

$$\begin{aligned} A^4 &= 12A^2 - 16A \\ &= 12(-2A + 8I) - 16A \\ &= -40A + 96I . \end{aligned} \tag{5.25}$$

Clearly we can continue this process and express any power of A as a linear combination of A and the identity matrix. Note also that using (5.23), (5.24), and (5.25) we deduce that

$$\begin{aligned} A^4 - A^3 - 13A^2 + 28A - 9I &= (-40A + 96I) - (12A - 16I) - 13(-2A + 8I) + 28A - 9I \\ &= 2A - I . \end{aligned} \tag{5.26}$$

We can see that given any polynomial $F(x)$ we can always find the coefficients α, β such that

$$F(A) = \alpha A + \beta I .$$

These coefficients can be calculated in several ways. We can of course use the method we have followed above for the case

$$F(x) = x^4 - x^3 - 13x^2 + 28x - 9 . \tag{5.27}$$

However, there are two methods that are worth discussing in detail.

5.2.1 The Division Method

Applying long division we derive the decomposition

$$x^4 - x^3 - 13x^2 + 28x - 9 = (x^2 + 2x - 8)(x^2 - 3x + 1) + 2x - 1 . \tag{5.28}$$

That means that when $F(x)$ is given by (5.27) we get

$$F(A) = m(A)(A^2 - 3A + I) + 2A - I .$$

Thus, whenever the matrix A satisfies the equation

$$m(A) = 0 ,$$

(in particular for the matrix of (5.18)), we have

$$F(A) = 2A - I$$

and this agrees with what we had previously obtained.

The general procedure can be described as follows. Suppose our matrix A satisfies the equation

$$A^p + c_1 A^{p-1} + \cdots + c_p I = 0 . \tag{5.29}$$

Set

$$m(x) = x^p + c_1 x^{p-1} + \cdots + c_p . \tag{5.30}$$

To evaluate a polynomial $F(x)$ of degree greater than p, we perform long division of $F(x)$ by $m(x)$ and obtain the decomposition

$$F(x) = m(x)Q(x) + R(x), \tag{5.31}$$

where $Q(x)$ and $R(x)$ denote the *quotient* and the *remainder* of the division. We know that $Q(x)$ and $R(x)$ are polynomials which are uniquely determined by (5.31) and the condition

$$\text{degree } R(x) < \text{ degree } m(x) . \tag{5.32}$$

This done, from (5.31) we get

$$F(A) = m(A)Q(A) + R(A)$$

and (5.29) gives

$$F(A) = R(A) .$$

We may thus summarize our findings in the following manner:

Theorem 5.2.1 *If A is any $n \times n$ matrix which satisfies the equation*

$$m(A) = A^p + c_1 A^{p-1} + \cdots + c_p I = 0,$$

then for any polynomial $F(x)$ we have the expression

$$F(A) = r_1 A^{p-1} + r_2 A^{p-2} + \cdots + r_{p-1} I,$$

where the polynomial

$$R(x) = r_1 x^{p-1} + r_2 x^{p-2} + \cdots + r_{p-1}$$

is the remainder of the division of $F(x)$ by $m(x)$.

In particular, by the Cayley–Hamilton theorem, every polynomial in A may be expressed in this manner as a linear combination of the matrices

$$I, A, A^2, \ldots, A^{n-1} .$$

5.2.2 The Interpolation Method

Long division to determine $R(x)$ may be quite tedious and indeed in practical situations it may involve a large number of intermediate steps. Also, for theoretical purposes, we may wish to have an explicit form for the resulting polynomial $R(x)$. It turns out that we can write down a formula for $R(x)$ in full generality. For simplicity we shall present it in the special case that $m(x)$ has distinct roots.

The result may be stated as follows:

Theorem 5.2.2 *If*

$$m(x) = (x - x_1)(x - x_2) \cdots (x - x_p) \tag{5.33}$$

with $x_1, x_2, \ldots, x_p$ distinct real or complex numbers, then the remainder $R(x)$ of the division of $F(x)$ by $m(x)$ is given by the formula

$$R(x) = \sum_{i=1}^{p} F(x_i) \chi_i(x), \tag{5.34}$$

where for $i = 1, 2, \ldots, n$,

$$\chi_i(x) = \frac{(x - x_1) \cdots (x - x_{i-1})(x - x_{i+1}) \cdots (x - x_n)}{(x_i - x_1) \cdots (x_i - x_{i-1})(x_i - x_{i+1}) \cdots (x_i - x_n)}. \tag{5.35}$$

Proof A polynomial $R(x)$ of degree at most $p - 1$ may be recovered from its values at p distinct points $x_1, x_2, \ldots, x_p$. Indeed, the interpolation formula of Lagrange (see Chapter 6, Section 6.5.3 for a proof) gives

$$R(x) = \sum_{i=1}^{p} R(x_i), \chi_i(x) \tag{5.36}$$

where $\chi_i(x)$ is as defined in (5.35).

This given, note that from (5.31) and (5.33) we get

$$F(x_i) = m(x_i)Q(x_i) + R(x_i) = R(x_i)$$

for $i = 1, 2, \ldots, p$. Substituting this in (5.36) gives (5.34) as desired.   □

Theorems 5.2.1 and 5.2.2 combined yield the following remarkable result:

Theorem 5.2.3 *If A satisfies the equation*

$$(A - x_1 I)(A - x_2 I) \cdots (A - x_p I) = 0 \tag{5.37}$$

with $x_1, x_2, \ldots, x_p$ distinct real or complex numbers, then for any polynomial $F(x)$ we have

$$F(A) = \sum_{i=1}^{p} F(x_i) \chi_i(A) \tag{5.38}$$

$$= \sum_{i=1}^{p} F(x_i) \frac{(A - x_1 I) \cdots (A - x_{i-1} I)(A - x_{i+1} I) \cdots (A - x_n I)}{(x_i - x_1) \cdots (x_i - x_{i-1})(x_i - x_{i+1}) \cdots (x_i - x_n)} .$$

Proof Theorem 5.2.1 gives that

$$F(A) = R(A),$$

where $R(x)$ is the remainder of the division of $F(x)$ by

$$(x - x_1)(x - x_2) \cdots (x - x_p)$$

and Theorem 5.2.2 gives that $R(A)$ has precisely the form given by the right hand side of (5.38). □

It may be good to illustrate this result by an example. Let us again take A to be the matrix of (5.18). We have then that

$$(A - 2I)(A + 4I) = A^2 + 2A - 8I = 0 . \tag{5.39}$$

Thus Theorem 5.2.3 applies with $x_1 = 2$, $x_2 = -4$. Formula (5.38) with $p = 2$ reduces to

$$F(A) = F(x_1) \frac{A - x_2 I}{x_1 - x_2} + F(x_2) \frac{A - x_1 I}{x_2 - x_1}.$$

Substituting $x_1 = 2$ and $x_2 = -4$ gives

$$F(A) = F(2) \frac{A + 4I}{6} + F(-4) \frac{A - 2I}{-6} . \tag{5.40}$$

For instance for

$$F(x) = x^4 - x^3 - 13x^2 + 28x - 9$$

we get

$$F(2) = 3, \quad F(-4) = -9 .$$

Thus in this case

$$F(A) = \frac{3}{6}(A + 4I) + \frac{9}{6}(A - 2I) = 2A - I ,$$

which again agrees with our previous findings.

Once in possession of formula (5.38) nothing prevents us from trying to use it for more general $F(x)$ than just polynomials. If we do this we discover that whenever it makes sense it is still valid!

For instance, setting $F(x) = 1/x$ in (5.40) gives

$$A^{-1} = \frac{1}{2}\frac{A+4I}{6} + \frac{1}{4}\frac{A-2I}{6} = \frac{1}{24}(3A+6I) = \frac{1}{8}\begin{bmatrix} 1 & 0 & 3 \\ 0 & 4 & 0 \\ 3 & 0 & 2 \end{bmatrix}$$

which agrees with our previous calculation.

However, formula (5.38) may be used in an even more surprising manner. To give an idea of the range of applications of the Cayley–Hamilton theorem we shall illustrate two of the most remarkable uses of (5.38).

5.2.3   Solutions of Differential Equations

Let us suppose we are to find the solutions of the system of differential equations

$$y_1'(t) = -y_1(t) + 3y_3(t)$$
$$y_2'(t) = 2y_2(t) \tag{5.41}$$
$$y_3'(t) = 3y_1(t) - y_3(t).$$

This system may be expressed using matrix notation in the form

$$\frac{d}{dt}\begin{bmatrix} y_1(t) \\ y_2(t) \\ y_3(t) \end{bmatrix} = \begin{bmatrix} -1 & 0 & 3 \\ 0 & 2 & 0 \\ 3 & 0 & -1 \end{bmatrix}\begin{bmatrix} y_1(t) \\ y_2(t) \\ y_3(t) \end{bmatrix} = A\begin{bmatrix} y_1(t) \\ y_2(t) \\ y_3(t) \end{bmatrix} \tag{5.42}$$

with A given by (5.18).

Now it is easy to see that if each of the column vectors

$$\begin{bmatrix} z_{11}(t) \\ z_{21}(t) \\ z_{31}(t) \end{bmatrix}, \quad \begin{bmatrix} z_{12}(t) \\ z_{22}(t) \\ z_{32}(t) \end{bmatrix}, \quad \begin{bmatrix} z_{13}(t) \\ z_{23}(t) \\ z_{33}(t) \end{bmatrix}$$

is a solution of (5.42), then the matrix

$$Z(t) = \begin{bmatrix} z_{11}(t) & z_{12}(t) & z_{13}(t) \\ z_{21}(t) & z_{22}(t) & z_{23}(t) \\ z_{31}(t) & z_{32}(t) & z_{33}(t) \end{bmatrix}$$

satisfies the equation

$$\frac{d}{dt} Z(t) = AZ(t) .$$ (5.43)

Forget for a moment that $Z(t)$ and A are matrices and interpret (5.43) as a differential equation involving a real valued function $Z(t)$ and a constant A.

Then from calculus we get that

$$Z(t) = Z(0)e^{At} .$$

In particular if

$$Z(0) = 1,$$

then

$$Z(t) = e^{At} .$$ (5.44)

Now, it is most remarkable that exactly the same formula holds even in the matrix case. More precisely, our formula (5.38) with

$$F(x) = e^{xt}$$ (5.45)

yields the matrix solution of (5.43) which satisfies the initial condition

$$Z(0) = I = \begin{bmatrix} 1 & 0 & 0 \\ 0 & 1 & 0 \\ 0 & 0 & 1 \end{bmatrix} .$$ (5.46)

We may verify this at once.

Recall that formula (5.38) for our matrix A reduces to (5.40). The latter for $F(x) = e^{xt}$ gives

$$Z(t) = e^{At} = e^{2t} \frac{A + 4I}{6} + e^{-4t} \frac{A - 2I}{-6} .$$ (5.47)

Now we see that

$$Z'(t) = e^{2t} \frac{A + 4I}{3} + 2e^{-4t} \frac{A - 2I}{3} .$$ (5.48)

On the other hand, from (5.39) we get

$$(A + 4I)A = A^2 + 4A = 2A + 8I = 2(A + 4I)$$

$$(A - 2I)A = A^2 - 2A = -4A + 8I = -4(A - 2I) .$$

Thus from (5.47)

$$Z(t)A = e^{2t}2\frac{A+4I}{6} + e^{-4t}(-4)\frac{(A-2I)}{-6} .$$

Comparing with (5.48) we see that we indeed have

$$Z'(t) = AZ(t)$$

as asserted. Moreover we see that setting $t = 0$ in (5.47) gives

$$Z(0) = \frac{A+4I}{6} - \frac{A-2I}{6} = I.$$

We leave it to the reader to verify that the general solution of the system in (5.41) is given by the formula

$$\begin{bmatrix} y_1(t) \\ y_2(t) \\ y_3(t) \end{bmatrix} = Z(t) \begin{bmatrix} y_1(0) \\ y_2(0) \\ y_3(0) \end{bmatrix} ,$$

with $Z(t)$ given by (5.47).

5.2.4   Solutions of Difference Equations

Formula (5.38) may also be used to obtain explicit solutions to difference equations. To see how this comes about, let us consider the recurrence

$$y_{n+3} = 2y_{n+2} + y_{n+1} - 2y_n . \tag{5.49}$$

This may be rewritten in the matrix form

$$\begin{bmatrix} y_{n+3} \\ y_{n+2} \\ y_{n+1} \end{bmatrix} = \begin{bmatrix} 2 & 1 & -2 \\ 1 & 0 & 0 \\ 0 & 1 & 0 \end{bmatrix} \begin{bmatrix} y_{n+2} \\ y_{n+1} \\ y_n \end{bmatrix} . \tag{5.50}$$

Thus, setting

$$Y_n = \begin{bmatrix} y_{n+3} \\ y_{n+2} \\ y_{n+1} \end{bmatrix} , \quad A = \begin{bmatrix} 2 & 1 & -2 \\ 1 & 0 & 0 \\ 0 & 1 & 0 \end{bmatrix} \tag{5.51}$$

we may rewrite (5.49) as

$$Y_n = AY_{n-1} .$$

Thus recursively we get

$$Y_1 = AY_0,$$
$$Y_2 = AY_1 = A^2 Y_0,$$
$$Y_3 = AY_2 = A^3 Y_0,$$
$$\vdots$$
$$Y_n = A^n Y_0$$
$$\vdots$$

Now, the characteristic polynomial of this matrix is

$$\det \begin{bmatrix} x-2 & -1 & 2 \\ -1 & x & 0 \\ 0 & -1 & x \end{bmatrix} = (x-2)(x-1)(x+1) .$$

Thus, by the Cayley–Hamilton theorem our matrix satisfies the equation

$$(A-2I)(A-I)(A+I) = 0 .$$

This is precisely of the form (5.37) with $p = 3$, $x_1 = 2$, $x_2 = 1$, $x_3 = -1$. Formula (5.38) then gives

$$A^n = 2^n \frac{(A-I)(A+I)}{(2-1)(2+1)} + 1^n \frac{(A-2I)(A+I)}{(1-2)(1+1)} + (-1)^n \frac{(A-2I)(A-I)}{(-1-2)(-1-1)} .$$

A simple calculation then yields that

$$A^n = \frac{2^n}{3} \begin{bmatrix} 4 & 0 & -4 \\ 2 & 0 & -2 \\ 1 & 0 & -1 \end{bmatrix} - \frac{1}{2} \begin{bmatrix} 1 & -1 & -2 \\ 1 & -1 & -2 \\ 1 & -1 & -2 \end{bmatrix} + \frac{(-1)^n}{6} \begin{bmatrix} 1 & -3 & 2 \\ -1 & 3 & -2 \\ 1 & -3 & 2 \end{bmatrix} .$$

The reader may verify that the general solution of the recurrence in (5.49) is given by the first component of the vector

$$A^n \begin{bmatrix} y_2 \\ y_1 \\ y_0 \end{bmatrix} ,$$

that is

$$y_{n+3} = \frac{2^n}{3}(4y_2 - 4y_0) - \frac{1}{2}(y_2 - y_1 - 2y_0) + \frac{(-1)^n}{6}(y_2 - 3y_1 + 2y_0).$$

Remark 5.2.1 Since most of the applications we have given hinge on the equation (5.37), it is worthwhile, before closing this section, to see what circumstances do guarantee that our matrix satisfies such a simpler equation.

We know that if A is a general $n \times n$ matrix, then its characteristic polynomial has the following factored form:

$$P_A(x) = (x - x_1)^{m_1}(x - x_2)^{m_2} \cdots (x - x_p)^{m_p}, \tag{5.52}$$

where $x_1, x_2, \ldots x_p$ are the *distinct* roots of the equation

$$P_A(x) = 0.$$

(Recall that $x_1, x_2, \ldots, x_p$ are usually referred to as the *eigenvalues* of A). In general, the only thing we can say about $m_1, m_2, \ldots, m_p$ is that they are positive integers and that

$$m_1 + m_2 + \cdots + m_p = n. \tag{5.53}$$

The Cayley–Hamilton theorem then assures that

$$(A - x_1 I)^{m_1}(A - x_2 I)^{m_2} \cdots (A - x_p I)^{m_p} = 0. \tag{5.54}$$

However, once the x_i are known, nothing prevents us from deciding whether or not we also have

$$m(A) = (A - x_1 I)(A - x_2 I) \cdots (A - x_p I) = 0. \tag{5.55}$$

In point of fact, there are a number of situations where we can be sure that our matrix A will satisfy such an equation.

Trivially, this will be the case if A has distinct eigenvalues since then from (5.53) we deduce that all the m_i are equal to one and (5.54) becomes precisely of the form (5.55).

It can be shown that for most matrices all the m_i are indeed equal to one. However, there are simple criteria which guarantee that our matrix will satisfy the equation in (5.55).

Recall that if A is a matrix, then its transpose A^T is the matrix obtained from A by moving the (i, j)th entry of A to the (j, i)th position. In other words A^T is obtained by reflecting A across the main diagonal. For instance

$$\begin{bmatrix} 1 & 0 & 2 \\ -1 & 0 & 1 \\ 5 & 0 & 1 \end{bmatrix}^T = \begin{bmatrix} 1 & -1 & 5 \\ 0 & 0 & 0 \\ 2 & 1 & 1 \end{bmatrix}.$$

If B is a matrix with complex entries, let $\overline{B}$ denote the matrix obtained by replacing each entry of B by its complex conjugate.

This given, we can be sure that (5.55) will hold true when

1. A is *symmetric*, that is

$$A^T = A,$$

2. A is *orthogonal*, that is

$$A^T = A^{-1},$$

or, when A has complex entries and
3. A is *Hermitian*, that is

$$A^T = \overline{A},$$

4. A is *unitary*, that is

$$\overline{A}^T = A^{-1}.$$

Finally, a condition which includes all of the previous ones as special cases is
A is *normal*, that is

$$A^T \overline{A} = \overline{A} A^T.$$

We see that the matrix in (5.18) is symmetric, thus the fact that it does satisfy an equation of type (5.55) is no accident.

It should be pointed out that the polynomial in (5.55) may be computed from $P_A(x)$ without the knowledge of the roots $x_1, x_2, \ldots, x_p$. Indeed, it is not difficult to show that this polynomial is given by the formula

$$m(x) = \frac{P_A(x)}{g(x)},$$

where $g(x)$ denotes the greatest common divisor of $P_A(x)$ and its derivative $\frac{d}{dx} P_A(x)$.

It should also be mentioned that equation (5.55) holds if and only if A is *diagonalizable,* that is if and only if for some invertible matrix P and constants $d_1, d_2, \ldots, d_n$ we have

$$A = P \begin{bmatrix} d_1 & 0 & \dots & 0 \\ 0 & d_2 & \dots & 0 \\ \vdots & \vdots & & \vdots \\ 0 & 0 & \dots & d_n \end{bmatrix} P^{-1} . \tag{5.56}$$

The *sufficiency* of this condition is easily verified. For instance, if

$$A = P \begin{bmatrix} 1 & 0 & 0 & 0 \\ 0 & 1 & 0 & 0 \\ 0 & 0 & 2 & 0 \\ 0 & 0 & 0 & 2 \end{bmatrix} P^{-1}$$

then

$$(A - I)(A - 2I) = P \begin{bmatrix} 0 & 0 & 0 & 0 \\ 0 & 0 & 0 & 0 \\ 0 & 0 & 1 & 0 \\ 0 & 0 & 0 & 1 \end{bmatrix} \begin{bmatrix} -1 & 0 & 0 & 0 \\ 0 & -1 & 0 & 0 \\ 0 & 0 & 0 & 0 \\ 0 & 0 & 0 & 0 \end{bmatrix} P^{-1} = 0 .$$

In general, we see that if A is of the form (5.56), then (5.55) will hold true with $x_1, x_2, \dots, x_p$ equal to the distinct values taken by $d_1, d_2, \dots, d_n$.

We find out that the *necessity* of the condition in (5.56) for (5.55) to hold true may be proved from our formula (5.38). We cannot go into details here without getting too much out of context. However, for the benefit of the inquisitive reader we give a sketch of the proof.

Define

$$Z_i(A) = (A - x_1 I) \cdots (A - x_{i-1} I)(A - x_{i+1} I) \cdots (A - x_n I) , \tag{5.57}$$

$$c_i = \frac{1}{(x_i - x_1) \cdots (x_i - x_{i-1})(x_i - x_{i+1}) \cdots (x_i - x_n)}$$

so that for $i = 1, 2, \dots, n$

$$\chi_i(A) = c_i Z_i(A) .$$

Note that with this notation, (5.38) with $F(x) = 1$ yields the formula

$$I = \sum_{i=1}^{p} c_i Z_i(A) . \tag{5.58}$$

We observe now that if A satisfies (5.55), then for each i we have

$$(A - x_i I) Z_i(A) = m(A) = 0 .$$ (5.59)

This given, from (5.58) we get that, for any n-vector

$$V = \begin{bmatrix} v_1 \\ v_2 \\ \vdots \\ v_n \end{bmatrix}$$

we have

$$V = \sum_{i=1}^{p} c_i V_i$$ (5.60)

with

$$V_i = Z_i(A) V .$$

Now, (5.59) implies that for each V_i we have

$$A V_i = x_i V_i ,$$

that is each V_i is an *eigenvector* of A. In other words (5.59) implies that every vector may be written as a linear combination of eigenvectors. Using this fact, it is not difficult to put together a matrix P yielding (5.56).

5.3 Exercises for Chapter 5

5.3.1 Show that for any square matrix A,

$$A^n A^m = A^{n+m} .$$

5.3.2 Use the Cayley–Hamilton theorem to calculate the inverses of the following matrices:

$$A = \begin{bmatrix} 5 & -6 & -6 \\ -1 & 4 & 2 \\ 3 & -6 & -4 \end{bmatrix}, \quad B = \begin{bmatrix} 2 & 1 & 1 \\ 2 & 0 & -1 \\ 1 & 1 & 2 \end{bmatrix}, \quad C = \begin{bmatrix} 0 & 0 & a \\ 1 & 0 & b \\ 0 & 1 & c \end{bmatrix} .$$

5.3.3 Verify that the matrix A in problem 5.3.2 satisfies

$$(A - I)(A - 2I) = 0.$$

5.3.4 Let

$$F_1(x) = x^4 + x^3 + x^2 + x + 1$$
$$F_2(x) = x^5 - x^4 + 2x + 3$$
$$F_3(x) = x^4 - 6x^3 + 13x^2 - 12x + 4.$$

Use the division algorithm to express $F_1(A)$, $F_2(A)$, and $F_3(A)$ as a linear combination of A and I where A is as in problem 5.3.2.

5.3.5 Redo problem 5.3.4 using the interpolation method.

5.3.6 The monic polynomial

$$m_A(x) = x^r + c_{r-1}x^{r-1} + \cdots + c_1 x + c_0$$

of least degree which vanishes when evaluated at A is called the *minimal polynomial* of A.

(a) Show that the minimal polynomial is unique.
(b) Show that if A is $n \times n$, then $\deg m_A(x) \le n$.

5.3.7 Suppose that for some polynomial p we have $p(A) = 0$. Show that $m_A(x)$ divides $p(x)$.

5.3.8 By problem 5.3.7, $m_A(x)$ divides $P_A(x)$. Prove the following stronger statement: Suppose

$$P_A(x) = (x - x_1)^{d_1}(x - x_2)^{d_2} \cdots (x - x_k)^{d_k}$$

with $d_i > 0$, $i = 1, 2, \ldots, k$. Then

$$m_A(x) = (x - x_1)^{r_1}(x - x_2)^{r_2} \cdots (x - x_k)^{r_k}$$

for some integers r_i with $0 < r_i \le d_i$ for $i = 1, 2, \ldots, k$.

5.3.9 Let C be the matrix of problem 5.3.2. Show that $P_C(x) = m_C(x)$.

5.3.10 Show that the characteristic polynomial of the matrix

$$A = \begin{bmatrix} 0 & 0 & 0 & . & . & -a_0 \\ 1 & 0 & 0 & . & . & -a_1 \\ 0 & 1 & 0 & . & . & -a_2 \\ . & . & . & . & . & . \\ 0 & 0 & 0 & . & 1 & -a_{n-1} \end{bmatrix}$$

is

$$P_A(x) = x^n + a_{n-1}x^{n-1} + \cdots + a_2x^2 + a_1x + a_0$$

and that this is also the minimal polynomial of A.

5.3.11 Let A be an $n \times n$ matrix such that $A^k = 0$ for some positive integer k. Show that $A^n = 0$.

5.3.12 Suppose A is an $n \times n$ matrix with $A^k = 0$ for some positive integer k. Prove that the matrix $I - A$ is invertible with

$$(I - A)^{-1} = I + A + A^2 + \cdots + A^n .$$

5.3.13 Let

$$A = \begin{bmatrix} 0 & 1 \\ 1 & 1 \end{bmatrix} .$$

Show that

$$A^n = f_n A + f_{n-1}I,$$

where f_n is the nth Fibonacci number.

5.3.14 Let

$$A_p = \begin{bmatrix} 0 & -p(p+1) \\ 1 & 2p+1 \end{bmatrix} .$$

Express $F_n(A_p)$ and $G_n(A_p)$ as a linear combination of I, where

$$F_n(x) = 1 + x + x^2 + \cdots + x^n$$
$$G_n(x) = ((2p+1)x - p(p+1))^n .$$

5.3.15 Suppose A satisfies

$$(A - x_1 I)(A - x_2 I) = 0$$

with $x_1 \neq x_2$. Use Theorem 5.2.3 to show that for any polynomial $F(x)$ one has

$$F(A) = \frac{F(x_1) - F(x_2)}{x_1 - x_2} A + \frac{F(x_2)x_1 - F(x_1)x_2}{x_1 - x_2} I$$

5.3.16 Suppose

$$A = \begin{bmatrix} 5 & -6 & -6 \\ -1 & 4 & 2 \\ 3 & -6 & -4 \end{bmatrix}, \quad B = \begin{bmatrix} -1 & -2 \\ -2 & 2 \end{bmatrix}.$$

(a) Use problem 5.3.15 to express

$$A^5 + 2A^4 + A^2 + A - I$$

as a linear combination of A and I.

(b) Similarly, express

$$B^n + B^{n-1} + \cdots + B + I$$

as a linear combination of B and I.

5.3.17 Suppose

$$(A - aI)^2 = 0$$

with $A \neq aI$. Use problem 5.3.15 to show that for any polynomial $F(x)$,

$$F(A) = F'(a)A + (F(a) - aF'(a))I .$$

5.3.18 Let $F_n(x) = x^n + x^{n-1} + \cdots + x + 1$ and

$$A = \begin{bmatrix} 0 & -1 \\ 1 & 2 \end{bmatrix}, \quad B = \begin{bmatrix} -3 & -1 \\ 4 & 1 \end{bmatrix}.$$

(a) Show that

$$F_n(A) = \frac{1}{2} \begin{bmatrix} (n+1)(2-n) & -n(n+1) \\ n(n+1) & (n+1)(n+2) \end{bmatrix} .$$

(b) Calculate $F_n(B)$.

5.3.19

(a) Verify that for any α the matrix

$$A(\alpha) = \begin{bmatrix} \cos\alpha & -\sin\alpha \\ \sin\alpha & \cos\alpha \end{bmatrix}$$

is orthogonal.

(b) Use the Cayley–Hamilton theorem to show that

$$A(\alpha)^n = \begin{bmatrix} \cos n\alpha & -\sin n\alpha \\ \sin n\alpha & \cos n\alpha \end{bmatrix} .$$

5.3.20 Suppose σ is a permutation of $\{1, 2, \ldots, n\}$. Let $M(\sigma)$ denote the $n \times n$ matrix $[\delta_{i,\sigma_j}]$. For instance for $\sigma = 312$ we have

$$M(\sigma) = \begin{bmatrix} 0 & 1 & 0 \\ 0 & 0 & 1 \\ 1 & 0 & 0 \end{bmatrix} .$$

(a) Show that $M(\sigma^{-1}) = M(\sigma)^{-1}$.
(b) Verify that if σ is an n-cycle, then

$$P_{M(\sigma)}(x) = x^n - 1 .$$

(c) Argue that if σ has p_i cycles of length i in its cycle decomposition, for $i = 1, 2, \ldots, n$, then

$$P_{M(\sigma)}(x) = (x - 1)^{p_1}(x^2 - 1)^{p_2} \cdots (x^n - 1)^{p_n} .$$

(d) Let $\sigma = 426513$ and put $A = M(\sigma)$. Verify directly that

$$M(\sigma^{-1}) = A^5 - A^4 - A^3 + A + I .$$

(e) For the following permutations σ use the Cayley–Hamilton theorem to express $M(\sigma^{-1})$ as a polynomial in $M(\sigma)$ as in part (d):

(a) $\sigma = 2143$,
(b) $\sigma = 32154$,
(c) $\sigma = 314256$.

5.3.21 Suppose $\sigma_i = n + 1 - i$ for $i = 1, 2, \ldots, n$. Prove that

$$\det M(\sigma) = (-1)^{\binom{n}{2}},$$

where $M(\sigma)$ is the matrix of σ defined in problem 5.3.20.

5.3.22 Let J denote the $n \times n$ matrix of all 1s.

(a) Show that for $n > 1$, the minimal polynomial of J is

$$M_J(x) = x(x - n) .$$

(b) By problem 5.3.8, the characteristic polynomial of J is of the form

$$P_J(x) = x_1^d (x - n)_2^d, \qquad d_1, d_2 \geq 1.$$

Show that $d_1 = n - 1$ and $d_2 = 1$.

(c) If $F(x) = a_0 + a_1 x + \cdots + a_m x^m$ show that

$$F(J) = \left(\sum_{k=1}^{m} a_k n^{k-1} \right) J + a_0 I.$$

5.3.23 Put $\tau = 2\,3 \cdots n\,1$ and let $A = M(\tau)$. Show that

$$I + A + A^2 + \cdots + A^{n-1} = J,$$

where J is the $n \times n$ matrix of all 1s defined in problem 5.3.22.

5.3.24 Put

$$A = \begin{bmatrix} 0 & 1 & 1 \\ 1 & 0 & 1 \\ 1 & 1 & 0 \end{bmatrix}.$$

(a) Calculate $P_A(x)$ and $m_A(x)$.

(b) Use problem 5.3.15 to verify that

$$A^n = \frac{1}{3} \left(2^n - (-1)^n \right) A + \frac{1}{3} \left(2^n + 2(-1)^n \right) I .$$

5.3.25 Calculate the number of paths of length n between the nodes 1 and 2 in the graph whose adjacency matrix is the matrix A of problem 5.3.24.

5.3.26 Let A be the matrix of the n-cycle $\tau = 2\,3 \cdots n\,1$ and suppose $p(x)$ is a polynomial of degree at most $n - 1$ satisfying

$$p(1) = np(0) .$$

(a) Prove that if n is odd, then the row sums, column sums, and principal diagonal sums of the matrix $p(A)$ are all equal.

(b) Prove that if n is even, $p(A)$ satisfies the conclusion of part (a) if and only if $p(x)$ has the additional property $p(-1) = 0$.

5.3.27 Suppose the sequence $\{Z_n\}$ satisfies the recursion

$$Z_n = a_1 Z_{n-1} - a_2 Z_{n-2} + \cdots + (-1)^{p-1} a_p Z_{n-p}.$$

(a) Show that for $n \geq 0$, Z_{n+p} can be written in the form

$$Z_{n+p} = c_{n,1} Z_{p-1} - c_{n,2} Z_{p-2} + \cdots + (-1)^{p-1} c_{n,p} Z_0,$$

where each $c_{n,k}$ is a polynomial in $a_1, a_2, \ldots, a_p$, for $k = 1, 2, \ldots, p$.

(b) Prove that $c_{n,k}$ is the determinant of the $(n+1) \times (n+1)$ matrix

$$
\begin{bmatrix}
a_1 & a_2 & a_3 & \cdots & \cdot & a_n & a_{k+n} \\
1 & a_1 & a_2 & \cdots & \cdot & \cdot & \cdot \\
0 & 1 & a_1 & \cdots & \cdot & \cdot & \cdot \\
0 & 0 & 1 & \cdots & \cdot & \cdot & \cdot \\
\cdot & \cdot & \cdots & a_2 & a_3 & & \cdot \\
\cdot & \cdot & \cdots & a_1 & a_2 & a_{k+2} \\
\cdot & \cdot & \cdots & 1 & a_1 & a_{k+1} \\
\cdot & \cdot & \cdots & 0 & 1 & a_k
\end{bmatrix}
$$

5.4 Sample Quiz for Chapter 5

1. Let

$$A = \begin{bmatrix} -1 & 2 \\ 3 & 4 \end{bmatrix}.$$

   (a) Calculate the characteristic polynomial $P_A(x)$.
   (b) Use the Cayley–Hamilton theorem to express the inverse of A as a polynomial in A.

2. Let A be as in problem 5.3.1. Given

$$F(x) = x^4 - 3x^3 - 9x^2 - 2x - 1 ,$$

   express $F(A)$ in the form $F(A) = \alpha A + \beta I$ by means of the

   (a) the division method,
   (b) the interpolation method.

3. Consider the difference equation

$$y_{n+2} = 4y_{n+1} - 2y_n$$

   for $n \geq 0$ with initial values y_0 and y_1.

   (a) Rewrite this recursion in matrix form.
   (b) Use the Cayley–Hamilton theorem and the interpolation method to find a formula for y_{n+2} in terms of y_0, y_1, and n.

Chapter 6
Exponential Structures and Polynomial Operators

6.1 More on Partitions and Permutations

We shall start by studying the generating function of the sequence of polynomials

$$\sigma_n(x) = \sum_{k=1}^{n} S_{n,k} x^k . \tag{6.1}$$

Here $S_{n,k}$ are the Stirling numbers of the second kind, and $\sigma_n(x)$ are the exponential polynomial introduced in Chapter 1, Section 1.8.

Let us for a moment denote by $B_{n,k}$ the class of *ordered* partitions of the set $[n] = \{1, 2, \ldots, n\}$ into k parts. The elements of $B_{n,k}$ can best be visualized as

placements of n distinguishable balls into k distinguishable boxes,

while as we have seen in Chapter 1, those of $\Pi_{n,k}$ (the k-partitions of the n-set) can be visualized as

placements of n distinguishable balls into k indistinguishable boxes.

Clearly, by ordering the parts of a k-partition of $[n]$ in all possible ways we obtain $k!$ distinct elements of $B_{n,k}$. This yields the relation

$$\#B_{n,k} = k! \, S_{n,k} . \tag{6.2}$$

It so happens that the collection $B_{n,k}$ is easier to study than $\Pi_{n,k}$. In fact, we have

Theorem 6.1.1 *The number of ordered partitions of $[n]$ with a total of k parts and whose ith part has p_i elements is given by the multinomial coefficient*

$$\frac{n!}{p_1! p_2! \cdots p_k!} . \tag{6.3}$$

© Springer Nature Switzerland AG 2021
Ö. Eğecioğlu, A. M. Garsia, *Lessons in Enumerative Combinatorics*, Graduate
Texts in Mathematics 290, https://doi.org/10.1007/978-3-030-71250-1_6

Proof An element $\beta \in B_{n,k}$ can be represented by a word $w(\beta)$ in the alphabet $X_k = \{x_1, x_2, \ldots, x_k\}$. Indeed, we simply put in the ith position of $w(\beta)$ the letter x_j if i belongs to the jth part of β.

This done, the elements of $B_{n,k}$ that are being counted here are represented by words in which the letter x_j occurs exactly p_j times. However, we know that these words are counted precisely by the multinomial coefficient in (6.3). $\square$

An immediate consequence of this result is a formula for $S_{n,k}$. Namely

Theorem 6.1.2

$$S_{n,k} = \frac{n!}{k!} \sum_{p_1 + p_2 \cdots + p_k = n} \frac{1}{p_1! p_2! \cdots p_k!} \qquad (p_i \geq 1) . \qquad (6.4)$$

Proof Just combine (6.2) and (6.3). $\square$

We can now derive the following result:

Theorem 6.1.3

$$1 + \sum_{n \geq 1} \frac{t^n}{n!} \sigma_n(x) = \exp(x(e^t - 1)) . \qquad (6.5)$$

Proof Since

$$e^t - 1 = \sum_{p \geq 1} \frac{t^p}{p!} ,$$

for any integer k we can write

$$(e^t - 1)^k = \sum_{p_1 \geq 1} \sum_{p_2 \geq 1} \cdots \sum_{p_k \geq 1} \frac{1}{p_1! p_2! \cdots p_k!} t^{p_1} t^{p_2} \cdots t^{p_k} .$$

This implies that the sum in the right hand side of (6.4) is simply the coefficient of t^n in

$$(e^t - 1)^k .$$

Thus formula (6.4) can be rewritten as

$$S_{n,k} = \frac{n!}{k!} (e^t - 1)^k \Big|_{t^n} . \qquad (6.6)$$

This gives (for $n \geq 1$)

$$\frac{1}{n!}\sigma_n(x) = \frac{1}{n!}\sum_{k=1}^n S_{n,k}x^k = \sum_{k\geq 1}\frac{(e^t-1)^k}{k!}x^k \Bigg|_{t^n}.$$

Multiplying by t^n and summing we obtain (6.5) as asserted.   □

The expression

$$B_n = \sum_{k=1}^n S_{n,k}$$

gives the total number of partitions of the set $[n]$. The B_n's are usually referred to as the *Bell numbers*. By setting $x = 1$ in (6.1) we get of course

$$B_n = \sigma_n(1).$$

Thus formula (6.5) yields the exponential generating function of the Bell numbers. Namely

Theorem 6.1.4

$$\sum_{n\geq 0}\frac{t^n}{n!}B_n = \exp(e^t-1), \qquad (here\ B_0 = 1). \tag{6.7}$$

We have shown combinatorially in Chapter 1, Theorem 1.8.2 that the Bell numbers satisfy the recursion

$$B_{n+1} = \sum_{k=0}^n \binom{n}{k}B_{n-k}. \tag{6.8}$$

Here we can obtain this result analytically from (6.7) upon differentiating both sides and equating coefficients of t^n.

There is a result concerning partitions that is somewhat analogous to Theorem 6.1.1. It can be stated as follows:

Theorem 6.1.5 *The number of partitions of the set $[n]$ which have p_i parts of cardinality i is given by the expression*

$$\frac{n!}{(1!)^{p_1}(2!)^{p_2}(3!)^{p_3}\cdots p_1!p_2!p_3!\cdots}. \tag{6.9}$$

Proof We shall present the argument in the special case $p_1 = 3$, $p_2 = 4$, $p_3 = 2$, and $n = 17$. The reader should have no difficulty in recognizing that the same argument applies in the general case.

We are going to enumerate all the partitions of [17] which have 3 one-element parts, 4 two-element parts, and 2 three-element parts. We may visualize the structure of such partitions by the diagram in Figure 6.1.

Fig. 6.1 The structure of a partition of [17] with three one-element, four two-element, and two three-element parts.

Note now that placing the successive elements of the permutation

$$12 \quad 2 \quad 11 \quad 5 \quad 14 \quad 17 \quad 3 \quad 7 \quad 6 \quad 15 \quad 9 \quad 10 \quad 1 \quad 4 \quad 8 \quad 13 \quad 16 \qquad (6.10)$$

into the circles of Figure 6.1 creates a configuration of balls in boxes depicted in Figure 6.2, which, as such, represents a partition with the desired distribution of part sizes. More precisely, the partition represented by Figure 6.2, with parts arranged in order of increasing least elements is

$$\{1, 4, 10\}, \quad \{2\}, \quad \{3, 17\}, \quad \{5, 14\}, \quad \{6, 7\}, \quad \{8, 13, 16\}, \quad \{9, 15\}, \quad \{11\}, \quad \{12\} . \qquad (6.11)$$

Fig. 6.2 Configuration of balls in boxes, labeled with the permutation in (6.10).

Note that if in this construction, we replace the permutation in (6.10) by the permutation

$$11 \quad 12 \quad 2 \quad 17 \quad 3 \quad 14 \quad 5 \quad 9 \quad 15 \quad 6 \quad 7 \quad 13 \quad 8 \quad 16 \quad 4 \quad 10 \quad 1 \qquad (6.12)$$

we obtain the *balls in boxes* configuration in Figure 6.3, which also represents the same partition. This is simply due to the fact that the permutation in (6.12) is obtained from the permutation in (6.10) by changing the order of the boxes and changing the order of the elements within some of the boxes. Clearly, neither of these orders counts in the construction of the resulting partition.

Fig. 6.3 Configuration of balls in boxes, labeled with the permutation in (6.12).

Now, the elements within a 2-element box may be rearranged in 2! ways and those within a 3-element box in 3! ways. While at the same time, the 1-element boxes can be rearranged in 3! ways among themselves, the 2-element boxes in 4! ways, and finally the 3-element boxes in 2! ways.

Since all these interchanges can be carried out simultaneously and independently of each other, we see that there is a total of

$$(2!)^4(3!)^2 3!4!2! = 165888$$

ways of altering the permutation in (6.10) without affecting the resulting partition. In other words, we have precisely 165888 permutations, all of which lead, by the above construction, to the partition in (6.11).

Clearly, the same reasoning applies regardless of what permutation we work with. Thus we must conclude that every permutation belongs to a class of 165888 permutations all leading to the same partition. Since every partition whose structure is given by the diagram in (6.1) can be obtained by our construction, we deduce that the number of such partitions multiplied by 165888 is equal to the total number of permutations of the numbers $1, 2, \ldots, 17$. In other words, we must have

$$N = \frac{17!}{(2!)^4(3!)^2 3!4!2!} \tag{6.13}$$

and this is formula (6.9) in the special case we are considering. $\qquad\square$

It is instructive to review this argument in terms of the Shepherd Principle: the "sheep" in question here are all partitions of the set [17] with 3 one-element, 4 two-element, and 2 three-element parts. We have a total number of 17! "legs" corresponding to the permutations of the set [17], and the remarks above show that each of our sheep has exactly

$$(2!)^4(3!)^2 3!4!2! = 165888$$

legs.

Thus the total number of sheep that we have is given by (6.13) as asserted.

Before closing this section, we should point out that a formula entirely analogous to (6.9) holds for permutations as well. This can be stated as follows:

Theorem 6.1.6 *The number of permutations of the set* [n] *which have* p_i *cycles of length i is given by the expression*

$$\frac{n!}{1^{p_1} 2^{p_2} 3^{p_3} \cdots p_1! p_2! p_3! \cdots}. \tag{6.14}$$

This result can be proved very much in the same way we proved formula (6.9). The reader should have no difficulty in making the appropriate changes in the arguments given at the start of this chapter and obtain a proof of (6.14).

6.2   Exponential Structures

There is a wide variety of combinatorial structures that can be put together by the same basic construction. Partitions and permutations are but the simplest examples of such structures. Before presenting the construction in the most general setting it will be good to go over two further examples.

We shall work first with composite objects that may be described as

distinguishable balls in indistinguishable tubes.

The collection of all configurations of n balls in k nonempty tubes will be denoted by $T_{n,k}$. For instance, an element of $T_{8,3}$ may be pictured as in Figure 6.4. Note that if we remove the balls from the tubes and put them in boxes we are led to the partition

$$\{1, 7, 8\}, \quad \{2, 3\}, \quad \{4, 5, 6\}. \tag{6.15}$$

Accordingly we have ordered the tubes in Figure 6.4 just as we would have ordered the parts of the partition in Figure 6.4. That is in order of increasing *least elements*. The least elements in this case are the balls 1, 2, and 4.

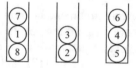

Fig. 6.4 Balls in tubes: an element of $T_{8,3}$.

Clearly, *balls in tubes* is simply a colorful way of picturing partitions whose parts are not subsets but ordered subsets. In other words, what we have here are partitions where the order of the parts does not count but the order of the elements within the parts does count.

This viewpoint leads us to a three-step procedure for constructing all the configurations of a given class $T_{n,k}$. For instance, to construct an element of $T_{8,3}$ we can proceed as follows:

We first select a partition $\Pi = (A_1, A_2, A_3)$ in $\Pi_{8,3}$. Secondly we select a total order for each of the parts A_1, A_2, and A_3. Finally, we place the balls with labels in A_1, A_2, and A_3 in the first, second, and third tubes respectively, according to the selected total orders.

In order to better understand the general mechanism that underlies this construction we must also give a pictorial description of these *total orders* we are selecting. To this end, let us assume that in the first step of the procedure we select the partition in (6.15). This given, before we fill the first tube with the elements of A_1 we must select one of the six different ways of filling a tube with three different balls. These may be represented by the list of patterns in Figure 6.5.

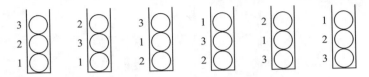

Fig. 6.5 List of ways of filling a tube with three balls.

Here the numbers on the side are to indicate where we must put the smallest, the next smallest, and the largest ball respectively. Thus, for instance, if we place the balls with labels in $A_1 = \{1, 7, 8\}$ in a tube according to the fifth pattern in Figure 6.5 we are led to the configuration in Figure 6.6.

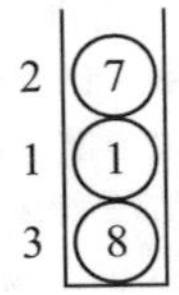

Fig. 6.6 Tube filled with $A_1 = \{1, 7, 8\}$ using the fifth pattern in Figure 6.5.

Thus we may interpret the first filled tube in Figure 6.4 as the result of *placing* the set A_1 into the fifth pattern of the list in Figure 6.5. Similarly, the last filled tube in Figure 6.4 can be obtained by placing the set $A_2 = \{4, 5, 6\}$ into the third pattern of the list in Figure 6.5. As for the second filled tube, the situation is entirely analogous. The list of all possible ways of filling a tube with two balls is shown in Figure 6.7.

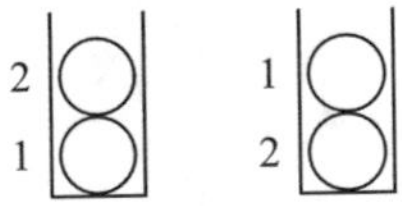

Fig. 6.7 List of ways of filling a tube with two balls.

Thus the second filled tube in Figure 6.4 is simply the result of placing the set $A_2 = \{2, 3\}$ into the first pattern of the list in Figure 6.7. To further illustrate this procedure, let us suppose that in the first step we select the partition

$$A_1 = \{1, 5, 8\}, \quad A_2 = \{2, 3, 7\}, \quad A_3 = \{4, 5\}$$

and that in the second step we select the second and fourth patterns in Figure 6.5 and the second pattern in Figure 6.7. Then placing A_1, A_2, and A_3 respectively into these three patterns yields the configuration in Figure 6.8.

Let now T_m denote the list of patterns representing all possible ways of placing m different balls in a tube. Let us suppose that we have been given these lists for all

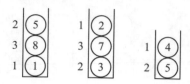

Fig. 6.8 Placing $A_1 = \{1, 5, 8\}$, $A_2 = \{2, 3, 7\}$, $A_3 = \{4, 5\}$ using the second and fourth patterns in Figure 6.5 and the second pattern in Figure 6.7.

m. This given, our three step procedure for constructing an element of $T_{n,k}$ may be described as follows. We first select a partition

$$\Pi = (A_1, A_2, \ldots, A_k)$$

in $\Pi_{n,k}$. Then, given that the number of elements in A_i is m_i, we select a pattern in the list T_{m_i}. Finally, in the third step of the procedure, we place the elements of A_i in this pattern. Of course, we do this for each $i = 1, 2, \ldots, k$.

But now that we have formalized in this manner the construction of the *balls in tubes* configurations, we are naturally led to the creation of all kinds of new families of composite objects. Indeed, all we have to do is use exactly the same three step procedure only with a different sequence of lists of patterns to be used in the second step of the procedure.

To illustrate this idea, let us suppose that the mth list of patterns is the collection C_m of Cayley trees on m nodes. Only here it is better to replace the nodes by circles. For instance the configurations in Figure 6.9 are elements of C_7, C_4, and C_5 respectively.

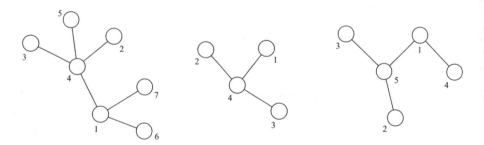

Fig. 6.9 Cayley trees as patterns.

If we now select the partition

$$A_1 = \{1, 2, 4, 7, 11, 12, 15\}, \quad A_2 = \{3, 5, 8, 16\}, \quad A_3 = \{6, 9, 10, 13, 14\}$$

in $\Pi_{16,3}$ and place A_1, A_2, A_3 in the above Cayley trees in the same exact manner we did for the *balls in tubes*, we obtain the composite configuration of Figure 6.10.

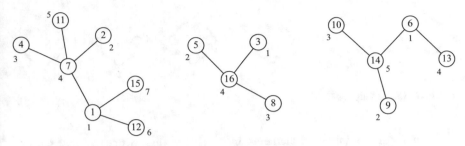

Fig. 6.10 Composite configuration in $F_{16,3}(C)$ after placing A_1, A_2, A_3.

It will be good to introduce some terminology here. First of all, a configuration consisting of a total of k trees will be referred to as a *k-forest*. Thus the composite object in Figure 6.10 is what we call a *Cayley 3-forest on 16 nodes*. Let us denote the collection of all Cayley k-forests on n nodes by the symbol $F_{n,k}(C)$.

Clearly, we can use the same procedure and introduce other kinds of forests by simply changing the lists of trees used as basic patterns. Thus if the mth list of patterns is the collection RC_m of all *rooted Cayley trees*, then the resulting composite objects will be what we may want to call *forests of rooted Cayley trees*. Let us denote the collection of such k-forests on n nodes by the symbol

$$F_{n,k}(RC) .$$

For instance, the configuration in Figure 6.11 is an element of $F_{16,3}(RC)$.

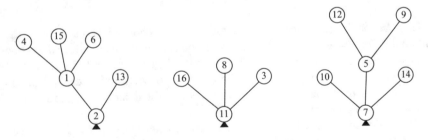

Fig. 6.11 An element of $F_{16,3}(RC)$.

The reader may construct further examples by using other lists of basic patterns.

Perhaps we should point out that we need not restrict ourselves to trees in our selection of basic patterns. Indeed, almost anything will do, as long as the mth list of patterns consists of configurations which have m circles in them labeled by the numbers $1, 2, \ldots, m$. For instance, suppose that the mth list is the collection Z_m consisting of all cycles with m nodes labeled $1, 2, \ldots, m$. We can refer to these as *m-cycles*. In particular, the collection of 4-cycles Z_4 is simply the list of patterns in Figure 6.12.

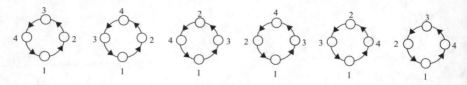

Fig. 6.12 List of 4-cycles Z_4.

Clearly Z_m has $(m - 1)!$ elements in it. This is due to the fact that we can always picture the cycle with the label 1 at the bottom and the arrows pointing in the counterclockwise direction. The remaining $m - 1$ labels can then be distributed in any one of the $(m - 1)!$ possible ways without creating any duplications.

This given, the reader should have no difficulty recognizing that the resulting composite objects, with this choice of basic patterns, are none other than the cycle diagrams of permutations. Thus the family $F_{n,k}(Z)$ can be identified with the class of permutations of $[n]$ whose cycle diagram consists of a total of k cycles.

Perhaps the best way to describe the general situation may be as follows. We are given a sequence $\{L_m\}$ of lists of patterns. We imagine that each pattern in L_m is actually a *playcard* displaying a certain picture which contains, among other things, m blank circles labeled $1, 2, \ldots, m$. A composite object can thus be thought of as a deck of filled cards. The family $F_{n,k}(L)$ is then the collection of all those decks which have k cards in them, containing all together n circles. More precisely, to construct an element of $F_{n,k}(L)$, we follow the three-step procedure we have described above. That is, first we select a partition

$$\Pi = (A_1, A_2, \ldots, A_k)$$

in $\Pi_{n,k}$. Secondly, given that A_i has m_i elements, we chose in the list L_{m_i} a card b_i. We do this for $i = 1, 2, \ldots, k$. We should emphasize that we are allowed to choose the same card more than once. Finally, in the third step, for $i = 1, 2, \ldots, k$, we fill the blank circles in the card b_i with the elements of A_i in the order indicated by the labels. The resulting deck then consists of these filled cards arranged in order of increasing least elements.

The collection of all families $F_{n,k}(L)$ constructed from a fixed sequence of patterns $\{L_m\}$ is referred to as the *exponential structure* associated with the sequence $\{L_m\}$. It will be denoted by the symbol $F(L)$. In other words symbolically we can write

$$F(L) = \sum_{n \geq 1} \sum_{k=1}^{n} F_{n,k}(L) .$$

Now there is a very general enumeration theorem which relates the cardinalities $\#F_{n,k}(L)$ of the families $F_{n,k}(L)$ to those of the cardinalities of the lists $\#L_m$. The purpose of the next section is to present this theorem and give some applications.

6.3 The Exponential Formula and Some Applications

Given an exponential structure $F(L)$ we associate to it a sequence of polynomials $\{\phi_n(x, L)\}$ by setting $\phi_0(x, L) = 1$ and

$$\phi_n(x, L) = \sum_{k=1}^{n} x^k \# F_{n,k}(L) \qquad \text{(for } n \geq 1\text{)} . \tag{6.16}$$

It turns out that a number of the classical sequences of polynomials can be associated in this manner to exponential structures.

We give a few examples. The first ones are the exponential polynomials we introduced in Section 6.1, namely

$$\sigma_n(x) = \sum_{k=1}^{n} x^k S_{n,k} .$$

Indeed we have

$$\sigma_n(x) = \phi_n(x, L)$$

whenever there is only one pattern in each list L_m. For then we shall have only one composite object for each partition, and the number of elements in $F_{n,k}(L)$ is simply equal to $\#\Pi_{n,k} = S_{n,k}$.

The next example is a bit more interesting. We recall that when we take L_m to be the list Z_m of all cycles, then the resulting exponential structure $F(Z)$ is the collection of all permutation diagrams. Thus we have

$$\phi_n(x, Z) = \sum_{k=1}^{n} x^k \# F_{n,k}(Z)$$

$$= \sum_{k=1}^{n} x^k p_{n,k} .$$

Note that our recursion in (1.83) gives

$$p_{n+1,k} = p_{n,k-1} + n p_{n,k}$$

so that

$$\phi_{n+1}(x, Z) = \sum_{k} p_{n+1,k} x^k$$

$$= x \sum_{k} p_{n,k-1} x^{k-1} + n \sum_{k} p_{n,k} x^k$$

$$= x \phi_n(x, Z) + n \phi_n(x, Z)$$

and this can be written in the form

$$\phi_{n+1}(x, Z) = \phi_n(x, Z)(x + n) .$$

Thus, since $\phi_1(x, Z) = x$, we recursively obtain

$$\phi_n(x, Z) = x(x + 1)(x + 2) \cdots (x + n - 1) . \qquad (6.17)$$

As we have had occasion to note before, these polynomials are referred to as the *upper factorial polynomials*.

We should also point out that a similar reasoning, based on the recursion (1.85) yields

$$\sum_{k=1}^{n} s_{n,k} x^k = x(x - 1)(x - 2) \cdots (x - n + 1) ,$$

where the $s_{n,k}$ are the Stirling numbers of the first kind, and the latter polynomials as we saw in Chapter 1 are the *lower factorial polynomials,* usually denoted by the symbol $(x)_n$.

Note that Newton's expansion

$$\sum_{n \geq 0} \binom{x + n - 1}{n} t^n = \frac{1}{(1 - t)^x}$$

may also be written in the form

$$1 + \sum_{n \geq 1} \frac{t^n}{n!} (x + n - 1)(x + n - 2) \cdots (x + 1)x = \exp\left(x \log \frac{1}{1 - t} \right) .$$

Comparing with (6.17) we see that we have

$$1 + \sum_{n \geq 1} \frac{t^n}{n!} \phi_n(x, Z) = \exp\left(x \log \frac{1}{1 - t} \right) \qquad (6.18)$$

which is a formula entirely analogous to (6.5).

Now as it turns out, both (6.5) and (6.18) are special cases of a general formula which holds for sequences of polynomials associated with an exponential structure. This may be stated as follows:

Theorem 6.3.1

$$1 + \sum_{n \geq 1} \frac{t^n}{n!} \phi_n(x, L) = \exp\left(x \sum_{m \geq 1} \frac{t^m}{m!} \#L_m \right) . \qquad (6.19)$$

Proof For convenience, let us set

$$s_m = \#L_m .$$

Our goal is to obtain first a formula for $\#F_{n,k}(L)$. To achieve this, suppose that in the first step of our procedure we choose a partition $\Pi \in \Pi_{n,k}$ which has p_i parts of cardinality i for $i = 1, 2, \dots, n$. A moment of thought should reveal that at the second step of the procedure we have a total of

$$(\#L_1)^{p_1} (\#L_2)^{p_2} (\#L_3)^{p_3} \cdots = s_1{}^{p_1} s_2{}^{p_2} s_3{}^{p_3} \cdots$$

choices available. From formula (6.9) we deduce that there is a total of

$$\frac{n!}{p_1! p_2! p_3! \cdots (1!)^{p_1} (2!)^{p_2} (3!)^{p_3} \cdots} s_1{}^{p_1} s_2{}^{p_2} s_3{}^{p_3} \cdots$$

choices yielding a composite object whose corresponding partition has p_i parts of cardinality i. Since the number of parts of such a partition is

$$p_1 + p_2 + \cdots + p_n ,$$

we see that we must have

$$\#F_{n,k}(L) = n! \sum_{\substack{k = p_1 + p_2 + p_3 + \cdots \\ n = p_1 + 2p_2 + 3p_3 + \cdots}} \frac{1}{p_1! p_2! p_3! \cdots} \left(\frac{s_1}{1!}\right)^{p_1} \left(\frac{s_2}{2!}\right)^{p_2} \left(\frac{s_3}{3!}\right)^{p_3} \cdots$$

This may be rewritten in the form

$$\#F_{n,k}(L) = \frac{n!}{k!} \sum_{k = p_1 + p_2 + \cdots + p_n} \frac{k!}{p_1! p_2! p_3! \cdots} \left(\frac{s_1 t}{1!}\right)^{p_1} \left(\frac{s_2 t^2}{2!}\right)^{p_2} \left(\frac{s_3 t^3}{3!}\right)^{p_3} \cdots \Bigg|_{t^n} .$$

Using the multinomial theorem, this may be simply expressed as

$$\#F_{n,k}(L) = \frac{n!}{k!} \left(\frac{s_1}{1!} t + \frac{s_2}{2!} t^2 + \frac{s_3}{3!} t^3 + \cdots\right)^k \Bigg|_{t^n} .$$

We thus obtain

$$\frac{1}{n!} \phi_n(x, L) = \frac{1}{n!} \sum_{k=1}^{n} x^k \#F_{n,k}(L) = \sum_{k \geq 1} \frac{x^k}{k!} \left(\sum_{m \geq 1} \frac{s_m}{m!} t^m\right)^k \Bigg|_{t^n} .$$

Multiplying by t^n and summing gives formula (6.19) as asserted. $\square$

Theorem 6.3.1 has many interesting applications. We can only illustrate a few. First of all, note that if we go back for a moment to our choice $L_m = Z_m$ and recall that $\#Z_m = (m-1)!$, we see that (6.19) in this case reduces to

$$1 + \sum_{n\geq 1} \frac{t^n}{n!} \phi_n(x, Z) = \exp\left(x \sum_{m\geq 1} \frac{t^m}{m} \right). \qquad (6.20)$$

Now, from calculus we know that

$$\sum_{m\geq 1} \frac{t^m}{m} = \log \frac{1}{1-t}.$$

Substituting this in (6.20) we obtain formula (6.18) again!

Let us next set $L_m = T_m$ (recall the *balls in tubes* configurations of Section 6.2). Since now $\#L_m = \#T_m = m!$ formula (6.19) gives

$$1 + \sum_{m\geq 1} \frac{t^n}{n!} \phi_n(x, T) = \exp\left(x \sum_{m\geq 1} t^m \right)$$

$$= \exp\left(x \frac{t}{1-t} \right) \qquad (6.21)$$

$$= \sum_{k\geq 0} \frac{x^k}{k!} \left(\frac{t}{1-t} \right)^k.$$

Using the expansion

$$\left(\frac{t}{1-t} \right)^k = \sum_{n\geq k} \binom{n-1}{k-1} t^n,$$

we can rewrite the identity in (6.21) as

$$1 + \sum_{n\geq 1} \frac{t^n}{n!} \phi_n(x, T) = \sum_{n\geq 0} t^n \sum_{k\leq n} \frac{x^k}{k!} \binom{n-1}{k-1}.$$

Equating coefficients of t^n in the two sides of this equation gives

$$\phi_n(x, T) = \sum_{k=1}^{n} x^k \frac{n!}{k!} \binom{n-1}{k-1}.$$

Since by definition

$$\phi_n(x, T) = \sum_{k=1}^{n} x^k \#T_{n,k} \, ,$$

we see that we must have

$$\#T_{n,k} = \frac{n!}{k!} \binom{n-1}{k-1} \, . \tag{6.22}$$

The reader may find it challenging to derive this fact by counting all possible configurations of n balls in k tubes directly.

As another example of the uses (6.19) may be put to, we shall derive from it yet another proof of the formula

$$\#C_n = n^{n-2} \tag{6.23}$$

giving the number of Cayley trees on n nodes. We shall derive this result by studying the exponential structure $F(RC)$ obtained by taking L_m to be the list of all rooted Cayley trees on n nodes. That is we set

$$L_m = RC_m \, .$$

Let us then suppose that we do not know (6.23), and let us set for convenience $\#C_n = c_n$.

Note that, since there are n different ways to make a Cayley tree with n nodes into a rooted Cayley tree, we must have

$$\#RC_n = nc_n \, . \tag{6.24}$$

Suppose now that we are given a Cayley tree with $n+1$ nodes labeled $0, 1, 2, \ldots, n$ and that we remove from it the node labeled "0" along with all the edges that emanate from it. Clearly the resulting object may be viewed as a forest of rooted Cayley trees. Indeed, we may take the roots of these trees to be the nodes which were originally attached to the node labeled 0 in the original tree we started with. This simple observation gives us the identity

$$c_{n+1} = \sum_{k=1}^{n} \#F_{n,k}(RC) \, . \tag{6.25}$$

Now, formula (6.19), written for $L_m = RC_m$ is

$$1 + \sum_{n \geq 1} \frac{t^n}{n!} \sum_{k=1}^{n} x^k \# F_{n,k}(RC) = \exp\left(x \sum_{m \geq 1} \frac{t^m}{m!} \# RC_m \right).$$

Setting $x = 1$ and using (6.24) and (6.25) gives

$$1 + \sum_{n \geq 1} \frac{t^n}{n!} c_{n+1} = \exp\left(\sum_{m \geq 1} \frac{t^m}{(m-1)!} c_m \right).$$

Multiplying by t and changing index of summation in the left hand side of this equation we get

$$\sum_{m \geq 1} \frac{t^m}{(m-1)!} c_m = t \exp\left(\sum_{m \geq 1} \frac{t^m}{(m-1)!} c_m \right).$$

In other words, we have obtained that the series

$$f(t) = \sum_{m \geq 1} \frac{t^m}{(m-1)!} c_m \tag{6.26}$$

satisfies the equation

$$f(t) = t \exp(f(t)) .$$

However, we can use the Lagrange inversion formula and obtain that

$$f(t) \bigg|_{t^m} = \frac{1}{m}(e^t)^m \bigg|_{t^{m-1}} = \frac{1}{m} \frac{m^{m-1}}{(m-1)!} .$$

Comparing with (6.26) gives

$$\frac{c_m}{(m-1)!} = \frac{m^{m-2}}{(m-1)!}$$

or

$$c_m = m^{m-2} ,$$

and this is what we wanted to show.

6.4 Polynomial Operators

In working with polynomials it is convenient to make use of *operator notation*. To this end we need to make some definitions. First of all a linear operator which sends polynomials into polynomials will be briefly referred to as a *polynomial operator*. One such operator is of course the ordinary derivative, which we shall denote here by the symbol D. Thus

$$Dx^n = nx^{n-1} \qquad (n \geq 0) . \tag{6.27}$$

Using D we can construct a whole class of polynomial operators, called *differential operators*. Indeed, for any given formal power series

$$F(t) = \sum_{k \geq 0} F_k t^k$$

we let

$$F(D) = \sum_{k \geq 0} F_k D^k$$

denote the operator which acts on polynomials in the obvious way, namely

$$F(D)P(x) = \sum_{k \geq 0} F_k D^k P(x), \qquad (D^0 P(x) = P(x)) . \tag{6.28}$$

Note that this makes perfectly good sense, since only a finite number of terms in the sum are different from zero. For instance, when $P(x)$ is of degree 5 all terms $D^k P(x)$ with $k > 5$ are equal to zero. In particular we get

$$F(D)x^n = \sum_{k=0}^{n} F_k n(n-1) \cdots (n-k+1)x^{n-k} . \tag{6.29}$$

Note that because of the linearity condition, to define a polynomial operator T we need only tell what T does to a sequence of polynomials forming a basis. In particular we know T as soon as we are given Tx^n for all $n \geq 0$. Thus (6.27) defines D and (6.29) defines $F(D)$.

A particular instance of (6.28) is

$$(D^2 + 3D)x^5 = 5 \cdot 4x^3 + 3 \cdot 5x^4 .$$

6.4.1 The Shift Operator

However there are considerably more surprising facts that we wish to illustrate. Indeed, let us define the *shift* operator E by setting for every polynomial $P(x)$

$$E P(x) = P(x + 1) .$$

Note that the binomial formula gives

$$Ex^n = (x + 1)^n = \sum_{k=0}^{n} \frac{1}{k!} n(n - 1) \cdots (n - k + 1) x^{n-k}$$

and this can be rewritten in the form

$$Ex^n = \sum_{k=0}^{n} \frac{D^k}{k!} x^n . \tag{6.30}$$

Comparing with (6.28) we discover that E *acts* exactly like the differential operator

$$e^D = \sum_{k \geq 0} \frac{D^k}{k!} ,$$

and thus here and after we can just as well write

$$E = e^D . \tag{6.31}$$

Actually, we can easily see that (6.30) is just an instance of Taylor's theorem. In fact, recall that for any polynomial $P(x)$, say of degree n, Taylor's theorem gives

$$P(x + y) = \sum_{k=0}^{n} \frac{y^k}{k!} P^{(k)}(x) . \tag{6.32}$$

Using our operator notation this result can be simply written as

$$P(x + y) = \left(\sum_{k \geq 0} \frac{y^k}{k!} D^k \right) P(x) .$$

In other words, we have

$$P(x + y) = e^{yD} P(x) , \tag{6.33}$$

and this reduces to (6.30) for $y = 1$ and $P(x) = x^n$.

6.4.2 The Evaluation Operator

It is useful to introduce also the "*letting* $x = a$" operators which we denote by the symbol L_a. More precisely, we set

$$L_a P(x) = P(a) .$$

Thus, for instance

$$L_2(x^3 + 5) = 2^3 + 5 = 13 .$$

Note that another instance of Taylor's theorem is the expansion

$$P(x) = P(0) + \frac{x}{1!} P'(0) + \frac{x^2}{2!} P''(0) + \frac{x^3}{3!} P'''(0) + \cdots \tag{6.34}$$

Using our notation, this can be written in the form

$$P(x) = \sum_{k \geq 0} \frac{x^k}{k!} L_0 D^k P(x) .$$

Thus, if we denote by I the *identity* operator (that is the operator which sends every polynomial back to itself), the expansion in (6.34) reduces to the operator identity

$$I = \sum_{k \geq 0} \frac{x^k}{k!} L_0 D^k . \tag{6.35}$$

Together with the derivative operator, we may also want to consider the integral operator $\int$ defined by setting

$$\int x^n = \frac{x^{n+1}}{n+1} \qquad (n \geq 0) . \tag{6.36}$$

Thus, the polynomial $P(x)$ which satisfies the equation

$$DP(x) = Q(x)$$

and the initial condition $P(0) = 0$, is simply given by the formula

$$P(x) = \int Q(x) .$$

In particular the solution of

$$DP(x) = 3x^2 + 2 \qquad (P(0) = 0) ,$$

is

$$P(x) = \int (3x^2 + 2) = x^3 + 2x .$$

It turns out that equation (6.36) is but a very special instance of a large family of equations we can solve by operator methods. We cannot give the general theory here, but we can present some examples to illustrate the basic ideas.

6.4.3 The Difference Operator

The operator

$$\Delta = E - I$$

is usually referred to as the *forward difference* or briefly as the *difference* operator. In other words, for any polynomial $P(x)$ we set

$$\Delta P(x) = P(x + 1) - P(x) .$$

In particular

$$\Delta x^3 = (x + 1)^3 - x^3 = 3x^2 + 3x + 1 .$$

Similarly the *backward difference operator* Δ_b is defined by setting

$$\Delta_b P(x) = P(x) - P(x - 1) .$$

Thus

$$\Delta_b x^3 = x^3 - (x - 1)^3 = 3x^2 - 3x + 1 .$$

Now it develops that the lower factorial polynomials $(x)_n$ are to Δ what the ordinary power polynomials are to D. More precisely we have the Δ-analogue of (6.27), namely

$$\Delta(x)_n = n(x)_{n-1} \qquad (n \geq 0) . \qquad (6.37)$$

This result is not difficult to verify for general n. For instance for $n = 4$ we see that

$$\Delta(x)_4 = (x + 1)x(x - 1)(x - 2) - x(x - 1)(x - 2)(x - 3)$$
$$= x(x - 1)(x - 2)(x + 1 - x + 3)$$
$$= 4x(x - 1)(x - 2) = 4(x)_3 .$$

Emulating what we have done in (6.36) we may define a new operator $\int$ by setting

$$\int (x)_n = \frac{(x)_{n+1}}{n+1} \qquad (n \geq 0) . \qquad (6.38)$$

We might refer to $\int$ as the *broken* integral operator. Using it we can solve for $P(x)$ in an equation of the form

$$\Delta P(x) = Q(x) \qquad (P(0) = 0) \qquad (6.39)$$

for any given polynomial $Q(x)$. In fact the solution of (6.39) is simply

$$P(x) = \int Q(x) . \qquad (6.40)$$

6.4.4 Applications of the Difference Operator

To show what this formula actually involves, we shall illustrate some of its special uses. Suppose we want to find the polynomial $P(x)$ which is such that

$$P(x + 1) - P(x) = x^2 \qquad (P(0) = 0) . \qquad (6.41)$$

We note that we have

$$x^2 = x(x - 1) + x = (x)_2 + (x)_1 .$$

Thus from (6.40) we get

$$P(x) = \int x^2 = \frac{(x)_3}{3} + \frac{(x)_2}{2} = \frac{x(x - 1)(2x - 1)}{6} . \qquad (6.42)$$

Note that (6.41) for $x = 0, 1, 2, 3$ gives

$$P(1) - P(0) = 0$$
$$P(2) - P(1) = 1^2$$
$$P(3) - P(2) = 2^2$$
$$P(4) - P(3) = 3^2 .$$

Summing these results (and recalling that $P(0) = 0$) we get

$$P(4) = 1^2 + 2^2 + 3^2 .$$

More generally (6.41) and (6.42) give

$$1^2 + 2^2 + \cdots + n^2 = P(n+1) = \frac{n(n+1)(2n+1)}{6} ,$$

which is a well known formula.

Let us next try to get a formula for the sum of the cubes of the first n integers. Proceeding in the same way we shall look for the solution of

$$P(x+1) - P(x) = x^3 \qquad (P(0) = 0) . \qquad (6.43)$$

To this end, we need to write x^3 as a linear combination of the lower factorial polynomials. That is we must find the coefficients c_1, c_2, c_3 in the expansion

$$x^3 = c_1(x)_1 + c_2(x)_2 + c_3(x)_3 . \qquad (6.44)$$

Now we find out that our table of Stirling numbers enables us to obtain this expansion without much further work. Indeed we have the following formula:

$$x^n = \sum_{k=1}^{n} S_{n,k}(x)_k \qquad (6.45)$$

that we first encountered in Chapter 1. In particular the coefficients in (6.44) are given by the third row in the table of Stirling numbers of the second kind. Thus we have

$$x^3 = (x)_1 + 3(x)_2 + (x)_3$$

and the solution of (6.43) is

$$P(x) = \int x^3 = \frac{(x)_2}{2} + (x)_3 + \frac{(x)_4}{4} = \frac{x^2(x-1)^2}{4} .$$

This gives the formula

$$1^3 + 2^3 + \cdots + n^3 = \frac{(n+1)^2 n^2}{4} .$$

To complete our argument here we need to establish (6.45). This identity was given a purely combinatorial proof in Chapter 1. However, in this context, it is best to proceed analytically for this will lead us to a very general result. The idea is to start from the formula (6.5) written in the form

$$1 + \sum_{n \geq 1} \frac{t^n}{n!} \sum_{k=1}^{n} S_{n,k} x^k = \sum_{k \geq 0} \frac{x^k}{k!} (e^t - 1)^k . \tag{6.46}$$

Note that since the coefficients of $t^n x^k$ on both sides of this equation are equal, we can replace x^k by $(x)_k$ and obtain

$$1 + \sum_{n \geq 1} \frac{t^n}{n!} \sum_{k=1}^{n} S_{n,k} (x)_k = \sum_{k \geq 0} \frac{(x)_k}{k!} (e^t - 1)^k .$$

On the other hand, the binomial theorem, written in the form

$$1 + \sum_{k \geq 1} \frac{(x)_k}{k!} t^k = \exp(x \log(1 + t)) \tag{6.47}$$

upon replacing t by $e^t - 1$, gives

$$1 + \sum_{k \geq 1} \frac{(x)_k}{k!} (e^t - 1)^k = \exp\left(x \log(1 + (e^t - 1))\right) = e^{xt} .$$

Thus we must conclude that

$$1 + \sum_{n \geq 1} \frac{t^n}{n!} \sum_{k=1}^{n} S_{n,k} (x)_k = \sum_{n \geq 0} \frac{x^n}{n!} t^n . \tag{6.48}$$

Equating coefficients of t^n here gives (6.45) as asserted.

6.4.5 Pairs of Polynomial Sequences

The preceding argument can be generalized to give a most remarkable result.

Theorem 6.4.1 *Let*

$$A_n(x) = \sum_{k=1}^{n} A_{n,k} x^k , \quad B_n(x) = \sum_{k=1}^{n} B_{n,k} x^k \tag{6.49}$$

be sequences of polynomials with exponential generating functions

$$(a) \quad 1 + \sum_{n \geq 1} \frac{t^n}{n!} A_n(x) = \exp(xa(t)) \,, \tag{6.50}$$

$$(b) \quad 1 + \sum_{n \geq 1} \frac{t^n}{n!} B_n(x) = \exp(xb(t)) \,.$$

Then the sequence of polynomials defined by setting

$$C_n(x) = \sum_{k=1}^{n} A_{n,k} B_k(x) \tag{6.51}$$

has the exponential generating function given by the formula

$$1 + \sum_{n \geq 1} \frac{t^n}{n!} C_n(x) = \exp(xb(a(t))) \,. \tag{6.52}$$

Proof We start with (6.50) (*a*) written in the form

$$1 + \sum_{n \geq 1} \frac{t^n}{n!} \sum_{k=1}^{n} A_{n,k} x^k = \sum_{k \geq 0} \frac{x^k}{k!} a^k(t) \,.$$

Replacing x^k by $B_k(x)$ gives

$$1 + \sum_{n \geq 1} \frac{t^n}{n!} \sum_{k=1}^{n} A_{n,k} B_k(x) = 1 + \sum_{k \geq 1} \frac{B_k(x)}{k!} a^k(t) \,. \tag{6.53}$$

Thus from (6.50) (*b*) with t replaced by $a(t)$ we get

$$1 + \sum_{n \geq 1} \frac{B_n(x)}{n!} a^n(t) = \exp(xb(a(t))) \,.$$

Combining with (6.53) we obtain (6.52) as asserted.   □

We can see that formula (6.48) is the particular case $a(t) = e^t - 1$, $b(t) = \log(1 + t)$ of (6.52). This leads us to the following generalization of formula (6.45):

Theorem 6.4.2 *Let the polynomials*

$$A_n(x) = \sum_{k=1}^{n} A_{n,k} x^k \,, \quad B_n(x) = \sum_{k=1}^{n} B_{n,k} x^k$$

have the exponential generating functions given in (6.50) and suppose further that $a(t)$ and $b(t)$ are inverse functions, that is

$$b(a(t)) = t . \tag{6.54}$$

Then the following identity holds for all n:

$$\sum_{k=1}^{n} A_{n,k} B_k(x) = x^n . \tag{6.55}$$

Proof This is an immediate corollary of Theorem 6.4.1. Indeed (6.55) follows from (6.54) upon equating coefficients of t in (6.52). □

However, we can get many other formulas from (6.51). For instance we know that the (Laguerre) polynomials

$$L_n(x) = \sum_{k=1}^{n} \binom{n-1}{k-1} \frac{n!}{k!} x^k$$

of *balls in tubes* exponential structure have the exponential generating function

$$1 + \sum_{n \geq 1} \frac{t^n}{n!} L_n(x) = \exp\left(x \frac{t}{1-t}\right) .$$

On the other hand, the polynomials $(x)_n$ have exponential generating function given by (6.47). This given, we can apply Theorem 6.4.1 with $A_n(x) = L_n(x)$ and $B_n(x) = (x)_n$. Formula (6.52) with $a(t) = t/(1-t)$ and $b(t) = \log(1+t)$ then gives

$$1 + \sum_{n \geq 1} \frac{t^n}{n!} \sum_{k=1}^{n} \binom{n-1}{k-1} \frac{n!}{k!} (x)_k = \exp\left(x \log(1 + \frac{t}{1-t})\right)$$

$$= \exp\left(x \log \frac{1}{1-t}\right) \tag{6.56}$$

$$= 1 + \sum_{n \geq 1} \frac{t^n}{n!} x(x+1) \cdots (x+n-1) ,$$

the latter identity being a consequence of formula (6.18). Equating coefficients of t^n in (6.56) yields the surprising identity

$$\sum_{n=1}^{n} \binom{n-1}{k-1} \frac{n!}{k!} (x)_k = x(x+1) \cdots (x+n-1) . \tag{6.57}$$

The reader may find it challenging to find a purely combinatorial proof of this identity. As a hint, we should say that (6.57) can be established by counting configurations consisting of n distinguishable balls in x distinguishable tubes, some of which may be left empty.

Polynomials with exponential generating functions of the type appearing in (6.50) have a number of properties very similar to those of the ordinary power sequence. We shall state a few.

First of all, note that we have a general result which includes (6.27) and (6.37) as very special cases. This can be stated as follows:

Theorem 6.4.3 *Let $P_n(x)$ be a sequence of polynomials with exponential generating function*

$$\sum_{n \geq 0} \frac{t^n}{n!} P_n(x) = \exp(x f(t)), \qquad (6.58)$$

where

$$f(t) = \sum_{m \geq 1} f_m t^m \qquad (f_1 \neq 0)$$

and let $F(t) = \sum_{m \geq 1} F_m t^m$ be the inverse function of $f(t)$, that is

$$f(F(t)) = t . \qquad (6.59)$$

Then for all $n \geq 1$ we have

$$F(D) P_n(x) = n P_{n-1}(x) \qquad (P_0 = 1) . \qquad (6.60)$$

Proof Differentiating (6.58) m times with respect to x gives

$$\sum_{n \geq 0} \frac{t^n}{n!} D^m P_n(x) = f(t)^m \exp(x f(t)) .$$

Multiplying by F_m and summing, we get (using (6.59))

$$\sum_{n \geq 0} \frac{t^n}{n!} F(D) P_n(x) = F(f(t)) \exp(x f(t)) = t \exp(x f(t)) .$$

Using (6.58) again, we deduce that

$$\sum_{n \geq 0} \frac{t^n}{n!} F(D) P_n(x) = t \sum_{n \geq 0} \frac{t^n}{n!} P_n(x) . \qquad (6.61)$$

The desired conclusion (6.60) now follows upon equating coefficients of t^n in this relation.

$\square$

We should note that since the polynomials $(x)_n$ have exponential generating function given by (6.47), Theorem 6.4.3 applies with $f(t) = \log(1 + t)$. Now the inverse of this function is

$$F(t) = e^t - 1 ,$$

and (6.60) in this case becomes

$$(e^D - 1)(x)_n = n(x)_{n-1} ,$$

which is (6.37) again.

The analogy with the power sequence case does not stop here; in fact we can even obtain a generalized version of Taylor's theorem. To see how this comes about, let us observe that (6.35) is a simple consequence of the fact that we have

$$\frac{1}{k!} L_0 D^k x^n = \begin{cases} 1 & \text{if } k = n \\ 0 & \text{if } k \neq n . \end{cases}$$

Indeed, from this we can immediately get that

$$\left(\sum_{k \geq 0} \frac{x^k}{k!} L_0 D^k \right) x^n = x^n$$

for all n. Thus this operator acts like the identity on the power sequence and therefore by linearity it must leave every polynomial unchanged. Thus (6.35) necessarily follows.

Now note that for Δ and $(x)_n$ we have as well

$$\frac{1}{k!} L_0 \Delta^k (x)_n = \begin{cases} 1 & \text{if } k = n \\ 0 & \text{if } k \neq n . \end{cases} \tag{6.62}$$

This is an immediate consequence of (6.37) and the fact that each $(x)_n$ (for $n \geq 1$) is equal to zero when $x = 0$.

This given, the same reasoning that gave us a proof of (6.35) will yield the Δ-analogue of Taylor's theorem. Namely

$$I = \sum_{n \geq 0} \frac{(x)_n}{n!} L_0 \Delta^n . \tag{6.63}$$

This identity gives another method for expanding a polynomial as a linear combination of lower factorial polynomials. We shall illustrate this for the polynomial

$$P(x) = 5x^3 - 2x^2 + x - 3 \, .$$

Clearly (6.63) in this case reduces to

$$P(x) = L_0 P(x) + \frac{(x)_1}{1!} L_0 \Delta P(x) + \frac{(x)_2}{2!} L_0 \Delta^2 P(x) + \frac{(x)_3}{3!} L_0 \Delta^3 P(x) \, . \quad (6.64)$$

Now, there is a very clever method for calculating the coefficients $L_0 \Delta^k P(x)$. The idea is to evaluate $P(x)$ for $x = 0, 1, 2, 3$ (in the general case of a polynomial of degree n we evaluate for $x = 0, 1, 2, \dots, n$) and construct the table in Figure 6.13.

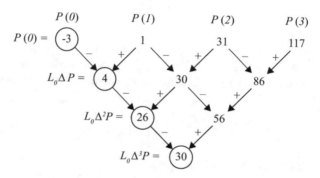

Fig. 6.13 The difference table to compute the coefficients $L_0 \Delta^k P(x)$.

The numbers in the first row here are the values of P as indicated. Each number on a lower row is then obtained by taking the difference of the two numbers right above it as indicated by the arrows. This done, the circled numbers give the desired coefficients. Thus (6.64) becomes here

$$5x^3 - 2x^2 + x - 3 = -3 + 4(x)_1 + \frac{26}{2!}(x)_2 + \frac{30}{3!}(x)_3$$

$$= -3 + 4(x)_1 + 13(x)_2 + 5(x)_3 \, .$$

Formulas (6.35) and (6.63) are special cases of the following general result:

Theorem 6.4.4 *Under the same hypotheses as in Theorem 6.4.3 we have the expansion*

$$I = \sum_{k \geq 0} \frac{P_k(x)}{k!} L_0(F(D))^k \, . \quad (6.65)$$

Proof Note that since $P_n(x)$ is of degree n (the coefficient of x^n in $P_n(x)$ is $(f_1)^n$) the sequence $\{P_n(x)\}$ forms a basis for the polynomials. Furthermore, by setting $x = 0$ in (6.58) and equating coefficients of t^n we deduce that

$$P_n(x) = \begin{cases} 1 & \text{if } n = 0 \\ 0 & \text{if } n \neq 0 . \end{cases}$$

This fact combined with (6.60) yields that

$$\frac{1}{k!} L_0(F(D))^k P_n(x) = \begin{cases} 1 & \text{if } k = n \\ 0 & \text{if } k \geq n . \end{cases} \tag{6.66}$$

Now, these identities immediately imply that the operator on the right hand side of (6.65) reproduces each $P_n(x)$. Since these polynomials are a basis, this operator must necessarily reproduce every polynomial. Thus (6.65) must hold as asserted. $\square$

In some applications, we are given $F(D)$ to start with and the $P_n(x)$ are to be found for which (6.60) as well as (6.65) do hold.

6.4.6 An Application of the Lagrange Inversion Formula

If we use Theorem 6.4.3 in reverse, we see that to find such a sequence $\{P_n\}$, we need only find the function $f(t)$ which is the functional inverse of $F(t)$. In some cases this function can be constructed by means of the Lagrange inversion formula. In point of fact, this idea leads to a useful formula for the $P_n(x)$ themselves. To this end we need a slightly different version of the Lagrange inversion formula than the one we gave in Section 3.3 of Chapter 3. It may be stated as follows:

Theorem 6.4.5 *Let $f(t)$ be the formal power series solution of the equation*

$$f(t) = t R(f(t)) . \tag{6.67}$$

Then for any formal power series $\Phi(t) = \sum_{n \geq 0} \Phi_k t^k$ we have

$$\Phi(f(t)) \Big|_{t^n} = \frac{1}{n} \Phi'(t) R^n(t) \Big|_{t^{n-1}} . \tag{6.68}$$

Proof Formula (3.44) of Chapter 3 can be rewritten in the form

$$(f(t))^k \Big|_{t^n} = \frac{k}{n} t^{k-1} R^n(t) \Big|_{t^{n-1}} . \tag{6.69}$$

Note that this is (6.68) in the special case $\Phi(t) = t^k$. To obtain (6.68) itself we simply multiply (6.69) by Φ_k and sum. $\square$

This given, we have the following remarkable result:

Theorem 6.4.6 *Let* $F(t) = t/R(t)$ *with*

$$R(t) = R_0 + R_1 t + R_2 t^2 + \cdots \qquad (R_0 \neq 0),$$

and let $f(t)$ *be the inverse of* $F(t)$, *that is*

$$\frac{f(t)}{R(f(t))} = t .$$

Then the sequence of polynomials $P_n(x)$ *with exponential generating function*

$$\sum_{n \geq 0} \frac{t^n}{n!} P_n(x) = e^{xf(t)} \qquad (6.70)$$

is given by the formula

$$P_n(x) = x R^n(D) x^{n-1} . \qquad (6.71)$$

Proof Note that from (6.70) we get that

$$\frac{1}{n!} P_n(x) = e^{xf(t)} \Big|_{t^n} .$$

Thus Theorem 6.4.5, or more specifically formula (6.68) with $\Phi(t) = e^{xt}$ gives

$$\frac{1}{n!} P_n(x) = \frac{1}{n} x e^{xt} R^n(t) \Big|_{t^{n-1}} . \qquad (6.72)$$

Now the Cauchy formula for the coefficient of a product of two series gives that

$$e^{xt} R^n(t) \Big|_{t^{n-1}} = \sum_{k=0}^{n-1} R^n(t) \Big|_{t^k} e^{xt} \Big|_{t^{n-1-k}} = \sum_{k=0}^{n-1} R^n(t) \Big|_{t^k} \frac{x^{n-1-k}}{(n-1-k)!} .$$

Substituting this in (6.72) and multiplying by $n!$ we get

$$P_n(x) = x \sum_{k=0}^{n-1} R^n(t) \Big|_{t^k} (n-1)(n-2) \cdots (n-k) x^{n-1-k}$$

$$= x \sum_{k=0}^{n-1} R^n(t) \Big|_{t^k} D^k x^{n-1} ,$$

and this is (6.71) in disguise. $\qquad \square$

It may be good to illustrate a use of this theorem. Let us take the case $F(t) = te^t$. Then $R(t) = e^{-t}$ and so formula (6.71) becomes (using (6.33))

$$P_n(x) = xe^{-nD}x^{n-1} = x(x - n)^{n-1} .$$

Note now that for a polynomial $P(x)$ we have in this case

$$F(D)P(x) = De^D P(x) = P'(x + 1)$$

and so

$$(F(D))^n P(x) = P^{(n)}(x + n) .$$

Theorem 6.4.4 then gives the remarkable expansion formula:

$$P(x) = P(0) + \sum_{n \geq 1} \frac{x(x - n)^{n-1}}{n!} P^{(n)}(n) . \tag{6.73}$$

6.5 Miscellaneous Applications and Examples

6.5.1 The Mullin–Rota Theory of Polynomial Sequences of Binomial Type

A sequence $\{b_n(x)\}$ of polynomials is said to be of *binomial type* [14, 22] if the identities

$$b_n(x + y) = \sum_{k=0}^{n} \binom{n}{k} b_k(x) b_{n-k}(y) \tag{6.74}$$

are satisfied for $n \geq 0$. It is often convenient to require the degree of $b_n(x)$ in such a sequence to be n.

The power sequence $\{x^n\}$ is the simplest example of a sequence of binomial type since in this case (6.74) becomes just a restatement of the binomial theorem.

It turns out that the theory of exponential structures together with the exponential formula furnish many nontrivial sequences of polynomials of binomial type.

Before we elaborate on this, it will be good to formulate the condition (6.74) in terms of the exponential generating function of a given sequence of polynomials.

Writing (6.74) in the form

$$\frac{b_n(x + y)}{n!} = \sum_{k=0}^{n} \frac{b_k(x)}{k!} \frac{b_{n-k}(y)}{(n - k)!} \tag{6.75}$$

and putting

$$B(x, t) = \sum_{n \geq 0} \frac{t^n}{n!} b_n(x) ,$$

we see that (6.74) is equivalent to the functional equation

$$B(x + y, t) = B(x, t)B(y, t) .$$

This is because the right hand side of (6.75) is the coefficient of t^n in the product $B(x, t)B(y, t)$ and the left hand side is the coefficient of t^n in $B(x + y, t)$.

Now recall that given an exponential structure $F(L)$ we have a sequence $\{\Psi_n(x, L)\}$ of polynomials associated to it which are defined by putting

$$\Psi_0(x, L) = 1$$

and

$$\Psi_n(x, L) = \sum_{k=1}^{n} x^k \# F_{n,k}(L) \qquad (n \geq 1) .$$

For notational convenience put $\Psi_n(x) = \Psi_n(x, L)$. The exponential generating function

$$\Psi(x, t) = 1 + \sum_{n \geq 1} \frac{t^n}{n!} \Psi_n(x)$$

of the sequence $\{\Psi_n(x)\}$ is given in closed form by the exponential formula (6.19) as

$$\Psi(x, t) = \exp\left(x \sum_{m \geq 1} \frac{t^m}{m!} \# L_m \right) .$$

If we set

$$l(t) = \sum_{m \geq 1} \frac{t^m}{m!} \# L_m ,$$

then

$$\Psi(x + y, t) = \exp((x + y)l(t)) = \exp(xl(t)) \exp(yl(t)) = \Psi(x, t)\Psi(y, t) .$$

We conclude that the sequence of polynomials $\{\Psi_n(x)\}$ satisfies the identity (6.74) and therefore these polynomials are of binomial type for any exponential structure $F(L)$. In this situation the degree of $\Psi_n(x)$ is n if and only if $\#F_{nn}(L) \neq 0$. The latter holds true for all of the exponential structures we have considered in the previous sections.

As examples we have

1. The *exponential* polynomials

$$\sigma_0(x) = 1, \quad \sigma_n(x) = \sum_{k=1}^{n} S_{n,k} x^k \quad (n \geq 1),$$

where $S_{n,k}$ are the Stirling numbers of the second kind.

2. The *upper factorial* polynomials

$$\phi_0(x) = 1, \quad \phi_n(x) = x(x+1) \cdots (x+n-1) \quad (n \geq 1),$$

where the underlying exponential structure is the collection of all permutations.

3. The *Laguerre* polynomials

$$\Psi_0(x) = 1, \quad \Psi_n(x) = \sum_{k=1}^{n} \frac{n!}{k!} \binom{n-1}{k-1} x^k \quad (n \geq 1),$$

where the underlying structure is *balls in tubes*.

Thus we see that all these sequences are of binomial type.

6.5.2 *Permutations with Even Cycles, Involutions Without Fixed Points*

We present two more applications of the exponential formula (6.19). We shall find generating functions for sequences of polynomials associated with two classes of permutations.

We first consider the class of permutations with even cycles.

Let us denote by E_{2n} the collection of all permutations of $\{1, 2, \ldots, 2n\}$ whose cycle decomposition consists of cycles of *even* length only. For example

$$E_2 = \{(12)\}$$

$$E_4 = \{(12)(34), (13)(24), (14)(23), (1234), (1243), (1324), (1342), (1423), (1432)\}.$$

Of course any permutation of a set with an odd number of elements necessarily contains an odd cycle in its cycle decomposition. Thus

$$E_{2n-1} = \emptyset \qquad (n \geq 1) .$$

For this particular problem, the list L of patterns which will give us the required exponential structure $F(L)$ consists of all cycles of even length. In other words, $L_{2m-1} = \emptyset$, and L_{2m} is the collection of all cycles of length $2m$, for $m \geq 1$. Thus

$$\#L_{2m-1} = 0$$

and, as we previously observed

$$\#L_{2m} = (2m - 1)! .$$

Set

$$\Psi_{2n}(x, L) = \sum_{k=1}^{2n} x^k \#F_{2n,k}(L) .$$

From the examples above we obtain the first two polynomials

$$\Psi_2(x, L) = x , \quad \Psi_4(x, L) = 6x + 3x^2$$

as special cases. In particular

$$\#E_2 = \Psi_2(1, L) = 1 , \quad \#E_4 = \Psi_4(1, L) = 9 .$$

For the general case, we make use of the exponential formula

$$1 + \sum_{n \geq 1} \frac{t^{2n}}{(2n)!} \Psi_{2n}(x, L) = \exp\left(x \sum_{m \geq 1} \frac{t^m}{m!} \#L_m \right)$$

$$= \exp\left(x \sum_{m \geq 1} \frac{t^{2m}}{2m} \right) . \tag{6.76}$$

It follows from calculus that

$$\sum_{m \geq 1} \frac{t^{2m}}{2m} = -\frac{1}{2} \log(1 - t^2) .$$

Substituting in (6.76) and simplifying gives

$$1 + \sum_{n \geq 1} \frac{t^{2n}}{(2n)!} \Psi_{2n}(x, L) = \left(1 - t^2 \right)^{-\frac{x}{2}} . \tag{6.77}$$

The right hand side of (6.77) can now be expanded by Newton's theorem as a power series

$$\left(1 - t^2\right)^{-\frac{x}{2}} = \sum_{n \geq 0} \binom{-\frac{x}{2}}{n}(-1)^n t^{2n} \, .$$

Therefore we have

$$\Psi_{2n}(x, L) = (2n)!(-1)^n \binom{-\frac{x}{2}}{n} \, .$$

This can be simplified further to give the formula

$$\Psi_{2n}(x, L) = 1 \cdot 3 \cdot 5 \cdots (2n - 1)x(x + 2) \cdots (x + (2n - 2)) \qquad (6.78)$$

for $\Psi_{2n}(x, L)$. In particular, setting $x = 1$ we get

$$\#E_{2n} = \Psi_{2n}(1, L) = 1 \cdot 3 \cdot 5 \cdots (2n - 1)1 \cdot 3 \cdot 5 \cdots (2n - 1) \, .$$

Thus the number of permutations of $\{1, 2, \ldots, 2n\}$ whose cycle factorization contains only even cycles is given by

$$\#E_{2n} = 1^2 \cdot 3^2 \cdot 5^2 \cdots (2n - 1)^2 \, .$$

Note that the expression

$$\Psi_4(x, L) = 6x + 3x^2$$

which was obtained directly by classifying the permutations in the set E_4 according to the number of cycles in their cycle factorization can be rewritten in the form

$$\Psi_4(x, L) = 1 \cdot 3x(x + 2) \, ,$$

which is in agreement with the formula (6.78).

Our next example is the class of *involutions without fixed points*.

We recall that an *involution* is a permutation whose cycle decomposition contains cycles of length 1 or 2 only. Note that if the cycle factorization of a permutation σ contains the cycle (i), then $\sigma_i = i$. In other words, σ fixes the point i. This means that involutions without fixed points are precisely those involutions whose cycle decomposition contains cycles of length 2 only.

Now let I_n denote the collection of involutions on $[n]$ without fixed points. Note that reasoning as in the previous example, we must have $I_{2n-1} = \emptyset$ for all $n \geq 1$. This is because, as we can easily see, any involution on a set with an odd number of elements must have a fixed point.

The list L in this instance is quite easy to describe: The only patterns of interest in the construction of involutions without fixed points are the 2-cycles. This means that all the L_m's are empty except for L_2 (which consists of a single 2-cycle).

Putting

$$\phi_{2n}(x, L) = \sum_{k=1}^{2n} x^k \#I_{2n,k}(L)$$

and using the exponential formula we have

$$1 + \sum_{n \geq 1} \frac{t^{2n}}{(2n)!} \phi_{2n}(x, L) = \exp\left(x \sum_{m \geq 1} \frac{t^m}{m!} \#L_m\right) = \exp\left(x \frac{t^2}{2}\right) . \tag{6.79}$$

This gives

$$\phi_{2n}(x, L) = (2n)! \frac{x^n}{n! 2^n} \tag{6.80}$$

if we make use of the expansion

$$\exp\left(x \frac{t^2}{2}\right) = \sum_{n \geq 0} \frac{1}{n!} \left(\frac{xt^2}{2}\right)^n$$

and equate the coefficients of t^{2n} in (6.79).

Formula (6.80) can be simplified further to

$$\phi_{2n}(x, L) = 1 \cdot 3 \cdot 5 \cdots (2n - 1) x^n . \tag{6.81}$$

In particular this gives that the number of involutions on $\{1, 2, \ldots, 2n\}$ without fixed points is

$$\#I_{2n} = 1 \cdot 3 \cdot 5 \cdots (2n - 1) .$$

6.5.3 Lagrange Interpolation

$X = \{x_0, x_1, \ldots, x_n\}$ is a set of $n + 1$ real numbers. It is often required to construct a polynomial $p(x)$ which takes on certain prescribed values $f_0, f_1, \ldots, f_n$ at these points: i.e.,

$$p(x_k) = f_k \qquad (k = 0, 1, \ldots, n) .$$

As an example if $x_0 = 1$, $x_1 = 2$, $x_2 = 3$ and $f_0 = -2$, $f_1 = 3$, $f_2 = 10$, the polynomial we wish to construct is required to satisfy

$$p(1) = -2, \quad p(2) = 3, \quad p(3) = 10.$$

To this end, let us first consider the following three polynomials:

$$\chi_0(x) = \frac{(x-2)(x-3)}{(1-2)(1-3)} = \frac{1}{2}(x-2)(x-3),$$

$$\chi_1(x) = \frac{(x-1)(x-3)}{(2-1)(2-3)} = -(x-1)(x-3),$$

$$\chi_2(x) = \frac{(x-1)(x-2)}{(3-1)(3-2)} = \frac{1}{2}(x-1)(x-2).$$

Observe that

$$\chi_0(x) = \begin{cases} 1 & \text{if } x = 1 \\ 0 & \text{if } x = 2 \text{ or } 3 \end{cases}$$

$$\chi_1(x) = \begin{cases} 1 & \text{if } x = 2 \\ 0 & \text{if } x = 1 \text{ or } 3 \end{cases}$$

$$\chi_2(x) = \begin{cases} 1 & \text{if } x = 3 \\ 0 & \text{if } x = 1 \text{ or } 2. \end{cases}$$

Note that if we put

$$p(x) = -2\chi_0(x) + 3\chi_1(x) + 10\chi_2(x),$$

then

$$p(1) = -2 \times 1 + 3 \times 0 + 10 \times 0 = -2$$
$$p(2) = -2 \times 0 + 3 \times 1 + 10 \times 0 = 3$$
$$p(3) = -2 \times 0 + 3 \times 0 + 10 \times 1 = 10.$$

Hence in this case we can take

$$p(x) = -(x-2)(x-3) - 3(x-1)(x-3) + 5(x-1)(x-2)$$
$$= x^2 + 2x - 5.$$

In general, if we set

$$p(x) = f_0 \chi_0(x) + f_1 \chi_1(x) + \cdots + f_n \chi_n(x), \tag{6.82}$$

where $\chi_i(x)$ are polynomials satisfying the conditions

$$\chi_i(x) = \begin{cases} 1 & \text{if } x = x_i \\ 0 & \text{if } x = x_j \text{ with } i \neq j \end{cases} \tag{6.83}$$

then $p(x)$ will indeed take on the value f_k at the point x_k.

Therefore the problem of constructing the desired polynomial is reduced to finding $n + 1$ polynomials $\chi_0, \chi_1, \ldots, \chi_n$ satisfying (6.83).

Observe that if we put

$$\chi_i(x) = \frac{(x - x_0) \cdots (x - x_{i-1})(x - x_{i+1}) \cdots (x - x_n)}{(x_i - x_0) \cdots (x_i - x_{i-1})(x_i - x_{i+1}) \cdots (x_i - x_n)},$$

then $\chi_i(x_i) = 1$ since the numerator is identical to the denominator for $x = x_i$. If $x_j \in X$ with $i \neq j$, then $\chi_i(x_j) = 0$ since the factor $(x - x_j)$ that appears in the numerator is zero for $x = x_j$. Therefore these $n + 1$ polynomials satisfy the required condition (6.83). Now using (6.82), we can easily construct the polynomial $p(x)$.

This method of obtaining $p(x)$ is due to Lagrange and it is usually referred to as the method of *Lagrange interpolation*.

It should be noted that the difference operator methods we have presented in Section 6.4 are applicable to this particular problem as well.

More precisely, suppose for the moment that $X = \{0, 1, \ldots, n\}$, and recall that the Δ-analogue of Taylor's theorem (formula (6.63)) is

$$p(x) = p(0) + \frac{(x)_1}{1!} L_0 \Delta p(x) + \frac{(x)_2}{2!} L_0 \Delta^2 p(x) + \cdots \tag{6.84}$$

The method we have presented for calculating the coefficients $L_0 \Delta^k p(x)$ can be reversed to find a polynomial $p(x)$ with

$$p(0) = f_0, \ p(1) = f_1, \ldots, \ p(n) = f_n$$

for arbitrarily prescribed numbers $f_0, f_1, \ldots, f_n$.

The idea is to start the table that we use to construct the coefficients $L_0 \Delta^k p$ by taking the first row to be

$$f_0, \ f_1, \ldots, \ f_n .$$

The numbers that appear on the left are then substituted into (6.84) to construct $p(x)$ as a linear combination of the lower factorial polynomials.

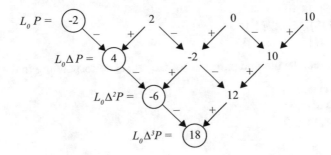

Fig. 6.14 The difference table for $f_0 = -2$, $f_1 = 2$, $f_2 = 0$, $f_3 = 10$.

For example, taking $f_0 = -2$, $f_1 = 2$, $f_2 = 0$, $f_3 = 10$ and constructing the table as developed in Section 6.4, namely as shown here in Figure 6.14, we obtain the polynomial

$$p(x) = -2 + \frac{4(x)_1}{1!} - \frac{6(x)_2}{2!} + \frac{18(x)_3}{3!} . \tag{6.85}$$

We can verify that

$$p(0) = -2$$
$$p(1) = -2 + 4 = 2$$
$$p(2) = -2 + 8 - \frac{12}{2} = 0$$
$$p(3) = -2 + 12 - \frac{36}{2} + \frac{108}{6} = 10 .$$

Of course, simplifying the expression (6.85), we can write

$$p(x) = x^3 - 2x^2 + x - 2$$

in the more familiar power basis.

With minor modifications, the above method can be used to obtain a polynomial $p(x)$ with $p(x_k) = f_k$, for $k = 0, 1, \ldots, n$, whenever the points $x_k \in X$ form an arithmetic progression: i.e., whenever $x_{i+1} - x_i$ is constant for all $i = 0, 1, \ldots, n-1$.

6.6   Exercises for Chapter 6

6.6.1 Making use of the correspondence given in the proof of Theorem 6.1.1, construct the words corresponding to the ordered partitions

(a) $(\{1, 2, 5\}, \{3, 4\})$

(b) $(\{2\}, \{1, 3\}, \{5, 6\}, \{4\})$

(c) $(\{2, 5, 7\}, \{1, 4\}, \{3, 6\})$

6.6.2 Construct the ordered partitions of the set [7] corresponding to the words

(a) $x_2 x_1 x_2 x_3 x_4 x_2 x_4$

(b) $x_1 x_2 x_3 x_1 x_2 x_2 x_3$

(c) $x_5 x_5 x_1 x_2 x_4 x_3 x_4$

6.6.3

(a) What is the number of ordered partitions of a 6 element set into 4 parts?

(b) What is the number of partitions of a 6 element set into 4 parts?

6.6.4 Calculate the number of partitions of a 9 element set with two 1-subsets, two 2-subsets, and one 3-subset.

6.6.5 Calculate the number of partitions of the set [16] into two 2-subsets and four 3-subsets.

6.6.6 Calculate the number of permutations of [16] having two 2-cycles and four 3-cycles.

6.6.7

(a) Calculate the number of partitions of [11] into two parts of size 1, three parts of size 2, and one part of size 3.

(b) Calculate the number of permutations of [11] which have two cycles of length 1, three cycles of length 2, and one cycle of length 3.

6.6.8 Calculate the number of permutations of [8] whose cycle decomposition consists of

(a) two cycles of length 1 and three cycles of length 2,

(b) one cycle of length 1, two cycles of length 2, and one cycle of length 3.

6.6.9 Calculate the number of ways of distributing 6 distinguishable balls into 4 indistinguishable boxes so that no box is empty.

6.6.10 Calculate the number of ways of placing 9 distinguishable balls into 3 identical tubes so that no tube is empty.

6.6.11 Calculate the number of ways of placing 6 distinguishable balls in 2 identical tubes if

(a) no tube is allowed to be empty,

(b) one of the tubes is allowed to be empty.

6.6.12 Show that for $n \geq 1$

$$\frac{(2n)!}{2^n n!} = 1 \cdot 3 \cdot 5 \cdots (2n - 1) \,.$$

6.6.13 Verify that

$$(2n)!(-1)^n \binom{-\frac{x}{2}}{n} = 1 \cdot 3 \cdot 5 \cdots (2n-1)x(x+2) \cdots (x+2n-2)$$

for $n \geq 1$.

6.6.14

(a) Show that

$$\exp(e^t) \Big|_{t^n} = \frac{1}{n!} \sum_{k \geq 0} \frac{k^n}{k!} .$$

(b) Show that multiplying equation (6.7) by e and equating the coefficients of t^n
gives

$$e \frac{B_n}{n!} = \exp(e^t) \Big|_{t^n} .$$

(c) Use (a) and (b) to deduce that

$$B_n = \frac{1}{e} \sum_{k \geq 0} \frac{k^n}{k!} .$$

This is known as Dobinski's formula for the Bell numbers.

6.6.15

(a) Put $x = -1$ in (6.5) and go through a computation along the lines of
problem 6.6.14 to show that

$$\sigma_n(-1) = e \sum_{k \geq 0} (-1)^k \frac{k^n}{k!} .$$

(b) As a special case of this find the sums

$$\sum_{k \geq 0} (-1)^k \frac{k^4}{k!} \quad \text{and} \quad \sum_{k \geq 0} (-1)^k \frac{k^5}{k!} .$$

6.6.16

(a) Use the multinomial theorem to show that

$$\sum \frac{n!}{p_1! p_2! \cdots p_k!} = k^n ,$$

where the summation is over all k-tuples $p_1 + p_2 + \ldots + p_k = n$ with $p_i \geq 0$.

(b) Use this result together with the formula (6.4) for the Stirling numbers $S_{n,k}$ to show that

$$S_{n,k} \leq \frac{k^n}{k!} .$$

6.6.17 By equation (6.6) we have

$$k!S_{n,k} = n!(e^t - 1)^k \Big|_{t^n} .$$

(a) Expand $(e^t - 1)^k$ using the binomial theorem to obtain

$$(e^t - 1)^k = \sum_{n \geq 0} \sum_{i=0}^{k} (-1)^i \binom{k}{i} (k - i)^n \frac{k^n}{n!} .$$

(b) Derive the formula

$$S_{n,k} = \frac{1}{k!} \sum_{i=0}^{k} (-1)^i \binom{k}{i} (k - i)^n .$$

(c) Use this formula to compute $S_{4,2}$ and $S_{4,3}$.

6.6.18 Put

$$Q_n(x_1, x_2, \ldots, x_k) = \sum x_1^{|A_1|} x_2^{|A_2|} \cdots x_k^{|A_k|},$$

where the summation is over all partitions $A_1 + A_2 + \cdots + A_k$ of $[n]$ into k parts.

(a) Show that

$$\sum_{n \geq 0} Q_n(x_1, x_2, \ldots, x_k)t^n =$$

$$\frac{x_1 x_2 \cdots x_k t^k}{(1 - tx_1)(1 - t(x_1 + x_2)) \cdots (1 - t(x_1 + x_2 + \cdots + x_k))} .$$

(b) Derive the identity

$$Q_n(x_1, x_2, \ldots, x_k) = x_k \sum_{i=k}^{n} Q_{i-1}(x_1, x_2, \ldots, x_{k-1})(x_1 + x_2 + \cdots + x_k)^{n-i} .$$

6.6.19

(a) Construct the language of partitions into k parts in which i is in the ith part for $i = 1, 2, 3$.

(b) Show that the number of partitions of $[n]$ into k parts in which 1 is in the first part, 2 is in the second part, and 3 is in the third part is

$$S_{n,k} - 3S_{n-1,k} + 2S_{n-2,k} \, .$$

6.6.20 Prove the identity

$$S_{n,k} = \sum_{j=k}^{n} S_{j-1,k-1} k^{n-j} \, .$$

6.6.21 Show that whenever $m < n$

$$\sum_{k=0}^{n} (-1)^k \binom{n}{k} (x + k)_m = 0 \, .$$

6.6.22 Show that

$$\sigma_n(x) e^x = \sum_{k \geq 0} k^n \frac{x^k}{k!} \, .$$

6.6.23 Calculate the coefficient of t^n in

$$\frac{t}{1-t} + \frac{t^2}{(1-t)(1-2t)} + \frac{t^3}{(1-t)(1-2t)(1-3t)} \, .$$

6.6.24 Let $X = \{x_1, x_2, \ldots\}$ be an infinite alphabet. Put $\mathcal{L}_1 = x_1^+$ and for $k > 1$, define the languages $\mathcal{L}_k$ recursively by setting

$$\mathcal{L}_k = \mathcal{L}_{k-1} x_k (x_1 + x_2 + \cdots + x_k)^* .$$

Give an interpretation of the language

$$\sum_{k \geq 1} \mathcal{L}_k$$

in terms of set partitions.

6.6.25 Show that the Bell numbers B_n have the ordinary generating function

$$\sum_{k \geq 1} \frac{t^k}{(1 - t)(1 - 2t) \cdots (1 - kt)}.$$

6.6.26 Call a partition of $[n]$ *separated* if 1 and n are not in the same part.

(a) Construct the language of all separated partitions into k parts.
(b) Find a formula for the number of separated partitions of $[n]$ into k parts.

6.6.27 Calculate the number of partitions of $[n]$ into k parts in which $2, 3, \ldots, i$ are all in the first part.

6.6.28 Calculate the number of partitions of $[n]$ into k parts in which n is in the kth part.

6.6.29

(a) Put $f_k(t) = \sum_{n \geq 0} S_{n+k,k} t^n$ for $k \geq 1$. Use the recursion (1.74) satisfied by the Stirling numbers $S_{n,k}$ to verify that

$$(1 - kt) f_k(t) = f_{k-1}(t) \qquad\qquad (k \geq 2).$$

(b) Deduce that

$$f_k(t) = \frac{1}{(1 - t)(1 - 2t) \cdots (1 - kt)}.$$

(c) Take $k = 2$ in part (b) and use partial fractions to expand

$$f_2(t) = \frac{1}{(1 - t)(1 - 2t)}$$

as a power series. Compare the coefficients of t^n on both sides to show that

$$S_{n,2} = 2^{n-1} - 1 \qquad\qquad (n \geq 2).$$

Verify that the numbers in the 2nd column of the Table 1.1 of Chapter 1 are indeed of this form.

(d) By a similar technique obtain the formula

$$S_{n,3} = \frac{1}{2} \left(3^{n-1} - 2^n + 1 \right) \qquad\qquad (n \geq 3).$$

Use this formula to compute $S_{5,3}$ and $S_{6,3}$.

6.6.30 Let A denote the permutations of $[7]$ whose cycle decomposition consists of exactly three cycles.

(a) Calculate $|A|$.
(b) How many permutations in A are derangements?

6.6.31 Show that any involution of a set with an odd number of elements must have at least one fixed point.

6.6.32 Calculate the number of involutions of $[2n + 1]$ having exactly one fixed point.

6.6.33 Let w_n denote the number of involutions of $[n]$. Show that $\{w_n\}$ satisfies

$$w_n = w_{n-1} + (n - 1)w_{n-2} \qquad (n \geq 3)$$

with the initial conditions $w_1 = 1, w_2 = 2$

(a) by using generating function methods,
(b) by giving a direct counting argument.
(c) Use this recursion to compute the numbers w_3 through w_6.

6.6.34 Let I denote the exponential structure of involutions.

(a) Find the generating function for the exponential polynomials associated to I.
(b) Find the coefficient of t^5 in this generating function.

6.6.35

(a) Find the exponential generating function of the sequence of polynomials associated with the exponential structure $F(L)$ whose lists L_m are empty for $m > 3$ and
(b) Give an example of an object in $F_{7,3}(L)$.

6.6.36 Suppose $P(L)$ is the exponential structure whose lists L_m are empty for $m \neq 1, 3$ and L_1 and L_3 are as given below:

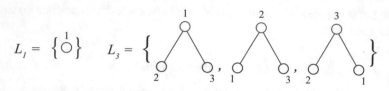

$$L_1 = \{\overset{1}{\circ}\} \qquad L_3 = \{\ \ ,\ \ ,\ \ \}$$

(a) Put together the exponential generating function of the polynomials $Q_n(x, L)$ associated to $P(L)$.

(b) Give an example of an object in $P_{8,4}(L)$.

6.6.37 Let a_n be the number of $\{0, 1, 2\}$-words of length n where 0 appears an odd number of times and 1 appears an even number of times.

(a) Put together the exponential generating function of the sequence $\{a_n\}$.

(b) Can you find a formula for a_n?

6.6.38 Let b_n be the number of words of length n in the letters A, B, C, D with A and B each appearing an even number of times.

(a) Find the exponential generating function of $\{b_n\}$.

(b) Show that

$$b_n = 4^{n-1} + 2^{n-1} \qquad\qquad (n \geq 1).$$

6.6.39 Let c_n denote the number of words of length n in the letters A, B, C with A appearing an odd number of times.

(a) Put together the exponential generating function of the sequence $\{c_n\}$.

(b) Find a formula for c_n.

6.6.40 Find a formula for the number of permutations of $[n]$ whose cycle diagram is made up of cycles of odd length only.

6.6.41 Let a_n denote the number of permutations of $[n]$ whose cycle factorization contains only cycles of length 1 and 4. Construct the exponential generating function of $\{a_n\}$.

6.6.42 Put $a_0 = 1$ and define

$$a_{2n} = 1 \cdot 3 \cdot 5 \cdots (2n - 1) \qquad\qquad (n \geq 1).$$

Show that

$$2^n a_{2n} = \sum_{k=0}^{n} \binom{2n}{2k} a_{2k} a_{2n-2k} \ .$$

(Hint: The sequence of polynomials given by (6.78) are of binomial type).

6.6.43 Put $p(x) = x^3 - 2x^2 + 3$. Compute the polynomials

(a) $(D^2 + D)p(x)$,
(b) $(\Delta^2 + \Delta)p(x)$,
(c) $e^D p(x)$,
(d) $e^\Delta p(x)$.

6.6.44 Let $p(x) = x^2 - 2x + 1$.

(a) Express $p(x)$ as a linear combination of lower factorial polynomials using the table of Stirling numbers of second kind.
(b) Find the polynomial $q(x)$ which solves the difference equation

$$q(x + 1) - q(x) = x^2 - 2x + 1 \qquad (q(0) = 0).$$

6.6.45 Find the solutions of the following difference equations. Assume that $p(0) = 0$.

(a) $\Delta p(x) = 2x$,
(b) $\Delta p(x) = x^3 - x$,
(c) $\Delta p(x) = (x)_3 + (x)_2$.

6.6.46 Express the polynomial

$$p(x) = x^3 - 2x^2 + x + 3$$

as a linear combination of lower factorial polynomials

(a) by means of the table of Stirling numbers $S_{n,k}$,
(b) by means of the Δ-analogue of Taylor's theorem (6.63).

6.6.47 Construct the polynomial $p(x)$ which solves the difference equation

$$p(x + 1) - p(x) = x^2 - 2x + 1 \qquad (p(0) = 2).$$

6.6.48

(a) Find the polynomial $p(x)$ which satisfies

$$p(x + 1) - p(x) = x^4 \qquad (p(0) = 0).$$

(b) Use the expression for $p(x)$ you have obtained in part (a) to find a formula for the sum

$$1^4 + 2^4 + 3^4 + \cdots + n^4 .$$

6.6.49 Suppose k is a fixed positive integer.

(a) Solve the difference equation

$$\Delta p(x) = \binom{x}{k} \qquad\qquad (p(0) = 0).$$

(b) Use part (a) to find a formula for the sum

$$\binom{k}{k} + \binom{k+1}{k} + \cdots + \binom{n}{k}.$$

6.6.50 Prove that if $p(x)$ is a polynomial of degree n, then

$$\Delta^{n+1} p(x) = 0 .$$

6.6.51 Suppose $p(x)$ satisfies the difference equation

$$\Delta^2 p(x) = 2x .$$

(a) Show that

$$\Delta p(x) = x + c$$

for some constant c.

(b) Find $p(x)$ if $p(0) = 1$ and $p(1) = 0$.

6.6.52 Use formula (6.33) to expand the polynomials

(a) x^3,

(b) $x^2 + 2x$.

6.6.53

(a) Express the polynomial $p(x) = x^3 - x^2 + 3x$ as a linear combination of lower factorial polynomials.

(b) Solve the difference equation

$$\Delta p(x) = x^3 - x^2 + 3x \qquad\qquad (p(0) = 0).$$

6.6.54 Use the Δ-analogue of Taylor's theorem to express

$$p(x) = x^3 - 4x^2 + x + 2$$

as a linear combination of lower factorial polynomials.

6.6.55 Construct the polynomial $p(x)$ of least degree such that

$$p(0) = 1, \ p(2) = 0, \ p(4) = -1, \ p(6) = 2.$$

6.6.56 Construct a polynomial $p(x)$ of degree 3 such that

$$p(0) = 1, \ p(1) = -1, \ p(2) = -3, \ p(3) = 1.$$

6.6.57 Suppose $X = \{x_0, x_1, \ldots, x_n\}$ is a set of real numbers with $x_{i+1} - x_i = d$ for $i = 0, 1, \ldots, n - 1$. Let $q(x)$ be a polynomial with

$$q(k) = a_k \qquad\qquad (k = 0, 1, \ldots, n),$$

where $f_0, f_1, \ldots, f_n$ are given real numbers.

(a) Show that the polynomial $p(x)$ defined by

$$p(x) = q\left(\frac{1}{d}(x - x_0)\right)$$

takes on the value f_k at the point x_k.

(b) Use this and the difference operator method described in Section 6.4 to construct a polynomial $p(x)$ with

$$p(-1) = 1, \ p(1) = -2, \ p(3) = 5.$$

6.6.58 Suppose $p(x)$ is a polynomial of degree n with

$$p(k) = f_k \qquad\qquad (k = 0, 1, \ldots, n)$$

for some integers $f_0, f_1, \ldots, f_n$. Show that $p(x)$ has integral coefficients if and only if $k!$ divides $L_0 \Delta^k p(x)$ for $k = 0, 1, \ldots, n$.

6.6.59

(a) Describe the collection of polynomials $p(x)$ for which

$$\Delta L_a p(x) = L_a \Delta p(x).$$

(b) Describe the collection of polynomials $p(x)$ for which

$$\Delta D p(x) = D \Delta p(x).$$

6.6.60 Suppose

$$p(x) = \sum_{k=0}^{m} c_k (x)_k .$$

Show that

$$\sum_{n\geq 0} p(n) 2^{-n} = 2 \sum_{k=0}^{m} k! c_k.$$

6.6.61 Use problem 6.6.60 to evaluate the following series:

(a) $\displaystyle\sum_{n\geq 0} n^2 2^{-n}$,

(b) $\displaystyle\sum_{n\geq 0} n^3 2^{-n}$,

(c) $\displaystyle\sum_{n\geq 0} (3 - n + 2n^2) 2^{-n}$,

(d) $\displaystyle\sum_{n\geq 0} n(n-1)(n-2) 2^{-n}$.

6.7 Sample Quiz for Chapter 6

1. Calculate the number of partitions of $\{1, 2, \ldots, 9\}$ into three 2-subsets and one 3-subset.
2. Calculate the number of permutations of $\{1, 2, \ldots, 9\}$ with three 2-cycles and one 3-cycle.
3. Calculate the number of ways of placing 12 balls in four identical tubes so that no tube is empty.
4. Let V_n denote the number of permutations of $\{1, 2, \ldots, n\}$ whose cycle diagram contains only cycles of length 3. Put together the exponential generating function of $\{V_n\}$.
5. Express the polynomial

$$P(x) = 3x^2 + 2x - 1$$

as a linear combination of lower factorial polynomials

a. by means of the table of Stirling numbers $S_{n,k}$,
b. by means of "Taylor's" formula for the difference operator.

6. Find the polynomial $P(x)$ which solves the difference equation

$$P(x + 1) - P(x) = 3x^2 + 2x - 1.$$

7. Find the generating function of the sequence of polynomials $\phi_n(x, L)$ associated with the exponential structure L whose lists L_m are all empty for $m > 3$ and L_1, L_2, L_3 are as given below:

Put together an example of a structure in $L_{8,3}$.

Chapter 7
The Inclusion-Exclusion Principle

7.1 A Special Case

Our goal here is to present a basic tool which can be used in a variety of enumeration problems. The need for such a tool is best illustrated by an example.

Let us suppose that we have six contiguous squares, as in the figure below

and we wish to color them using twice each of the following three colors: blue, red, and yellow. For instance, indicating each color by its first letter, one of the possibilities may be

$$\text{Y Y R B R B} \tag{7.1}$$

It is easy to see that there are all together

$$\frac{6!}{2!2!2!} = 90$$

different ways of doing this.

Difficulties arise if we try counting colorings satisfying certain restrictions.

For instance, let us say that a coloring is *sharp* if no two adjacent squares are given the same color (this would exclude the coloring given in (7.1)), and suppose that our task is to calculate the number of sharp colorings. It turns out that this is but a very special case of a general class of counting problems for which there is a powerful method of solution.

We illustrate the method by working on this example.

© Springer Nature Switzerland AG 2021
Ö. Eğecioğlu, A. M. Garsia, *Lessons in Enumerative Combinatorics*, Graduate
Texts in Mathematics 290, https://doi.org/10.1007/978-3-030-71250-1_7

First of all, we have a space, let us call it Ω, consisting of all possible configurations. In this case Ω is the collection of all unrestricted ways we can color the six boxes with the given colors. As we have observed

$$|\Omega| = 90 .$$

Next, we have a certain number of subsets of Ω which represent the *properties* that we <u>do not</u> want our configurations to have. For instance, in our case we have three such properties. Namely, let us say that a coloring has property 1 if two adjacent squares are colored blue, property 2 if it has two adjacent red squares, and property 3 if two yellow adjacent squares occur. Let us denote by A_1, A_2, A_3 the collections of colorings having respectively properties 1, 2, 3.

Indicating Ω by a rectangle and the subsets A_1, A_2, A_3 by circles, we can represent the general situation schematically by the diagram in Figure 7.1.

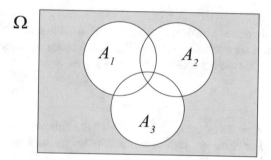

Fig. 7.1 Schematic representation of A_1, A_2, A_3.

In particular, the portion of Ω common to the circles A_1 and A_2 represents the collection of configurations which have two blue and two red adjacent squares.

Our problem is to count the number of configurations which have none of these properties. In the figure above this is represented precisely by the portion of Ω which lies outside of the three circles.

Given a subset A, let us denote by χ_A the function

$$\chi_A(x) = \begin{cases} 1 & \text{if } x \in A, \\ 0 & \text{otherwise.} \end{cases}$$

This function is usually referred to as the *indicator* of the set A. Indeed, knowing it is equivalent to knowing the set A itself. Note that we could just as well write $\chi_A(x) = \chi(x \in A)$, since the right hand side of this is 1 exactly when x is in A and 0 otherwise.

For convenience, let D denote the set of sharp configurations. Then what we are interested in is simply the sum

$$d = \sum_{x \in \Omega} \chi_D(x) . \tag{7.2}$$

It so happens that there is a simple way to express χ_D in terms of $\chi_{A_1}, \chi_{A_2}, \chi_{A_3}$. Namely, for every $x \in \Omega$,

$$\chi_D(x) = \left(1 - \chi_{A_1}(x)\right)\left(1 - \chi_{A_2}(x)\right)\left(1 - \chi_{A_3}(x)\right) . \tag{7.3}$$

In fact, if x belongs to any of the sets A_1, A_2, A_3, then the right hand side of (7.3) is equal to zero. On the other hand, if x belongs to none of them, then all the factors in (7.3) are equal to one and their product is equal to one as well. Thus, (7.3) must hold true as asserted.

Carrying out the multiplication in (7.3) and omitting the dependence on x for notational simplicity gives

$$\chi_D = 1 - \chi_{A_1} - \chi_{A_2} - \chi_{A_3} + \chi_{A_1}\chi_{A_2} + \chi_{A_1}\chi_{A_3} + \chi_{A_2}\chi_{A_3} - \chi_{A_1}\chi_{A_2}\chi_{A_3} .$$

Substituting in (7.2) and carrying out the summation we derive that

$$\begin{aligned}
d = \sum_{x \in \Omega} 1 - \sum_{x \in \Omega} \chi_{A_1} - \sum_{x \in \Omega} \chi_{A_2} - \sum_{x \in \Omega} \chi_{A_3} \\
+ \sum_{x \in \Omega} \chi_{A_1}\chi_{A_2} + \sum_{x \in \Omega} \chi_{A_1}\chi_{A_3} + \sum_{x \in \Omega} \chi_{A_2}\chi_{A_3} \\
- \sum_{x \in \Omega} \chi_{A_1}\chi_{A_2}\chi_{A_3} .
\end{aligned} \tag{7.4}$$

Note in particular that the sum

$$\sum_{x \in \Omega} \chi_{A_1}\chi_{A_2}$$

gives the number of elements common to A_1 and A_2, that is the number of elements of the set $A_1 \cap A_2$. Thus (7.4) may be rewritten in the form

$$d = |\Omega| - |A_1| - |A_2| - |A_3| + |A_1 \cap A_2| + |A_1 \cap A_3| + |A_2 \cap A_3| - |A_1 \cap A_2 \cap A_3| . \tag{7.5}$$

For a given subset T of $\{1, 2, 3\}$ let us denote by a_T the number of configurations that have property i for each i in T. For instance, $a_{\{2,3\}}$ denotes the number of configurations that have at least properties 2 and 3. Note, we do not exclude those configurations that have also property 1. In other words

$$a_{\{2,3\}} = |A_2 \cap A_3| .$$

For convenience let us set $a_\emptyset = |\Omega|$. Using this notation, formula (7.5) may be rewritten in the form

$$d = a_\emptyset - a_{\{1\}} - a_{\{2\}} - a_{\{3\}} + a_{\{1,2\}} + a_{\{1,3\}} + a_{\{2,3\}} - a_{\{1,2,3\}} \tag{7.6}$$

or, using summation notation

$$d = \sum_{T \subseteq \{1,2,3\}} (-1)^{|T|} a_T .$$

The significance of formula (7.6) derives from the fact that, although d itself may be difficult to calculate directly, the numbers a_T are almost immediately available to us.

For instance, we can easily put together all the configurations having at least properties 2 and 3. These are the configurations having 2 adjacent red squares and two adjacent yellow squares. One of them is

$$\boxed{Y}\,\boxed{Y}\,\boxed{B}\,\boxed{R}\,\boxed{R}\,\boxed{B}$$

We see that in these configurations the *blocks* RR and YY occur interspersed with two squares B, B.

We may thus interpret them as words in the four letters:

$$RR,\ YY,\ B,\ B .$$

We can easily deduce that there are a total of

$$\frac{4!}{1!1!2!} = 12$$

such words, and thus $a_{\{1,2\}} = 12$.

The reader should have no difficulty applying the same reasoning to the other cases and conclude that

$$a_{\{1\}} = a_{\{2\}} = a_{\{3\}} = \frac{5!}{1!2!2!} = 30$$

$$a_{\{1,2\}} = a_{\{1,3\}} = a_{\{2,3\}} = 12$$

$$a_{\{1,2,3\}} = 6 .$$

Substituting these values in (7.6) and recalling that $|\Omega| = 90$ we finally obtain that there are

$$d = 90 - 3 \times 30 + 3 \times 12 - 6 = 30$$

sharp colorings.

7.2   The General Formulation

In the general case we shall have again a collection Ω of unrestricted configurations x, and a certain number, say n, of unwanted properties. We indicate as before by $A_1, A_2, \ldots, A_n$, the collections of configurations having properties $1, 2, \ldots, n$ respectively.

For a subset

$$T \subseteq \{1, 2, \ldots, n\}$$

we let as before a_T denote the number of configurations having property i, at least for each i in T. Thus if

$$T = \{i_1, i_2, \ldots, i_k\},$$

we have

$$a_T = |A_{i_1} \cap A_{i_2} \cap \cdots \cap A_{i_k}| .$$

In shorthand notation this may be written in the form

$$a_T = \left| \bigcap_{i \in T} A_i \right| .$$

This given the general formula may be stated as follows:

Theorem 7.2.1 (Inclusion-Exclusion Principle) *The number d of configurations that have none of the properties is given by*

$$d = \sum_{T \subseteq \{1,2,\ldots,n\}} (-1)^{|T|} a_T . \tag{7.7}$$

Proof The argument follows almost step by step what we have done in the special case previously considered. We let D denote the set of configurations having none of the properties. In set theoretical notation this is

$$D = \Omega - (A_1 \cup A_2 \cup \cdots \cup A_n) .$$

Our desired number is

$$d = \sum_{x \in \Omega} \chi_D(x) . \tag{7.8}$$

We then observe that

$$\chi_D = \prod_{i=1}^{n} (1 - \chi_{A_i}) .$$

Expanding the product on the right hand side of this formula we get

$$\chi_D = \sum_{T \subseteq [n]} (-1)^{|T|} \prod_{i \in T} \chi_{A_i} .$$

Substituting this in (7.8) gives

$$d = \sum_{x \in \Omega} \sum_{T \subseteq [n]} (-1)^{|T|} \prod_{i \in T} \chi_{A_i}(x)$$

and interchanging order of summation we obtain

$$d = \sum_{T \subseteq [n]} (-1)^{|T|} \sum_{x \in \Omega} \prod_{i \in T} \chi_{A_i}(x) . \qquad (7.9)$$

Note that the product

$$\prod_{i \in T} \chi_{A_i}(x)$$

is equal to one if and only if x has property i for each $i \in T$. Thus

$$\sum_{x \in \Omega} \prod_{i \in T} \chi_{A_i}(x) = a_T . \qquad (7.10)$$

Substituting this in (7.9) gives the desired formula (7.7).   □

7.3   Two Classical Examples

There are two classical applications of the inclusion-exclusion principle that are worth including in any treatment of the subject. The first one is as follows:

7.3.1   Divisibility Properties

We are given to calculate the number of integers less than 120 that are relatively prime to 120. Since the prime factors of 120 are 2, 3, and 5, the integers we are to count are those which are not divisible by any of these primes.

We can easily recognize that this is an instance of the inclusion-exclusion setting. Here Ω consists of the integers

$$1, 2, 3, 4, \ldots, 120 .$$

The unwanted properties are *divisibility* by $2, 3$, and 5. Accordingly, let A_1, A_2, A_3 denote the subsets consisting of those integers of Ω that are divisible by $2, 3$, and 5, respectively.

We see that

$$|A_1| = \frac{120}{2}, \quad |A_2| = \frac{120}{3}, \quad |A_3| = \frac{120}{5} .$$

Clearly we have also

$$|A_1 \cap A_2| = \frac{120}{2 \cdot 3}, \quad |A_1 \cap A_3| = \frac{120}{2 \cdot 5}, \quad |A_2 \cap A_3| = \frac{120}{3 \cdot 5}$$

and

$$|A_1 \cap A_2 \cap A_3| = \frac{120}{2 \cdot 3 \cdot 5} .$$

Thus from Theorem 7.2.1 we obtain that our desired number d is given by the expression

$$d = 120 - \frac{120}{2} - \frac{120}{3} - \frac{120}{5} + \frac{120}{2 \cdot 3} + \frac{120}{2 \cdot 5} + \frac{120}{3 \cdot 5} - \frac{120}{2 \cdot 3 \cdot 5}$$

$$= 120 \left(1 - \frac{1}{2} \right) \left(1 - \frac{1}{3} \right) \left(1 - \frac{1}{5} \right) = 32 .$$

This simple calculation suggests the following general result:

Theorem 7.3.1 *Let N be an integer and let $p_1, p_2, \ldots, p_n$ be its prime divisors. Then the number d of integers less than or equal to N, relatively prime to N is*

$$d = N \left(1 - \frac{1}{p_1} \right) \left(1 - \frac{1}{p_2} \right) \cdots \left(1 - \frac{1}{p_n} \right) . \tag{7.11}$$

Proof We follow step by step the argument given in the special case we have just considered. Let Ω denote the set

$$1, 2, 3, \ldots, N .$$

Let A_i denote the set of integers in Ω that are divisible by p_i. We see that if $T = \{i_1, i_2, \ldots, i_k\}$, then

$$a_T = |A_{i_1} \cap A_{i_2} \cap \cdots \cap A_{i_k}|$$

is equal to the number of integers in Ω that are divisible by the product

$$p_{i_1} p_{i_2} \cdots p_{i_k} \, .$$

This gives

$$a_T = \frac{N}{p_{i_1} p_{i_2} \cdots p_{i_k}} = N \prod_{i \in T} \frac{1}{p_i} \, .$$

Thus from Theorem 7.2.1 we obtain

$$d = \sum_{T \subseteq [n]} (-1)^{|T|} N \prod_{i \in T} \frac{1}{p_i} = N \sum_{T \subseteq [n]} (-1)^{|T|} \prod_{i \in T} \frac{1}{p_i} \, .$$

It is easy to see that the last expression is an expanded version of that appearing on the right hand side of (7.11). □

7.3.2   Permutations Without Fixed Points

The next application may be given a frivolous setting yet it leads to a rather surprising mathematical development.

It may be stated as follows:

A secretary is requested to send 25 letters to 25 different people. The letters are all the same except for the name and address of the recipient. The secretary gets the letters and corresponding envelopes printed out by a computer in the same order and into two separate piles. Now it happens that in transporting the material this man, being rather clumsy, spills the pile of envelopes. As a result the envelopes get out of order and in final assembly some letters get into the wrong envelope.

Assuming that the envelopes end up in a completely random rearrangement, the question is:

What is the probability that every *letter is sent to the wrong recipient?*

In a less frivolous setting, what we have to count here is the number of permutations

$$\sigma = \begin{pmatrix} 1 & 2 & \ldots & 25 \\ \sigma_1 & \sigma_2 & \ldots & \sigma_{25} \end{pmatrix} \tag{7.12}$$

which satisfy the condition

$$\sigma_i \neq i, \quad \text{for } i = 1, 2, \ldots, 25 \, .$$

Indeed, if σ_i denotes the position of the ith envelope in the rearranged pile, then $\sigma_i = i$ corresponds to the event that the ith letter is placed in its proper envelope.

We can again recognize the inclusion-exclusion setting. Here Ω is taken to be the set of all permutations of [25] and A_i for $i = 1, 2, \ldots, 25$, denotes the set of σ for which $\sigma_i = i$.

It is easy to see that

$$|\Omega| = 25! \, ,$$

and

$$|A_1| = |A_2| = \cdots = |A_{25}| = 24! \, .$$

For instance, to construct all permutations in A_5, all we have to do is place 5 in its right position (that is set $\sigma_5 = 5$) and then place the remaining 24 integers in the remaining positions in any of the 24! possible ways. A similar reasoning applies in the calculation of a_T for any T. In particular we get

$$a_{\{3,6,8\}} = |A_3 \cap A_6 \cap A_8| = 22! \, .$$

Indeed, to construct the permutations in $A_3 \cap A_6 \cap A_8$ we simply set $\sigma_3 = 3$, $\sigma_6 = 6$, $\sigma_8 = 8$ and place the remaining integers into the 22 available positions in any of the 22! possible ways.

We thus see that for any k-element set T we have

$$a_T = (25 - k)! \, .$$

Since there are a total of

$$\binom{25}{k}$$

k-element subsets of $\{1, 2, \ldots, 25\}$, formula (7.7) in this case reduces to

$$d = \sum_{k=0}^{25} (-1)^k \binom{25}{k} (25 - k)! = \sum_{k=0}^{25} (-1)^k \frac{25!}{k!} = 25! \sum_{k=0}^{25} \frac{(-1)^k}{k!} \, .$$

Thus the desired probability is

$$\frac{d}{25!} = \sum_{k=0}^{25} \frac{(-1)^k}{k!} \approx \frac{1}{e} = 0.3678794411\ldots$$

It is to be noted that the sum above approximates $1/e$ remarkably well. Indeed, it can be shown that the error involved is less than $1/26!$.

It is quite remarkable that an elementary problem such as this can lead to a rather esoteric number such as e.

It is worthwhile stating our result in full generality. First of all, given a permutation as in (7.12), an element $i \in [n]$ is said to be a *fixed point* of σ if $\sigma_i = i$. Permutations without fixed points are called *derangements*. The arguments we have presented in the special case $n = 25$ can be easily generalized to yield the following result:

Theorem 7.3.2 *The number D_n of derangements of $\{1, 2, \ldots, n\}$ is given by the formula*

$$D_n = n! \sum_{k=0}^{n} \frac{(-1)^k}{k!} . \tag{7.13}$$

There are two recursions satisfied by the number D_n that are worth pointing out. Indeed, they may be used to calculate the first few values of D_n a bit quicker than by means of (7.13).

They may be stated as follows:

Theorem 7.3.3

$$(a) \quad D_n = n D_{n-1} + (-1)^n , \tag{7.14}$$

$$(b) \quad D_{n+1} = n D_n + n D_{n-1} .$$

Proof The recursion (7.14) (*a*) is obtained by isolating the last term in the sum (7.13). More precisely, we can rewrite (7.13) in the form

$$D_n = n! \sum_{k=0}^{n-1} \frac{(-1)^k}{k!} + n! \frac{(-1)^n}{n!} . \tag{7.15}$$

On the other hand, again from (7.13) we get

$$(n-1)! \sum_{k=0}^{n-1} \frac{(-1)^k}{k!} = D_{n-1} .$$

Combining this with (7.15), we obtain (7.14) (*a*) as asserted.

This given, formula (7.14) (*b*) follows immediately from (7.14) (*a*). Indeed, using formula (7.14) (*a*) twice in succession (once for $n + 1$ and once for n) we obtain

$$D_{n+1} = (n+1) D_n + (-1)^{n+1} = n D_n + D_n - (-1)^n = n D_n + n D_{n-1} .$$

Thus (7.14) (*b*) holds precisely as asserted. □

Remark 7.3.1 The reader may find it amusing to derive formula (7.14) (*b*) by a purely combinatorial reasoning. We give a brief sketch of the idea.

Suppose we represent the derangements of $[n + 1]$ by their cycle diagrams. Imagine that we remove $n + 1$ from one of these diagrams. There are two cases to be considered according as $n + 1$ is in a 2-cycle or in a larger cycle. In the former case, by removing the 2-cycle we are left with the cycle diagram of a derangement of the remaining $n - 1$ integers. We can see that this case accounts for $n D_{n-1}$ different possibilities, leading to the second term in (7.14) (b). When $n + 1$ is in a larger cycle, by removing $n + 1$ we get the cycle diagram of a derangement of $[n]$. Since there are n different positions that $n + 1$ can occupy in the cycle diagram of such a derangement, we see that this case accounts for $n D_n$ different possibilities, leading to the first term in (7.14) (b).

7.4 Further Identities

The method of proof that yielded the identity in (7.7) can be used, with only minor modifications, to obtain further useful identities.

For convenience, if A is a subset of Ω let $\overline{A}$ denote the *complement* of A, that is the set of elements of Ω that are not in A. We can easily see that for any x

$$\chi_{\overline{A}}(x) = 1 - \chi_A(x) . \tag{7.16}$$

This given, the idea is to replace the product

$$\prod_{i=1}^{n}(1 - \chi_{A_i}(x)) = \prod_{i=1}^{n}\chi_{\overline{A_i}}(x) \tag{7.17}$$

by the expression

$$\prod_{i=1}^{n}\left(\chi_{\overline{A_i}}(x) + t\chi_{A_i}(x)\right) , \tag{7.18}$$

where t is an auxiliary variable. We see that (7.18) reduces to (7.17) upon setting $t = 0$.

Note further that the expression in (7.18), for fixed x, is a polynomial in t. Indeed, we have

$$\prod_{i=1}^{n}(\chi_{\overline{A_i}}(x) + t\chi_{A_i}(x)) = \sum_{T \subseteq [n]} t^{|T|} \prod_{i \in T}\chi_{A_i}(x) \prod_{j \notin T}\chi_{\overline{A_j}}(x) . \tag{7.19}$$

The reader should have no difficulty understanding the general case of this identity by verifying the special cases $n = 2$ or 3.

Now observe that the function

$$\prod_{i \in T} \chi_{A_i}(x) \prod_{j \notin T} \chi_{\overline{A}_j}(x)$$

is none other than the indicator of the subset E_T of Ω consisting of the elements x which belong to the set A_i precisely for each i in T and no others. In other words E_T consists of those elements of Ω which have only the properties in T. Set for the moment

$$e_T = |E_T| = \sum_{x \in \Omega} \chi_{E_T}(x).$$

We see that if we sum the expression in (7.19) for all x in Ω we end up with the polynomial

$$P(t) = \sum_{T \subseteq [n]} t^{|T|} \sum_{x \in \Omega} \chi_{E_T}(x) = \sum_{T \subseteq [n]} t^{|T|} e_T.$$

Grouping together the terms corresponding to sets T having the same number (say k) of elements, we end up with the expression

$$P(t) = \sum_{k=0}^{n} t^k \sum_{\substack{T \subseteq [n] \\ |T|=k}} e_T. \tag{7.20}$$

It is easy to see that the sum

$$\sum_{|T|=k} e_T \tag{7.21}$$

gives the number of elements of Ω which have exactly k of the properties. Indeed, e_T itself counts those elements which have the properties in T and no others and in (7.21) we are summing these counts for all the subsets of cardinality k. For convenience let us denote the integer in (7.21) by the symbol e_k. This given, we can write (7.20) in the form

$$P(t) = \sum_{k=0}^{n} e_k t^k.$$

We may interpret this as the generating function of the numbers e_k.

Now it develops that we can find a new expression for the same polynomial $P(t)$ in terms of the numbers a_T used in Sections 7.1 and 7.2.

This result may be stated as follows:

Theorem 7.4.1 *Let us set for a given* $k = 0, 1, \ldots, n,$

$$a_k = \sum_{|T|=k} a_T . \tag{7.22}$$

Then the following identity holds:

$$P(t) = \sum_{k=0}^{n} a_k (t-1)^k . \tag{7.23}$$

Proof Note that, in view of (7.16), we may write the product in (7.18) in the form

$$\prod_{i=1}^{n} \left(\chi_{\overline{A_i}}(x) + t \chi_{A_i}(x) \right) = \prod_{i=1}^{n} \left(1 + (t-1) \chi_{A_i}(x) \right) .$$

We then also get that

$$P(t) = \sum_{x \in \Omega} \prod_{i=1}^{n} \left(1 + (t-1) \chi_{A_i}(x) \right)$$

$$= \sum_{x \in \Omega} \sum_{T \subseteq [n]} (t-1)^{|T|} \prod_{i \in T} \chi_{A_i}(x).$$

Thus interchanging order of summation and using formula (7.10) we finally obtain

$$P(t) = \sum_{T \subseteq [n]} (t-1)^{|T|} \sum_{x \in \Omega} \prod_{i \in T} \chi_{A_i}(x)$$

$$= \sum_{T \subseteq [n]} (t-1)^{|T|} a_T .$$

Following this, formula (7.23) is obtained by grouping together terms for which T has cardinality k and using (7.22). □

It may be good to illustrate the significance of this identity by an example. Let us go back to our initial coloring problem of Section 7.1. In that case we have

$$a_0 = a_\emptyset = 90$$

$$a_1 = a_{\{1\}} + a_{\{2\}} + a_{\{3\}} = 90$$

$$a_2 = a_{\{12\}} + a_{\{13\}} + a_{\{23\}} = 36$$

$$a_3 = a_{\{123\}} = 6 .$$

Thus from formula (7.23) we obtain

$$P(t) = e_0 + e_1 t + e_2 t^2 + e_3 t^3 = 90 + 90(t-1) + 36(t-1)^2 + 6(t-1)^3 \,.$$

Expanding the powers of $t-1$ and collecting equal powers of t yield

$$e_0 + e_1 t + e_2 t^2 + e_3 t^3 = 30 + 36t + 18t^2 + 6t^3 \,,$$

that is

$$e_0 = 30, \ e_1 = 36, \ e_2 = 18, \ e_3 = 6 \,.$$

The first of these equalities says what we already know, namely that there are 30 colorings with no adjacent squares of the same color. The last says something which in this case is almost obvious. Namely that there are 6 colorings in which 3 pairs of adjacent squares are of the same color. However, we get two additional conclusions that are not entirely obvious even in this simple situation. That is

1. *There are 36 colorings which have exactly 1 pair of adjacent squares of the same color,*
2. *There are 18 colorings which have exactly 2 pairs of adjacent squares of the same color.*

It is clear from this example that by expanding the powers of $t-1$ occurring in (7.23) we may derive formulas expressing the es in terms of the as.

More precisely we obtain the following identities:

Theorem 7.4.2 *For each $m = 0, 1, \ldots, n$ we have*

$$e_m = a_m - \binom{m+1}{m} a_{m+1} + \binom{m+2}{m} a_{m+2} - \cdots \tag{7.24}$$

Moreover, setting

$$l_m = e_m + e_{m+1} + \cdots + e_n \,, \tag{7.25}$$

we have also

$$l_m = a_m - \binom{m}{m-1} a_{m+1} + \binom{m+1}{m-1} a_{m+2} - \cdots \tag{7.26}$$

Proof Using the binomial theorem we may rewrite (7.23) in the form

$$P(t) = \sum_{k=0}^{n} a_k \sum_{m=0}^{k} \binom{k}{m} t^m (-1)^{k-m}$$

$$= \sum_{m=0}^{n} t^m \sum_{k=m}^{n} \binom{k}{m}(-1)^{k-m} a_k \, .$$

Equating coefficients of t^m gives

$$e_m = \sum_{k=m}^{n} \binom{k}{m}(-1)^{k-m} a_k \, ,$$

and this is (7.24) written in mathematical shorthand.

To prove (7.26) note first that

$$\sum_{m=0}^{n} l_m t^m = \sum_{m=0}^{n} t^m \sum_{k=m}^{n} e_k$$

$$= \sum_{k=0}^{n} e_k \sum_{m=0}^{k} t^k = \sum_{k=0}^{n} e_k \frac{t^{k+1} - 1}{t - 1} \, .$$

In other words, from our original definition of $P(t)$, we have

$$\sum_{m=0}^{n} l_m t^m = \frac{t P(t) - P(1)}{t - 1} \, .$$

Substituting for $P(t)$ as given by (7.23), we get

$$\sum_{m=0}^{n} l_m t^m = \frac{t \sum_{k=0}^{n} a_k (t - 1)^k - a_0}{t - 1}$$

$$= a_0 + t \sum_{k=1}^{n} a_k (t - 1)^{k-1} \, .$$

Expanding the powers of $(t - 1)$ by the binomial theorem gives

$$\sum_{m=0}^{n} l_m t^m = a_0 + t \sum_{k=1}^{n} a_k \sum_{m=0}^{k-1} \binom{k-1}{m} t^m (-1)^{k-1-m}$$

$$= a_0 + \sum_{m=0}^{n-1} t^{m+1} \sum_{k=m+1}^{n} a_k \binom{k-1}{m} (-1)^{k-1-m} \, .$$

Equating the coefficients of t^m we finally obtain

$$l_m = \sum_{k=m}^{n} a_k \binom{k-1}{m-1}(-1)^{k-m} ,$$

and this is (7.26).

$\square$

We should point out that the quantity in (7.25) gives the number of objects which have *at least* m of the unwanted properties. In particular, for $n = 3$ and $m = 2$, (7.26) reduces to

$$l_2 = a_2 - \binom{2}{1}a_3 .$$

For our sharp coloring problem this gives

$$l_2 = 36 - 2 \cdot 6 = 24 ,$$

and indeed we do have

$$l_2 = e_2 + e_3 = 18 + 6 = 24 .$$

7.5 Miscellaneous Applications and Examples

The inclusion-exclusion principle has many applications and ramifications, we can only indicate a few.

7.5.1 The Ménage Problem

We are asked to seat 4 married couples at a round table so that women and men alternate and no husband is seated next to his wife. The question is in how many ways can this be done.

We may seat the wives first and label them W_1, W_2, W_3, W_4 clockwise as in Figure 7.2.

The possible positions to be occupied by the husbands are indicated by chairs and labeled P_1, P_2, P_3, P_4. Let H_1, H_2, H_3, H_4 denote the husbands of W_1, W_2, W_3, W_4, respectively.

According to our requirements we *may not* sit

$$H_1 \text{ in } P_1, P_4$$

$$H_2 \text{ in } P_1, P_2$$

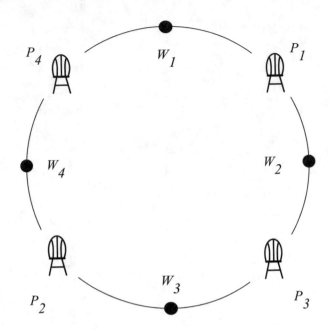

Fig. 7.2 The placement of W_1, W_2, W_3, W_4.

$$H_3 \text{ in } P_2, P_3$$

$$H_4 \text{ in } P_3, P_4 \, .$$

Each way to seat the husbands may be identified with a permutation of $1, 2, 3, 4$. Indeed, the permutation

$$\begin{pmatrix} 1 \ 2 \ 3 \ 4 \\ 3 \ 1 \ 4 \ 2 \end{pmatrix} \tag{7.27}$$

may be taken to represent seating H_1 in P_3, H_2 in P_1, H_3 in P_4 and H_4 in P_2. This gives the seating arrangement in Figure 7.3. Of course this arrangement does not satisfy our requirements.

Our original seating problem is thus reduced to counting permutations with restricted entries. That is, we are to count the permutations which satisfy the four conditions

$$1) \ \sigma_1 \neq 1, 4$$

$$2) \ \sigma_2 \neq 1, 2 \tag{7.28}$$

$$3) \ \sigma_3 \neq 2, 3$$

$$4) \ \sigma_4 \neq 3, 4 \, .$$

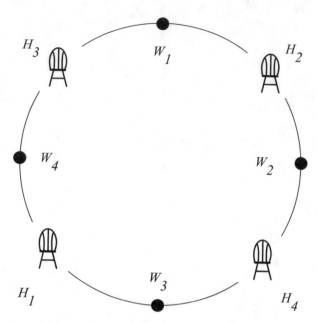

Fig. 7.3 A seating arrangement that does not satisfy the requirements.

We should now be able to recognize the inclusion-exclusion setting. Clearly, the set Ω consists of the 24 permutations of $\{1, 2, 3, 4\}$ and we have 4 unwanted properties, namely those respectively corresponding to the 4 conditions in (7.28). We set

$$A_1 = \{\sigma \mid \sigma_1 = 1 \text{ or } 4\} \; (H_1 \text{ is seated next to } W_1)$$

$$A_2 = \{\sigma \mid \sigma_2 = 1 \text{ or } 2\} \; (H_2 \text{ is seated next to } W_2)$$

$$A_3 = \{\sigma \mid \sigma_3 = 2 \text{ or } 3\} \; (H_3 \text{ is seated next to } W_3)$$

$$A_4 = \{\sigma \mid \sigma_4 = 3 \text{ or } 4\} \; (H_4 \text{ is seated next to } W_4).$$

To facilitate our calculation and for our later considerations we shall make use of *permutation diagrams*. These are graphical representations of permutations which are constructed as follows. To represent

$$\sigma = \begin{pmatrix} 1 & 2 & \dots & n \\ \sigma_1 & \sigma_2 & \dots & \sigma_n \end{pmatrix}$$

we draw an $n \times n$ chessboard and in the ith column we place a cross in the square that is at height σ_i. This is best communicated by an example. For instance the diagram of the permutation in (7.27) is simply as shown in Figure 7.4.

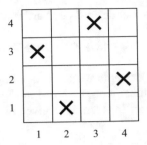

Fig. 7.4 The diagram of the permutation in (7.27).

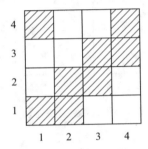

Fig. 7.5 To place nonattacking rooks. The shaded squares are forbidden.

Note that in these diagrams there is only one cross in each row and in each column. For this reason they are colorfully referred to as placements of *nonattacking rooks* on a chessboard.

We see then that our problem is to count the number of ways we can place 4 nonattacking rooks in the 4×4 board in Figure 7.5 with the additional condition that the positions corresponding to the shaded squares are *forbidden*. Our subsets A_1, A_2, A_3, A_4 correspond to the collections of configurations where a cross falls in the darkened area of the 1st, 2nd, 3rd, and 4th columns respectively. This given it is not difficult to derive that

$$a_{\{1\}} = a_{\{2\}} = a_{\{3\}} = a_{\{4\}} = 2 \cdot 3! = 12$$

$$a_{\{1,2\}} = a_{\{1,4\}} = a_{\{2,3\}} = a_{\{3,4\}} = 3 \cdot 2 = 6$$

$$a_{\{1,3\}} = a_{\{2,4\}} = 8 \tag{7.29}$$

$$a_{\{1,2,3\}} = a_{\{1,2,4\}} = a_{\{1,3,4\}} = a_{\{2,3,4\}} = 4$$

$$a_{\{1,2,3,4\}} = 2 .$$

Thus formula (7.7) gives that our desired number is

$$d = 24 - 48 + 40 - 16 + 2 = 2 \, .$$

Of course, for such a small example we could obtain the answer in a more direct manner. Indeed, reasoning directly on the board in Figure 7.5 we quickly discover that the only possible configurations are those given in Figure 7.6.

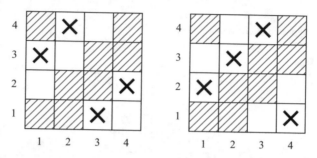

Fig. 7.6 Placement of nonattacking rooks on the board in Figure 7.5.

Note however, that our formulas combined with the counts in (7.29) yield some additional information.

In fact, for a permutation σ let $n(\sigma)$ denote the number of crosses of the diagram of σ that *hit* the forbidden region, and let

$$Q(t) = \sum_{\sigma} t^{n(\sigma)} \, .$$

The coefficient of t^k in this polynomial gives the number of permutation diagrams which hit the forbidden region in exactly k places. In other words this coefficient gives precisely the number of configurations that have exactly k of the unwanted properties. In other words

$$Q(t) = P(t) = \sum_{k=0}^{n} e_k t^k \, .$$

Since here

$$a_0 = 24,$$
$$a_1 = 48,$$
$$a_2 = 40,$$
$$a_3 = 16,$$

$$a_4 = 2$$

formula (7.23) gives

$$Q(t) = 24 + 48(t-1) + 40(t-1)^2 + 16(t-1)^3 + 2(t-1)^4$$
$$= 2 + 8t + 4t^2 + 8t^3 + 2t^4 \ .$$

The reader may verify that the values of e_0, e_1, e_2, e_3, e_4 thus obtained are in agreement with those given by formula (7.24).

We should note that the diagrams in Figure 7.6 correspond to the seating arrangements shown in Figure 7.7.

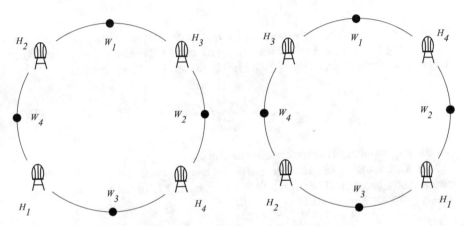

Fig. 7.7 The seating arrangements corresponding to the diagrams in Figure 7.6.

Clearly each of these arrangements (by permuting husbands and wives in the same way) produces a total of 24 different arrangements. Moreover, by interchanging the positions of W_i and H_i for each i we can double the total. We may thus conclude that there are all together 96 seating arrangements satisfying our original requirements.

The reader may rightly wonder if we can obtain the analogous count for the case of n couples, for an arbitrary n. It turns out that there is a beautiful formula for this that is due to Touchard [28]. We shall present it next since it leads to several further interesting developments.

7.5.2   Rook Numbers and Hit Polynomials

Let B denote an $n \times n$ board and let A be a subset of its squares. For instance, for $n = 6$ we may have the configuration depicted in Figure 7.8.

Here the subset A is represented by the darkened squares.

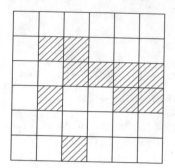

Fig. 7.8 A is the set of shaded squares.

For each $k = 1, 2, \ldots, n$ we denote by $r_k(A)$ the number of ways of placing k nonattacking rooks in A. We shall refer to $r_k(A)$ as the kth *rook number* of the *board* A. It is convenient to set $r_0(A) = 1$. With this convention the expression

$$R(x, A) = \sum_{k=0}^{n} r_k(A)x^k \tag{7.30}$$

is usually referred to as the *rook polynomial* of A.

Note that some of the rook numbers may be equal to zero, so that $R(x, A)$ may actually be of smaller degree than indicated in (7.30). In fact, for the board in Figure 7.8 we have

$$r_5(A) = r_6(A) = 0.$$

This is due to the fact that all of the squares of A are contained in four rows of B and we cannot place more than one rook in any given row. It also is easy to see that

$$r_1(A) = 10$$

since the first rook number of any board A is always equal to the cardinality of A.

The remaining rook numbers require a little more effort. To calculate $r_2(A)$ we first chose the two rows in which we are to place the two rooks, then count the number of ways the rooks may be placed in those two rows. For instance, if we chose the 2nd and 3rd rows of the board in Figure 7.8, we have a total of

$$4 + 3 = 7$$

possibilities (4 if we place the rook in the first darkened square of the 2nd row and 3 if we place it in the second).

Repeating this argument for each of the 6 pairs of rows and adding the counts we get

$$r_2(A) = 7 + 5 + 1 + 10 + 3 + 3 = 29.$$

Proceeding in a similar manner for the other rook numbers we obtain

$$r_3(A) = 25, \ r_4(A) = 4 .$$

This gives

$$R(x, A) = 1 + 10x + 29x^2 + 25x^3 + 4x^4 . \tag{7.31}$$

As we can see, rook numbers may be quite tedious to calculate. However, some simplifications do occur for special boards.

Let us say that a subset A of B is a *product* board, if we can divide the squares of A into two parts A_1 and A_2 so that no rook placed in A_1 can attack any rook placed in A_2. We may briefly express this by saying that A breaks up into the *product* of A_1 and A_2. For instance the board in Figure 7.9 breaks up into the product of the subsets A_1, A_2 consisting of the vertically and horizontally shaded squares respectively.

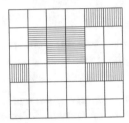

Fig. 7.9 A product board.

This given, note that to construct a configuration of k nonattacking rooks on A we may place a number of them, say i, in A_1 and then the remaining $k - i$ in A_2. Since the latter cannot be attacked by any of the former, these two placements can be carried out entirely independently of each other. Doing this for each $i = 0, 1, \ldots, k$ and adding the resulting counts yield the identity

$$r_k(A) = \sum_{i=0}^{k} r_i(A_1) r_{k-i}(A_2) . \tag{7.32}$$

This result is best expressed in the following form:

Theorem 7.5.1 *If the board A breaks up into the product of A_1 and A_2, then*

$$R(x, A) = R(x, A_1) \cdot R(x, A_2) . \tag{7.33}$$

Proof The right hand side of (7.32) is (by Cauchy's formula) the coefficient of x^k in the product of the polynomials $R(x, A_1)$ and $R(x, A_2)$. $\square$

This result may yield some dramatic simplifications. Indeed, applying it to the board in Figure 7.9 we immediately get the rook polynomial as calculated in Figure 7.10.

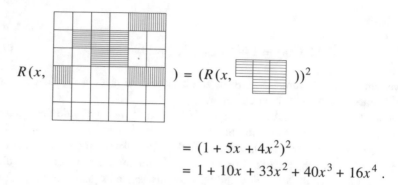

$$= (1 + 5x + 4x^2)^2$$
$$= 1 + 10x + 33x^2 + 40x^3 + 16x^4 .$$

Fig. 7.10 Calculation of the rook polynomial of the board in Figure 7.9.

Remark 7.5.1 It should be noted that the rook polynomial of a board A only depends on the shape of A and not on the positions that the squares of A occupy on the chessboard. In fact, a yet stronger rule holds, namely

> *The rook polynomial does not change if we replace the board A by a board A' which is obtained from A by any number of row or column interchanges.*

The reason for this is clear, the interchanges that send A into A' send nonattacking rook configurations of A into nonattacking rook configurations of A'. Thus $r_k(A) = r_k(A')$ must hold for all k and the assertion follows.

Applying this rule we deduce that we can proceed as shown in Figure 7.11 and this accounts for the result in Figure 7.10.

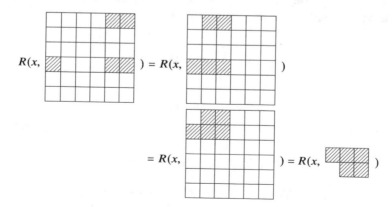

Fig. 7.11 Using row and column interchanges to compute the rook polynomial.

Rook polynomials may be calculated recursively by removing squares from the board one at a time. To see how this comes about let us consider an example and suppose that A is the *staircase* board in Figure 7.12.

Fig. 7.12 A staircase board A.

Note that although A itself is not a product board, the board A' obtained upon removing the circled square is. Thus it clearly pays off to relate the rook polynomials of A and A'.

For convenience let us denote the board obtained by removing from A the square s by $A - s$. Moreover, let us denote by A/s the board obtained upon removing from A all the squares that lie in the same row or column of s. For the particular board in Figure 7.12 we have these two boards as shown in Figure 7.13.

$A - s =$ 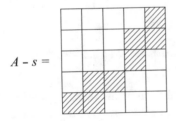 $A / s =$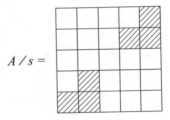

Fig. 7.13 The boards $A - s$ and A/s.

Our recursive algorithm is based upon the following simple fact:

Theorem 7.5.2 *For any board A and any square s of A we have*

$$R(x, A) = R(x, A - s) + x\, R(x, A/s) . \qquad (7.34)$$

Proof If a rook configuration of A does not have a rook on s, then it clearly pertains to the board $A-s$ as well. On the other hand, if the configuration does have a rook on s and it has k rooks all together, then the remaining $k - 1$ rooks form a configuration of nonattacking rooks on A/s. We can thus divide the k-rook configurations of A into two classes C_1, C_2; those of C_1 corresponding to k-rook configurations of $A - s$ and those of C_2 corresponding to $(k - 1)$-rook configurations of A/s. This observation yields that for each $k = 1, 2, \ldots, n$ we have

$$r_k(A) = r_k(A - s) + r_{k-1}(A/s) .$$

We may consider this valid for $k = 0$ as well by setting $r_{-1} = 0$. With this, multiplying both sides of the equation by x^k and summing for $k = 0, 1, \ldots, n$ we obtain

$$R(x, A) = \sum_{k=0}^{n} r_k(A - s)x^k + x \sum_{k=1}^{n} r_{k-1}(A/s)x^{k-1} \,,$$

and this is clearly another way of writing (7.34). □

This result applied to the staircase board of Figure 7.12, in view of Figure 7.13, yields the calculation in Figure 7.14.

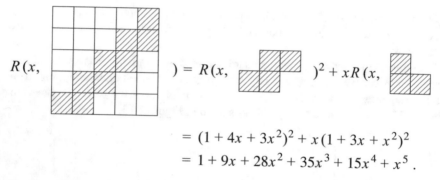

$$= (1 + 4x + 3x^2)^2 + x(1 + 3x + x^2)^2$$
$$= 1 + 9x + 28x^2 + 35x^3 + 15x^4 + x^5 \,.$$

Fig. 7.14 Calculation of the rook polynomial of the staircase board in Figure 7.12.

We are now ready to consider the general form of the ménage problem. Let us study a slightly larger example, say the case of 6 couples. Seating the wives first as we did for 4 couples, we see that each way of seating the husbands corresponds to a configuration of 6 nonattacking rooks on the 6×6 board avoiding the set A of darkened squares in Figure 7.15.

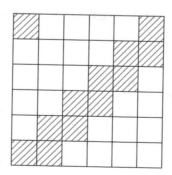

Fig. 7.15 Board M_6 for the ménage problem with six couples.

Similarly, we see that the general case of the ménage problem requires counting n-rook configurations on an $n \times n$ board which avoid a set A of squares consisting of a staircase board and an additional square at the upper left corner.

It develops that with the same effort we can solve a considerably more general problem. Indeed, let A be any subset of the $n \times n$ chessboard and for each permutation σ let

$$A \cap \sigma$$

denote the set of squares common to A and the diagram of σ.

Set

$$Q(t) = \sum_\sigma t^{|A \cap \sigma|} .$$

This is usually referred to as the *hit polynomial* of the board A. Note that the coefficient of t^k in $Q(t)$ is equal to the number of nonattacking n-rook configurations which *hit* A in precisely k places. In particular $Q(0)$ gives the number of permutation diagrams which entirely avoid the set A.

The relevancy of our study of rook numbers to the ménage problem and to the general problem of counting permutations with restricted entries is expressed by the following remarkable result:

Theorem 7.5.3 *For any subset A of the $n \times n$ board we have*

$$Q(t) = \sum_\sigma t^{|A \cap \sigma|} = \sum_{k=0}^{n} (n-k)! \, r_k(A)(t-1)^k . \tag{7.35}$$

In particular the number of permutation diagrams which avoid A is given by the expression

$$Q(0) = \sum_{k=0}^{n} (n-k)! \, r_k(A)(-1)^k . \tag{7.36}$$

Proof We have here another instance of the inclusion-exclusion setting. The set Ω consists of all configurations of n nonattacking rooks. The unwanted properties correspond to the sets $A_1, A_2, \ldots, A_n$, where A_i consists of all configurations which hit A on some square of the ith column.

This given, the coefficient of t^k in $Q(t)$ may be viewed as the number of configurations which have exactly k of the unwanted properties. In short, using the notation of Section 7.4, we must have

$$Q(t) = \sum_{k=0}^{n} e_k t^k = P(t) .$$

From Theorem 7.4.1 we then get

$$Q(t) = \sum_{k=0}^{n} a_k(t-1)^k . \tag{7.37}$$

We aim to show that in this case we have

$$a_k = (n-k)!\, r_k(A) . \tag{7.38}$$

This formula is best understood if we work with a special board. Let us take the ménage board of Figure 7.15 and suppose we want to calculate a_2. From our definitions we get

$$a_2 = a_{\{1,2\}} + a_{\{1,3\}} + a_{\{1,4\}} + \cdots + a_{\{5,6\}} . \tag{7.39}$$

Recall for instance that

$$a_{\{1,2\}} = |A_1 \cap A_2| .$$

A configuration σ in $A_1 \cap A_2$ is one which hits A in the first and second columns *at least*. For instance such a σ could be as shown in Figure 7.16.

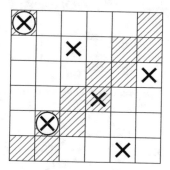

Fig. 7.16 An example σ which hits A in the first two columns.

Note that we may group the configurations in $A_1 \cap A_2$ according to which 2-rook configuration they have in the first 2 columns of the board (the circled rooks in Figure 7.16). In this case we have three groups corresponding to the three 2-rook configurations given in Figure 7.17.

Each of these groups has a total of 4! configurations. For instance, the group corresponding to the first of the 2-rook configurations in Figure 7.17 consists of the diagrams of the permutations

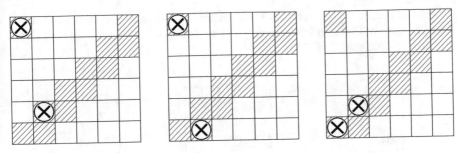

Fig. 7.17 Classification by the hits in first two columns of A.

$$\begin{pmatrix} 1 & 2 & 3 & 4 & 5 & 6 \\ \sigma_1 & \sigma_2 & \sigma_3 & \sigma_4 & \sigma_5 & \sigma_6 \end{pmatrix}$$

for which $\sigma_1 = 6$, $\sigma_2 = 2$ and the other entries are entirely unrestricted. That is, $(\sigma_3, \sigma_4, \sigma_5, \sigma_6)$ can be any permutation of the numbers $1, 3, 4, 5$. More generally, given any configuration C of k nonattacking rooks on A, there are $(n - k)!$ permutation diagrams which hit A *at least* in C.

In particular we have

$$a_{\{1,2\}} = |A_1 \cap A_2| = 3 \cdot 4! \, .$$

We thus see that each configuration of 2 nonattacking rooks on A contributes $4!$ to the sum in (7.39). We must therefore conclude that

$$a_2 = 4! r_2(A) = (6 - 2)! r_2(A) \, ,$$

and this is the special case $n = 6$, $k = 2$ of (7.38).

It is not difficult to see that the same reasoning applies in the general case and (7.38) must necessarily hold as asserted.

Substituting (7.38) in (7.37) yields our desired formula (7.35). □

It is good to illustrate our results by an example. Indeed, let us solve the ménage problem for 5 couples. To do this we need to calculate the rook numbers of the ménage board M_5 in Figure 7.18.

$$M_5 =$$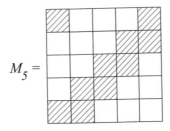

Fig. 7.18 The ménage board M_5.

Now we may use Theorem 7.5.2 and reduce this board to two staircase boards. More precisely, Theorem 7.5.2 (taking s to be the upper left corner square) yields the identity in Figure 7.19.

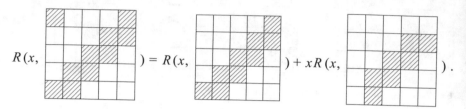

Fig. 7.19 The decomposition of M_5 for the calculation of $R(x, M_5)$.

In other words we have

$$R(x, M_5) = R(x, S_5) + x R(x, S_4) ,\qquad(7.40)$$

where by S_n we denote the $n \times n$ staircase board.

Another simple use of Theorem 7.5.2 yields the rook polynomial in Figure 7.20.

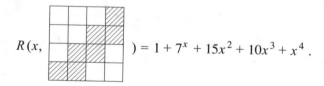

Fig. 7.20 The rook polynomial of the staircase board S_4.

Substituting this and Figure 7.14 in (7.40) we get

$$R(x, M_5) = 1 + 10x + 35x^2 + 50x^3 + 25x^4 + 2x^5 .$$

Finally, (7.36) gives

$$Q(0) = 5! - 4! \cdot 10 + 3! \cdot 35 - 2! \cdot 50 + 1! \cdot 25 - 0! \cdot 2 = 13 .$$

Thus we may conclude that there are total of $2 \cdot 5! \times 13 = 3120$ ways of seating these 5 couples!

To solve the general problem we need to calculate the rook numbers of the nth ménage board M_n. As in the case $n = 5$, Theorem 7.5.2 yields

$$R(x, M_n) = R(x, S_n) + x R(x, S_{n-1}) .\qquad(7.41)$$

This leads us to calculate the rook numbers of the staircase boards.

It turns out that we have the following very simple formula:

$$r_k(S_n) = \binom{2n-k}{k}. \tag{7.42}$$

Using this in (7.41) gives

$$R(x, M_n) = \sum_{k=0}^{n} \binom{2n-k}{k} x^k + \sum_{k=0}^{n-1} \binom{2n-2-k}{k} x^{k+1}.$$

Now the second sum may be rewritten in the form

$$\sum_{k=1}^{n} \binom{2n-1-k}{k-1} x^k.$$

We thus obtain

$$R(x, M_n) = 1 + \sum_{k=1}^{n} \left[\binom{2n-k}{k} + \binom{2n-1-k}{k-1} \right] x^k.$$

In other words we must have that

$$r_k(M_n) = \binom{2n-k}{k} + \binom{2n-1-k}{k-1} = \frac{2n}{2n-k} \binom{2n-k}{k}.$$

We can combine our results into the following beautiful formula:

Theorem 7.5.4 (Touchard) *The hit polynomial of the ménage board is*

$$\sum_{k=0}^{n} (n-k)! \frac{2n}{2n-k} \binom{2n-k}{k} (t-1)^k.$$

In particular, the number of ways of seating n couples at a round table so that no husband is next to his wife is

$$2 \cdot n! \sum_{k=0}^{n} (n-k)! \frac{2n}{2n-k} \binom{2n-k}{k} (-1)^k.$$

We see that for $n = 5$ the sum reduces to

$$5! - 4! \frac{10}{9} \binom{10-1}{1} + 3! \frac{10}{8} \binom{10-2}{2} - 2! \frac{10}{7} \binom{10-3}{3} + 1! \frac{10}{6} \binom{10-4}{4} - 0! \frac{10}{5} \binom{10-5}{5} = 13,$$

which is in perfect agreement with our previous calculation.

To complete our argument we need to establish formula (7.42). Now, there is a rather amusing way to derive this result. We shall illustrate it in the case $n = 7$, $k = 4$.

We want to count the number of ways of placing 4 nonattacking rooks in the staircase board in Figure 7.21.

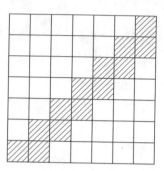

Fig. 7.21 Staircase board S_7.

It is easy to see that 2 rooks placed in the staircase are nonattacking if and only if they are not in adjacent squares. Thus our desired number is obtained by counting 4-subsets of the 13 squares laid out as in Figure 7.22 with no adjacent elements.

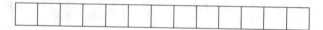

Fig. 7.22 The shaded squares of S_7 laid out in a row.

To find this number imagine that each of the squares represents a bar stool and we wish to seat 4 people at the bar so that the first 3 from left to right are in a position to set their right foot on the next bar stool. A possible configuration might be as in Figure 7.23.

Fig. 7.23 Forbidding the next bar stool.

Note that if we replace each of the patterns ◎⊖ by a P (for *person*) and do the same for the last ◎ , then replace each ○ by an E (for *empty*) we get the word

$$EPPEEPPEEE.$$

Clearly each of our configurations can be represented in this manner by a 10 letter word with 4 Ps and 6 Es. Thus the number of such configurations is

$$\binom{10}{4}.$$

This gives that

$$r_4(S_7) = \binom{14-4}{4},$$

which is the special case $n = 7, k = 4$ of (7.42).

In the general case we have $2n - 1$ bar stools and k people to sit. The first $k - 1$ from left to right place their right foot on the next bar stool. That leaves $2n - 1 - (k - 1) = 2n - k$ places. Proceeding as we did for $n = 7$, we represent our seating configurations by $2n-k$-letter words with k Ps and $2n-2k$ Es. Each of these words yields one of our configurations upon replacing the first $k - 1$ Ps by the pattern ◍◒ the last P by ◍ and each Es by ○

There are

$$\binom{2n-k}{k}$$

such words, thus we must conclude that (7.42) does hold as asserted.

7.5.3   *Rises of Permutations and Ferrers Boards*

Let

$$\sigma = \begin{pmatrix} 1 & 2 & \dots & n \\ \sigma_1 & \sigma_2 & \dots & \sigma_n \end{pmatrix}$$

be a permutation. We say that i is a *rise* of σ if

$$i \leq \sigma_i . \tag{7.43}$$

Similarly we say that i is a *fall* of σ if

$$i > \sigma_i . \tag{7.44}$$

We also let $r(\sigma)$ and $f(\sigma)$ respectively denote the number of rises and the number of falls of σ.

For instance, for the permutation

$$\sigma = \begin{pmatrix} 1\ 2\ 3\ 4\ 5\ 6\ 7 \\ 3\ 2\ 5\ 1\ 7\ 6\ 4 \end{pmatrix} \tag{7.45}$$

the indices $1, 2, 3, 5, 6$ are rises and $4, 7$ are falls. Thus in this case

$$r(\sigma) = 5, \quad f(\sigma) = 2 .$$

Since each index is either a rise or a fall, for any permutation σ of $[n]$ we must have

$$r(\sigma) + f(\sigma) = n . \tag{7.46}$$

It will be convenient to refer to an index i for which

$$i < \sigma_i$$

as a *strict* rise of σ and let $sr(\sigma)$ denote the total number of strict rises of σ. This given, we set

$$E_n(t) = \sum_{\sigma \in P_n} t^{r(\sigma)} .$$

Clearly, we have

$$E_n(t) = \sum_{k=0}^{n} e_{n,k} t^k , \tag{7.47}$$

where $e_{n,k}$ denotes the number of permutations of $[n]$ which have exactly k rises.
    The $\{E_n(t)\}$ are usually referred to as the *Eulerian* polynomials, and the $e_{n,k}$ as the *Euler numbers* [7].
    Note that since

$$r \begin{pmatrix} 1\ 2\ 3 \\ 1\ 2\ 3 \end{pmatrix} = 3, \quad r \begin{pmatrix} 1\ 2\ 3 \\ 1\ 3\ 2 \end{pmatrix} = 2, \quad r \begin{pmatrix} 1\ 2\ 3 \\ 2\ 1\ 3 \end{pmatrix} = 2,$$

$$r \begin{pmatrix} 1\ 2\ 3 \\ 2\ 3\ 1 \end{pmatrix} = 2, \quad r \begin{pmatrix} 1\ 2\ 3 \\ 3\ 1\ 2 \end{pmatrix} = 1, \quad r \begin{pmatrix} 1\ 2\ 3 \\ 3\ 2\ 1 \end{pmatrix} = 2,$$

we have

$$E_3(t) = t + 4t^2 + t^3 .$$

Studies on the average performance of certain data structures require the knowledge of the average behavior of the number of rises as the permutation varies among all permutations of a given set. Such information can be obtained from the Eulerian polynomials.

As it turns out, the results of this chapter are precisely what is needed for the study of these polynomials. This can be seen as follows. Note that the rises of the permutation in (7.45) can be quickly detected upon drawing the diagram. Indeed, the rises are precisely the crosses that hit the region above or on the diagonal squares of the board (see Figure 7.24).

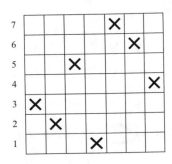

Fig. 7.24   Locating the rises and the falls.

We can also see that the falls are given by the crosses that lie strictly below the diagonal squares.

For convenience, let SS_n denote the region below the diagonal. For instance for $n = 7$, this is represented by the shaded area in Figure 7.25.

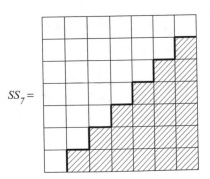

Fig. 7.25 SS_7 is the set of shaded squares.

We shall refer to SS_n as the *solid staircase* to distinguish it from the staircases considered in the previous subsection.

We see then that the Eulerian polynomials are none other than the hit polynomials of complements of solid staircase boards. To be precise, recall that if A is a subboard of the $n \times n$ board and P_n denotes the collection of all permutations of $[n]$, then the hit polynomial is defined by the formula

$$\sum_{\sigma \in P_n} t^{|A \cap \sigma|} .$$

It will be convenient to denote this polynomial using the notation

$$H(t, A) .$$

Our observation is simply that

$$(a) \quad r(\sigma) = |\overline{SS}_n \cap \sigma| \tag{7.48}$$
$$(b) \quad f(\sigma) = |SS_n \cap \sigma|$$

and thus

$$E_n(t) = \sum_{\sigma \in P_n} t^{r(\sigma)} = H(t, \overline{SS}_n) . \tag{7.49}$$

Moreover

$$\sum_{\sigma \in P_n} t^{f(\sigma)} = H(t, SS_n) . \tag{7.50}$$

Clearly for any subset of the $n \times n$ board we have that the number of times σ hits A and the number of times σ hits the complement of A add up to n. That is

$$|A \cap \sigma| + |\overline{A} \cap \sigma| = n .$$

This gives that

$$\sum_{\sigma \in P_n} t^{|A \cap \sigma|} = \sum_{\sigma \in P_n} t^{n - |\overline{A} \cap \sigma|} = t^n \sum_{\sigma \in P_n} \left(\frac{1}{t}\right)^{|\overline{A} \cap \sigma|} .$$

In other words

$$H(t, A) = t^n H\left(\frac{1}{t}, \overline{A}\right) . \tag{7.51}$$

Rewriting this for $A = SS_n$ and combining with (7.49) and (7.50) we get

$$\sum_{\sigma \in P_n} t^{f(\sigma)} = H(t, SS_n) = t^n H\left(\frac{1}{t}, \overline{SS}_n\right) = t^n E_n\left(\frac{1}{t}\right) . \tag{7.52}$$

Looking again at the diagram in Figure 7.24 we note that the strict rises are the crosses that hit the region above the diagonal squares. For $n = 7$ this is given by the shaded area in Figure 7.26.

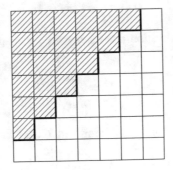

Fig. 7.26 The region for the strict rises.

We have observed (see Remark 7.5.1) that the rook numbers of a board depend only on the shape and not on the position of the board. Comparing Figure 7.26 and Figure 7.27 we see that these two boards have the same rook numbers and thus (by Theorem 7.5.3) they must also have the same hit polynomials. In other words

$$\sum_{\sigma \in P_n} t^{f(\sigma)} = \sum_{\sigma \in P_n} t^{sr(\sigma)} . \tag{7.53}$$

We shall see that this very simple observation yields a surprising fact.

Given a permutation σ, let us denote by $\uparrow \sigma$ the permutation obtained by increasing each σ_i by 1 (modulo n). This is the same as moving all the crosses of the permutation diagram of σ up one square, except the highest cross which is moved down to the bottom of its column. For instance, for the diagram in Figure 7.24 the crosses move as indicated in Figure 7.27.

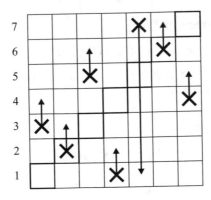

Fig. 7.27 Moving the crosses of σ with wraparound.

From this picture we can easily observe that the rises of σ become strict rises of $\uparrow \sigma$ except for the rise corresponding to the index i for which $\sigma_i = n$, that is lost all together. We may thus conclude that quite generally we have

$$sr(\uparrow \sigma) = r(\sigma) - 1 . \tag{7.54}$$

It is quite clear that the operation that takes σ to $\uparrow \sigma$ maps P_n onto itself in a one-to-one fashion, thus

$$\sum_{\sigma \in P_n} t^{sr(\sigma)} = \sum_{\sigma \in P_n} t^{sr(\uparrow \sigma)} .$$

Substituting (7.54) into this and combining with (7.53) we get

$$\sum_{\sigma \in P_n} t^{f(\sigma)} = \sum_{\sigma \in P_n} t^{r(\sigma)-1} = \frac{1}{t} E_n(t) . \tag{7.55}$$

Combining (7.55) with (7.52) we finally obtain that the Eulerian polynomials satisfy the identity

$$E_n(t) = t^{n+1} E_n\left(\frac{1}{t}\right) . \tag{7.56}$$

To understand the significance of this relation it is best to look at a special case. For instance for $n = 6$ (using (7.47)) what we have in (7.56) may be written in the form

$$e_{6,1}t + e_{6,2}t^2 + e_{6,3}t^3 + e_{6,4}t^4 + e_{6,5}t^5 + e_{6,6}t^6 =$$
$$e_{6,1}t^{7-1} + e_{6,2}t^{7-2} + e_{6,3}t^{7-3} + e_{6,4}t^{7-4} + e_{6,5}t^{7-5} + e_{6,6}t^{7-6} .$$

Equating coefficients of t^2 yields

$$e_{6,2} = e_{6,5} = e_{6,7-2} .$$

Thus (7.56) implies in particular that

There are as many permutations of [6] with 2 rises as there are with 5 rises!

In general, by equating coefficients of t^k in (7.56) we obtain a remarkable symmetry of Eulerian numbers, namely

$$e_{n,k} = e_{n,n+1-k} \qquad \text{for } k = 1, 2, \ldots, n .$$

Another consequence of (7.55) and (7.52) is that

$$E_n(t) = t \, H(t, SS_n) .$$

Thus Theorem 7.5.3 written for $A = SS_n$ gives the formula

$$E_n(t) = t \sum_{k=0}^{n} (n - k)! \, r_k(SS_n)(t - 1)^k . \qquad (7.57)$$

This leads us to study the rook numbers of the solid staircase boards. However, another surprise awaits us here. Namely, the latter are none other than the Stirling numbers of the second kind!

More precisely, for any $n \geq k$ we have

$$r_k(SS_n) = S_{n,n-k} . \qquad (7.58)$$

To see how this comes about let us proceed a bit more generally and consider boards which like SS_n consist of n piles of squares whose lengths increase (weakly) from left to right. Since these boards resemble Ferrers diagrams, we shall call them Ferrers boards.

It will be convenient to call the lengths of the piles $a_1, a_2, \ldots, a_n$ from left to right. For instance a Ferrers board in a 7×7 square could be as given by the shaded area in Figure 7.28.

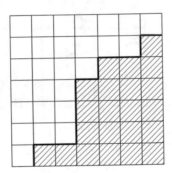

Fig. 7.28 A Ferrers board.

For the example in Figure 7.28 we have

$$a_1 = 0, \; a_2 = a_3 = 1, \; a_4 = 4, \; a_5 = a_6 = 5, \; a_7 = 6 .$$

It has been found that there is a rather surprising and speedy method for calculating the rook numbers of such a board. This follows from a clever observation of Goldman–Joichi–White [15].

We illustrate the idea on the board of Figure 7.28. To begin with let us denote by $F + x$ the figure obtained by adjoining x additional rows of n squares each to the given Ferrers board F. We have depicted in Figure 7.29 the resulting figure for the board of Figure 7.28 when $x = 9$ (here $n = 7$):

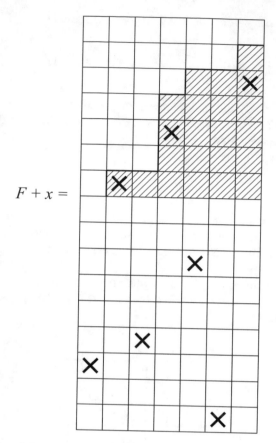

$$F + x =$$

Fig. 7.29 Ferrers board F with $x = 9$ additional rows.

The solid line at the bottom of the original figure will be referred to as the *ground*.

We have depicted in Figure 7.29 also a configuration of $n = 7$ nonattacking rooks in $F + x$. For convenience we shall refer to these configurations as *complete* (one in each column) rook configurations of $F + x$, or briefly *complete configurations*.

Note that the configuration depicted in Figure 7.29 has 4 rooks below ground and 3 rooks above ground. Moreover the partial configuration consisting of the rooks above ground is one accounted for in the calculation of $r_3(F)$.

The Goldman–Joichi–White idea consists in counting complete configurations in two different ways and thereby obtaining an equation involving the rook numbers of the original Ferrers board F.

The first count is obtained from the following construction. We imagine putting together complete configurations by placing the rooks one at a time. First rook in first column, second rook in second column (from left to right), third rook in third column, etc.

Note that (see Figure 7.29) we have $x + a_1 (= 9 + 0 = 9)$ squares where we can place the first rook. After this rook is in place, his row is no longer available to any of the remaining rooks. In particular we have only $x + a_2 - 1 (= 9 + 1 - 1 = 9)$ squares where we can place the second rook. After this rook is placed, we have two rows unavailable to the other rooks. Thus we are left with only $x+a_3-2(=9+1-2=8)$ squares for the third rook. Reasoning in this manner the number of choices is

$$x + a_4 - 3 \ (= 9+4-3 = 10) \text{ for the 4th rook,}$$

$$x + a_5 - 4 \ (= 9+5-4 = 10) \text{ for the 5th rook,}$$

$$x + a_6 - 5 \ (= 9+5-5 = 9) \text{ for the 6th rook,}$$

$$x + a_7 - 6 \ (= 9+6-6 = 9) \text{ for the 7th rook.}$$

We thus have all together

$$(x+a_1)(x+a_2-1)(x+a_3-2)(x+a_4-3)(x+a_5-4)(x+a_6-5)(x+a_7-6)$$
$$= 9 \cdot 9 \cdot 8 \cdot 10 \cdot 10 \cdot 9 \cdot 9 = 5248800$$

different choices for placing our 7 rooks, each leading to a different complete configuration.

The same reasoning applied in the general case of a subboard of the $n \times n$ board yields that the total number of complete configurations in $F + x$ is given by the product

$$(x + a_1)(x + a_2 - 1)(x + a_3 - 2) \cdots (x + a_n - n + 1) . \tag{7.59}$$

Our second count is based on a different procedure for putting together complete configurations. To be precise, now we first chose the partial configuration of rooks above the ground and then place the remaining rooks below the ground column by column as before.

Let us illustrate this procedure on the board of Figure 7.29. Say we chose to place 3 rooks above the ground as indicated in Figure 7.30.

This given, the remaining 4 rooks are to be placed in the 1st, 3rd, 6th, and 7th columns respectively.

Note that the number of squares where we may place the first (below ground) rook is x, following the same reasoning as before, we see that we have $x-1$ squares available for the second (below ground) rook. Similarly, for the 3rd and 4th (below ground) rooks we have respectively $x-2$ and $x-3$ choices. Thus the total number of ways of placing the 4 below ground rooks is given by the lower factorial polynomial evaluated at $x = 9$. That is

$$(x)_4 = x(x - 1)(x - 2)(x - 3) = 9 \cdot 8 \cdot 7 \cdot 6 = 3024 .$$

Note that whatever the partial configuration of above ground rooks is chosen to be, as long as it consists of 3 nonattacking rooks, the number of ways to place the remaining 4 rooks below ground in $F + x$ is always given by $(x)_4$.

$$F + x = $$

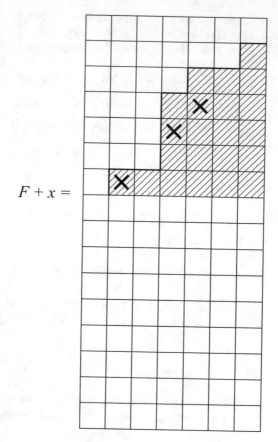

Fig. 7.30 Three rooks placed above the ground.

Since there are $r_3(F)$ ways of placing our 3 rooks above ground, the total number of complete configurations in $F + x$ having exactly 4 rooks below ground is given by the product

$$r_3(F)(x)_4 \,.$$

This reasoning can be carried out in full generality. We may conclude that the total number of complete configurations in $F + x$ with k rooks below ground is given by the product

$$r_{n-k}(F)(x)_k \,.$$

Summing this for $k = 0, 1, 2, \ldots, n$ we get that the total number of complete configurations in $F + x$ is also given by the formula

$$\sum_{k=0}^{n} r_{n-k}(F)(x)_k \, . \tag{7.60}$$

Note that for some boards it may be impossible to place too few rooks below ground. However the formula is still correct since in these cases the corresponding rook numbers are equal to zero. For instance for the board in Figure 7.28 we have

$$r_7(F) = 0 \, .$$

So in this case (7.60) reduces to

$$r_6(F)(x)_1 + r_5(F)(x)_2 + r_4(F)(x)_3 + r_3(F)(x)_4 + r_2(F)(x)_5 + r_1(F)(x)_6 + (x)_7 \, .$$

Of course, our two counts must agree. That is for any value of x the expressions in (7.59) and (7.60) must yield the same number. This observation yields the equation

$$(x+a_1)(x+a_2-1)(x+a_3-2)\cdots(x+a_n-n+1) = \sum_{k=0}^{n} r_{n-k}(F)(x)_k \, . \tag{7.61}$$

Note that for $F = SS_4$ we have

$$a_1 = 0, \ a_2 = 1, \ a_3 = 2, \ a_4 = 3 \, .$$

Thus in this case (7.61) reduces to

$$x^4 = r_3(SS_4)(x)_1 + r_2(SS_4)(x)_2 + r_1(SS_4)(x)_3 + (x)_4 \, .$$

In general, for $F = SS_n$ we obtain the identity

$$x^n = \sum_{k=1}^{n} r_{n-k}(SS_n)(x)_k \, .$$

Comparing with (6.45) of Chapter 6 we see that (7.58) must indeed hold true as asserted.

We have thus established the following two remarkable results:

Theorem 7.5.5 *For a Ferrers board F with piles of heights $a_1, a_2, \ldots, a_n$ we have*

$$\prod_{i=1}^{n}(x + a_i - i + 1) = \sum_{k=0}^{n} r_{n-k}(F)(x)_k \, . \tag{7.62}$$

Theorem 7.5.6 *The Eulerian polynomials and the Stirling numbers of the second kind are related by the formula*

$$E_n(t) = t \sum_{k=1}^{n} k! \, S_{n,k} \, (t-1)^{n-k} \, . \tag{7.63}$$

Proof Just substitute (7.58) in (7.57) and reverse direction of summation. □

Remark 7.5.2 Equation (7.63) may also be rewritten in the form

$$\frac{E_n(t)}{(1-t)^n} = \sum_{k=1}^{n} k! \, S_{n,k} \, \frac{t^k}{(1-t)^k} \, . \tag{7.64}$$

In this form it is usually referred to as *Frobenius* formula.

To obtain (7.64) from (7.63) simply divide by $(t-1)^n$, replace t by $1/t$ and use the identity in (7.56).

It is to be noted that a formula similar to (7.64) holds for any Ferrers board F. To see this, start with the identity (Theorem 7.5.3)

$$H(t, F) = \sum_{k=0}^{n} (n-k)! \, r_k(F)(t-1)^k \, ,$$

divide by $(t-1)^n$, and reverse summation to obtain

$$\frac{H(t, F)}{(t-1)^n} = \sum_{k=0}^{n} k! \, r_{n-k}(F) \frac{1}{(t-1)^k} \, .$$

Upon replacing t by $1/t$ and using (7.51) we get

$$\frac{H(t, \overline{F})}{(1-t)^n} = \sum_{k=0}^{n} k! \, r_{n-k}(F) \frac{t^k}{(1-t)^k} \, . \tag{7.65}$$

and this is what (7.64) looks like for general Ferrers boards.

At this point we have all the tools we need to obtain the exponential generating function of the Eulerian polynomials.

Our point of departure is the following identity:

Theorem 7.5.7 *For any Ferrers board F with piles of heights $a_1, a_2, \ldots, a_n$, we have*

$$\frac{H(t, \overline{F})}{(1-t)^{n+1}} = \sum_{m \geq 0} t^m \prod_{i=1}^{n} (m + a_i - i + 1) \, . \tag{7.66}$$

Proof Formula (7.62) with x replaced by an integer $m \geq 0$ may be written in the form

$$\prod_{i=1}^{n}(m + a_i - i + 1) = \sum_{k=0}^{n} k! \, r_{n-k}(F)\binom{m}{k} . \tag{7.67}$$

Note that from the binomial theorem we get

$$\sum_{m \geq 0}\binom{m}{k} t^m = \frac{t^k}{(1-t)^{k+1}} .$$

This given, if we multiply (7.67) by t^m and sum, we do get

$$\sum_{m \geq 0} t^m \prod_{i=1}^{n}(m + a_i - i + 1) = \sum_{k=0}^{n} k! \, r_{n-k}(F)\frac{t^k}{(1-t)^{k+1}} . \tag{7.68}$$

Thus (7.66) is obtained by combining the right hand sides of (7.68) and (7.65). $\square$

Our final result here may be stated as follows:

Theorem 7.5.8

$$1 + \sum_{n \geq 1}\frac{x^n}{n!}E_n(t) = \frac{1-t}{1 - te^{(1-t)x}} . \tag{7.69}$$

Proof Writing (7.66) with $F = SS_n$ and using (7.49) we get

$$\frac{E_n(t)}{(1-t)^{n+1}} = \sum_{m \geq 1} t^m m^n . \tag{7.70}$$

Setting for convenience $E_0(t) = t$, we see that this formula remains true even for $n = 0$. This given, multiplying (7.70) by $\frac{y^n}{n!}$ and summing, we get

$$\sum_{n \geq 0}\frac{y^n}{n!}\frac{E_n(t)}{(1-t)^{n+1}} = \sum_{m \geq 1} t^m e^{my} = \frac{te^y}{1 - te^y} .$$

This gives

$$1 + \frac{t}{1-t} + \sum_{n \geq 1}\frac{y^n}{n!}\frac{E_n(t)}{(1-t)^{n+1}} = 1 + \frac{te^y}{1 - te^y} ,$$

or better

$$\frac{1}{1-t} + \sum_{n \geq 1} \frac{y^n}{n!} \frac{E_n(t)}{(1-t)^{n+1}} = \frac{1}{1-te^y} ,$$

and (7.69) follows upon making the substitution $y/1-t = x$ and multiplying both sides by $1-t$. $\square$

7.5.4   Descents and Ascents of Permutations

Let

$$\sigma = \begin{pmatrix} 1 & 2 & \dots & n \\ \sigma_1 & \sigma_2 & \dots & \sigma_n \end{pmatrix}$$

be a permutation. An index $i \leq n-1$ is said to be a *descent* of σ if

$$\sigma_i > \sigma_{i+1} .$$

Similarly we say that i is an *ascent* of σ if

$$\sigma_i < \sigma_{i+1} .$$

Descents and ascents behave very much like falls and rises. Indeed if $d(\sigma)$ denotes the number of descents of σ we have

$$\sum_{\sigma \in P_n} t^{d(\sigma)} = \frac{1}{t} E_n(t) . \tag{7.71}$$

A completely analogous result holds for the number of ascents of σ.

The identity in (7.71) is proved by constructing a transformation $\sigma \to \sigma^*$ with maps P_n onto itself in a one-to-one fashion and converts falls of σ into descents of σ^*. We give a brief sketch of the idea.

Let

$$\sigma = \begin{pmatrix} 1 & 2 & 3 & 4 & 5 & 6 \\ 4 & 2 & 6 & 1 & 3 & 5 \end{pmatrix}$$

and construct the cycle diagram of σ as shown in Figure 7.31.

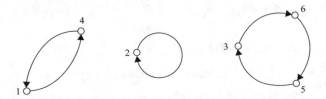

Fig. 7.31 The cycle diagram of σ.

Now write the individual cycles of σ starting from the highest element and arrange them in order of increasing largest elements. This gives the cycle factorization

$$\sigma = (2)\,(41)\,(653)\,.$$

Let now σ be the permutation of [6] obtained by *removing the parentheses*. That is

$$\sigma^* = 2\,4\,1\,6\,5\,3\,.$$

It is easy to see that the falls of σ are precisely the descents of σ^*. This is due to the fact that from this construction the only pairs i, σ_i which do not occur as adjacent elements of σ^* are those for which σ_i is the highest element of a cycle. Thus in going from σ to σ^* we lose only rises.

We should be concerned that in removing parentheses we might produce descents of σ^* which are not falls of σ. However, this can never happen since the first element of a cycle is bigger (by the way we order the cycles) than every element of a previous cycle (including the last). This means that removal of parentheses always produces ascents of σ^*, and never descents.

We thus get

$$f(\sigma) = d(\sigma^*)\,. \tag{7.72}$$

It is not difficult to show that the map $\sigma \rightarrow \sigma^*$ is reversible and thus obtain that it is a one-to-one map of P_n onto itself. This gives

$$\sum_{\sigma \in P_n} t^{d(\sigma^*)} = \sum_{\sigma \in P_n} t^{d(\sigma)}$$

and (7.72) yields

$$\sum_{\sigma \in P_n} t^{f(\sigma)} = \sum_{\sigma \in P_n} t^{d(\sigma)}$$

and this proves (7.71).

7.6 Exercises for Chapter 7

7.6.1 List all derangements of [4].

7.6.2 Use Theorem 7.3.3 to calculate the numbers D_n up to $n = 9$.

7.6.3 Calculate the number of integers less than or equal to 315 and relatively prime to it.

7.6.4 How many integers between 100 and 1000 inclusive are relatively prime to 100?

7.6.5 How many integers between 1 and 10000 inclusive are divisible by none of 3, 5, or 7?

7.6.6 Calculate the number of integers from 1 to 1000 inclusive which are neither perfect squares nor perfect cubes.

7.6.7 How many integers between 1 and 10000 inclusive are there which are neither perfect squares, perfect cubes, nor perfect fourth powers?

7.6.8 Show that the number of positive integers less than or equal to $30n$ $(n \geq 1)$ which are divisible by none of 2, 3, 5 is $8n$.

7.6.9 Find a formula for the number of integers between 1 and $210n$ inclusive which are divisible by none of the primes 2, 3, 5, 7.

7.6.10 Calculate the number of permutations of [6] which have

(a) exactly one fixed point,
(b) at least 3 fixed points,
(c) no more than 4 fixed points.

7.6.11 Show that for $n > 1$, at least a third of all permutations of [n] are derangements.

7.6.12 Suppose two decks of cards are individually shuffled. If we compare the decks one by one, what is the probability that we shall go right through the decks without obtaining a single identical pair of cards?

7.6.13 Give a formula for the number of permutations of [n] which have exactly k fixed points.

7.6.14 Calculate the number of permutations of [n] which have

(a) at least k fixed points,
(b) at most k fixed points.

7.6.15 Show that

$$1 + \sum_{n \geq 1} \frac{t^n}{n!} D_n = \frac{e^{-t}}{1 - t} .$$

7.6.16 Prove the identity

$$n! = \sum_{k=0}^{n} \binom{n}{k} D_k \qquad\qquad (D_0 = 1)$$

in two different ways:

(a) by a counting argument,
(b) by generating function methods.

7.6.17 Suppose $w \in \{x_1, x_2, \ldots, x_n\}^*$. Call x_i a *fixed point* of w if the ith letter of w is x_i.

(a) What is the number of words of length m with no fixed points?
(b) How many words of length m are there with exactly k fixed points?

7.6.18 Let Ω denote the collection of all rearrangements of the word *AAABB-BCCC*.

(a) Calculate the number of words in Ω in which none of *AAA, BBB, CCC* appears.
(b) How many words in Ω are there in which no two adjacent letters are identical?

7.6.19 Calculate the number of lattice paths (East and North unit steps) from A to B in the following mesh:

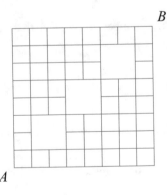

7.6.20 Calculate the number of integral solutions to

$$x_1 + x_2 + x_3 + x_4 = 50$$

with

(a) $0 \le x_i \le 15$,
(b) $5 \le x_i \le 15$,

for $i = 1, 2, 3, 4$.

7.6.21 Find a general formula for the number of integral solutions to

$$x_1 + x_2 + \cdots + x_k = n,$$

where

(a) $0 \le x_i \le s$,
(b) $r \le x_i \le s$,

for $i = 1, 2, \ldots, k$.

7.6.22 Calculate the number of permutations σ of $[4]$ such that $\sigma_{i+1} \ne 1 + \sigma_i$ for $i = 1, 2, 3$.

7.6.23 Find a formula for the number of permutations σ of $[n]$ $(n \ge 4)$ in which $\sigma_{i+1} \ne 1 + \sigma_i$, for $i = 1, 2, 3$.

7.6.24 Let $X = \{x_1, x_2, \ldots, x_m, y_1, y_2, \ldots, y_m\}$. Calculate the number of words length n over X in which none of the patterns $x_i y_i$ occurs, $i = 1, 2, \ldots, m$.

7.6.25 In the general setting of inclusion-exclusion,

(a) show that the number O of objects satisfying an odd number of properties is

$$O = \sum_{k=1}^{n} (-2)^{k-1} a_k .$$

(b) Similarly show that the number of objects that satisfy an even number of properties is

$$E = a_0 - \sum_{k=1}^{n} (-2)^{k-1} a_k \qquad (= |\Omega| - O)) .$$

(c) For the coloring problem presented in Section 7.1, show that $O = 42$ and $E = 48$.

7.6.26 Show that the number of permutations of $[n]$ with an odd number of fixed points is

$$\sum_{k=1}^{n} (-2)^{k-1} \binom{n}{k} (n - k)! .$$

Conclude that

$$\sum_{k \ge 0} \binom{n}{2k + 1} D_{n-2k-1} = \sum_{k=1}^{n} (-2)^{k-1} \frac{n!}{k!} .$$

7.6.27

(a) Calculate the number of rearrangements of $w = AABBCCDD$ in which no two adjacent letters are identical.

(b) How many rearrangements of w are there with exactly one pair of adjacent identical letters? With exactly 2 pairs of adjacent identical letters?

(c) Calculate the number of rearrangements of w with an even number of pairs of adjacent identical letters.

7.6.28 Put $A = \{a_1, a_2, \ldots, a_n\}$ and let $\mathcal{L} \subseteq A^*$ denote all words with no repeated letters. Set

$$p_n = |\mathcal{L}| .$$

Show that

$$\lim_{n \to \infty} \frac{p_n}{D_n} = e^2.$$

7.6.29 Show that the Bell numbers B_n and the derangement numbers D_n ($D_0 = 1$) are related by

$$B_n = \frac{1}{n!} \sum_{k=0}^{n} (n - k)^n \binom{n}{k} D_k.$$

7.6.30 Let $X = \{x_1, x_2, \ldots, x_k\}$ and suppose $\mathcal{L} \subseteq X^*$ consists of all words in which each x_i appears at least once. Show that for $n \geq k$,

$$|\mathcal{L}_n| = \sum_{i=1}^{k} (-1)^i \binom{k}{i} (k - i)^n .$$

7.6.31 Take X as in problem 7.6.30. Calculate $|\mathcal{L}_n|$ where $\mathcal{L} \subseteq X^*$ consists of all words in which each x_i appears at least twice.

7.6.32 Calculate the number of functions mapping $[n]$ onto $[k]$.

7.6.33 Let O_n and E_n denote the number of derangements of $[n]$ with an odd and an even number of cycles, respectively.

(a) Use the recursion (7.14) (b) for D_n and remark 7.3.1 to derive the recursions

$$O_{n+1} = nE_{n-1} + nO_n$$

$$E_{n+1} = nE_n + nO_{n-1}.$$

(b) Put $g_n = O_n - E_n$. Show that

$$g_{n+1} = ng_n - ng_{n-1} .$$

Conclude that $g_n = n - 1$ for $n \geq 1$.

(c) Derive the formulas

$$O_n = \frac{1}{2}(D_n + n - 1)$$

$$E_n = \frac{1}{2}(D_n + 1 - n).$$

(d) Use part (c) to show that D_n is odd if and only if n is even.

7.6.34 Let $\phi(N)$ denote the number of integers less than or equal to N, relatively prime to N as given by Theorem 7.3.1. (This function is referred to as the Euler ϕ-function or the Euler *totient* function).

(a) Verify that if

$$N = p_1^{d_1} p_2^{d_2} \cdots p_n^{d_n},$$

where $p_1, p_2, \ldots, p_n$ are distinct primes, then

$$\phi(N) = \phi(p_1^{d_1})\phi(p_2^{d_2}) \cdots \phi(p_n^{d_n}).$$

(b) Show that

$$\sum_{d \mid N} \phi(d) = N,$$

where the summation is over all divisors of N.

7.6.35 Use Theorem 7.3.1 to verify that for $m > 0$

$$\phi(n^m) = n^{m-1}\phi(n).$$

7.6.36 Show that for $p, q > 0$,

$$\phi(n^p(n+1)^q) = n^{p-1}(n+1)^{q-1}\phi(n)\phi(n+1).$$

7.6.37 Let $p_1, p_2, \ldots$ be the sequence of prime numbers. Show that the number of positive integers less than or equal to n which are products of distinct primes is

$$\sum_T (-1)^{|T|} \left\lfloor \frac{n}{\prod_{i \in T} p_i^2} \right\rfloor,$$

where the summation is over all finite subsets T of $\{1, 2, 3, \ldots\}$ and the brackets denote the greatest integer function.

7.6.38 Let F_n denote all partitions of n. Put $m = \lfloor \frac{n}{2} \rfloor$ and for $i = 1, 2, \ldots, m$ set

$$A_i = \{\pi \in F_n \mid \pi \text{ has at least 2 parts of size } i\}$$
$$B_i = \{\pi \in F_n \mid \pi \text{ has at least one part of size } i\}$$

(a) For any $T \subseteq [m]$ construct a bijection between the sets

$$\bigcap_{i \in T} A_i \quad \text{and} \quad \bigcap_{i \in T} B_i .$$

(b) Use the inclusion-exclusion principle to show that the partitions of n into distinct parts is equinumerous with partitions of n into odd parts.

7.6.39 Use inclusion-exclusion to prove that the number of partitions of n where no part is divisible by d is equal to the number of partitions where each part appears at most $d - 1$ times.

7.6.40

(a) Calculate the number of permutations σ of $[4]$ which satisfy

$$\sigma_1 \neq 2$$
$$\sigma_2 \neq 1$$
$$\sigma_3 \neq 2$$
$$\sigma_4 \neq 4.$$

(b) Use Theorem 7.4.1 to construct the polynomial $P(t)$ defined in (7.20).

7.6.41 How many permutations σ of $[4]$ are there with

$$\sigma_1 \neq 2$$
$$\sigma_2 \neq 1$$
$$\sigma_3 \neq 4$$
$$\sigma_4 \neq 3 ?$$

7.6.42 Let τ be a fixed permutation of $[n]$. Show that the number of permutations σ such that $\sigma_i \neq \tau_i$ for $i = 1, 2, \ldots, n$ is D_n.

7.6.43 Let $a_0(x_1, x_2, \ldots, x_n) = 1$ and for $0 < k \leq n$ put

$$a_k(x_1, x_2, \ldots, x_n) = \sum_T \prod_{i \in T} x_i,$$

where the summation is over all $T \subseteq [n]$ with $|T| = k$. For instance for $n = 3$ we have

$$a_1(x_1, x_2, x_3) = x_1 + x_2 + x_3$$

$$a_2(x_1, x_2, x_3) = x_1x_2 + x_1x_3 + x_2x_3$$

$$a_3(x_1, x_2, x_3) = x_1x_2x_3 .$$

Show that a_k is a symmetric function: i.e., for each permutation σ of $[n]$

$$a_k(x_{\sigma_1}, x_{\sigma_2}, \ldots, x_{\sigma_n}) = a_k(x_1, x_2, \ldots, x_n) .$$

The a_ks are called the *elementary* symmetric functions in $x_1, x_2, \ldots, x_n$.

7.6.44 Verify the following properties of the elementary symmetric functions:

(a) $a_k(1, 1, \ldots, 1) = \binom{n}{k}$

(b) $a_k(x_1, x_2, \ldots, x_{n-1}, 0) = a_k(x_1, x_2, \ldots, x_{n-1})$ for $k = 0, 1, \ldots, n - 1$

(c) $\displaystyle\sum_{k=0}^{n} a_k(x_1, x_2, \ldots, x_n)t^k = (1 + x_1t)(1 + x_2t) \cdots (1 + x_nt)$

(d) $a_k(x_1, x_2, \ldots, x_n) = a_k(x_1, x_2, \ldots, x_{n-1}) + x_n a_{k-1}(x_1, x_2, \ldots, x_{n-1})$.

7.6.45 Suppose $w = w_1w_2 \cdots w_n$ is an alphabetic word over the alphabet $[n]$ in which the letter i appears x_i times, $i = 1, 2, \ldots, n$. Let $D_n(x_1, x_2, \ldots, x_n)$ denote the number of permutations σ of $[n]$ with $\sigma_i \neq w_i$ for all $i = 1, 2, \ldots, n$.

(a) Show that

$$D_n(x_1, x_2, \ldots, x_n) = \sum_{k=0}^{n}(-1)^k a_k(x_1, x_2, \ldots, x_n)(n - k)!$$

where a_k is the kth elementary symmetric function in the variables $x_1, x_2, \ldots, x_n$.

(b) Verify that $D_n(1, 1, \ldots, 1)$ is the derangement number D_n.

7.6.46 Show that the assumption that w is alphabetic is not necessary in problem 7.6.45.

7.6.47 Calculate the number of permutations $\sigma = \sigma_1\sigma_2 \cdots \sigma_n$ with $\sigma_i \neq w_i$ ($i = 1, 2, \ldots, n$) where $w = w_1w_2 \cdots w_n$ is given as below:

(a) $w = 1133$
(b) $w = 2244666$
(c) $w = 7554754$

7.6.48 Calculate the number of falls, rises, and strict rises of the following permutations:

(a) $\sigma = 3567214$
(b) $\sigma = 7536421$
(c) $\sigma = 4761532$

7.6.49 For each of the permutations σ in problem 7.6.48, construct the corresponding permutation σ^* such that $f(\sigma) = d(\sigma^*)$.

7.6.50 Calculate the number of descents and the number of ascents of the following permutations:

(a) $\sigma^* = 3251764$
(b) $\sigma^* = 4321756$
(c) $\sigma^* = 1726354$.

7.6.51 For each of the permutations σ^* in problem 7.6.50, construct the corresponding permutation σ such that $d(\sigma^*) = f(\sigma)$.

7.6.52 Calculate the number of ways of seating 6 couples at a round table so that no husband is next to his wife.

7.6.53 How many rearrangements of the word *AAABBCC* are there in which none of *AAA, BB, CC* appears?

7.6.54 Calculate the number of words of length m over the alphabet $\{x_1, x_2, \ldots, x_n\}$ in which no x_i appears exactly once, $i = 1, 2, \ldots, n$.

7.6.55 The following dartboard is to be painted with three colors B, R, and Y, using each color twice:

(a) Calculate the number of colorings where no two adjacent wedges are given the same color.
(b) What are the numbers a_k for this problem, $k = 0, 1, 2, 3$?
(c) Construct the generating function of the numbers e_k, $k = 0, 1, 2, 3$.
(d) How many colorings are there in which exactly one pair of adjacent wedges is colored the same?
(e) How many colorings are there in which exactly two pairs of adjacent wedges are given the same color?

7.6.56

(a) Show that

$$\sum_{k=0}^{n} l_k = P(1) + P'(1),$$

where $l_k = e_k + e_{k+1} + \cdots + e_n$ and $P(t)$ is the polynomial defined in (7.20) (see Theorem 7.4.2).

(b) Use Theorem 7.4.1 to obtain the formulas

$$a_0 + a_1 = \sum_{k=0}^{n} (k+1)e_k$$

$$a_1 = \sum_{k=1}^{n} k e_k .$$

7.6.57

(a) Use Theorem 7.4.1 to show that

$$\sum_{k=0}^{n} a_k t^k = \sum_{k=0}^{n} e_k (t+1)^k .$$

(b) Derive the formula

$$a_m = \sum_{k=m}^{n} \binom{k}{m} e_k .$$

7.6.58 Construct the rook polynomials and the hit polynomials of the following boards:

(a) 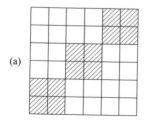   (b) 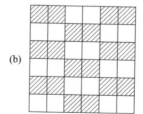   (c)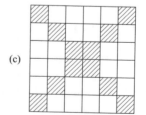

7.6.59 Calculate the number of permutations of [6] which avoid the shaded regions in problem 7.6.58.

7.6.60

(a) Calculate the number of ways of placing k nonattacking rooks on the black squares of an 8×8 chessboard.

(b) Construct $R(x, A)$ where A consists of the black squares of an 8×8 chessboard as in part (a).

(c) Use Theorem 7.5.3 to construct the hit polynomial of A.

(d) How many permutations of [8] are there which avoid A?

7.6.61 Calculate the number of permutations $\sigma = \sigma_1 \sigma_2 \sigma_3 \sigma_4 \sigma_5 \sigma_6$ in which

$$\sigma_1 \neq 2 \text{ or } 4,$$
$$\sigma_2 \neq 1 \text{ or } 5,$$
$$\sigma_4 \neq 2 \text{ or } 4,$$
$$\sigma_5 \neq 6,$$
$$\sigma_6 \neq 1 .$$

7.6.62

(a) Construct the hit polynomial of the following $n \times n$ board:

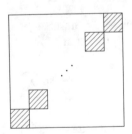

(b) Use Theorem 7.5.3 to derive the formula (7.13) for the number of derangements.

7.6.63 Calculate the number of rearrangements of *AABBCC* such that A is not in the 1st position, B is not in the 4th or the 5th position, and C is not in the 3rd position.

7.6.64 Calculate the number of rearrangements of the word *ABCDEFG* in which *BAD* and *DEC* do not appear.

7.6.65 Let $X = \{a, b\}$ and $\mathcal{L} \subseteq X^*$ consist of all words in which *aaa* and *bbb* do not appear. Use inclusion-exclusion to calculate $|\mathcal{L}_n|$.

7.6.66 Let $\mathcal{L} \subseteq \{x_1, x_2, x_3\}^*$ denote the language of all words in which x_1 appears an even number of times. Use the inclusion-exclusion principle to show that

$$|\mathcal{L}_n| = \frac{3^n + 1}{2}.$$

7.6.67 Let p_n denote the number of words in $\{x_1, x_2, \ldots x_n\}$ with no repeated letters. Show that

$$\sum_{k=0}^{n} \binom{n}{k} p_k D_{n-k} = (n+1)!$$

where $D_0 = 1$.

7.6.68 Let X be the alphabet $\{a, b\}$ and $\mathcal{L} \subseteq X^*$ the language of words in which the pattern aa does not appear.

(a) Show that $\mathcal{L}$ satisfies

$$\mathcal{L} = 1 + a + b\mathcal{L} + ab\mathcal{L}.$$

(b) Put

$$f(x, t) = \sum_{w \in \mathcal{L}} t^{|w|_a} x^{|w|_b},$$

where $|w|_a$ and $|w|_b$ denote the number of occurrences of a and b in w, respectively. Use part (a) to prove that

$$f(x, t) = \frac{1 + t}{1 - x - xt} = \sum_{m \geq 0} x^m (1 + t)^{m+1}.$$

(c) Argue that the coefficient of the term $t^k x^{n-k}$ in $f(x, t)$ is the number of k-element subsets of $[n]$ which contain no consecutive integers.

(d) Derive the formula for $r_k(S_n)$ given in (7.42).

7.6.69 Let X and $\mathcal{L}$ be as in problem 7.6.68.

(a) Verify that

$$g(t) = \sum_{w \in \mathcal{L}} t^{|w|} = \frac{1 + t}{1 - t - t^2}.$$

(b) Use generating function methods to show that

$$g(t)\bigg|_{t^n} = f_{n+1},$$

where f_n in the nth Fibonacci number.

(c) Derive the identity

$$f_{n+1} = \sum_{k=0}^{n} \binom{n-k+1}{k}.$$

7.6.70 Show that the number of k-element subsets of $[n]$ in which no two elements are consecutive or differ by 2 is

$$\binom{n-2k+2}{k}.$$

7.6.71 Show that the number of permutations of $[n]$ which hit a forbidden region A at exactly k places is

$$\sum_{i=k}^{n} (n-i)! r_i(A)(-1)^{i-k} \binom{i}{k}.$$

7.6.72

(a) Verify directly that the rook polynomial of a full $n \times n$ board is

$$\sum_{k=0}^{n} (n)_k^2 \frac{x^k}{k!}.$$

where $(n)_k = n(n-1)\cdots(n-k+1)$.

(b) What is the hit polynomial of an $n \times n$ board?

7.6.73 Use problems 7.6.71 and 7.6.72 to prove the identity

$$\sum_{i=k}^{n} \binom{n}{i}\binom{i}{k}(-1)^i = 0$$

for $k < n$.

7.6.74 Show directly that the number of ways of placing n nonattacking rooks in a Ferrers board with piles of heights $a_1, a_2, \ldots, a_n$ is

$$a_1(a_2 - 1)(a_3 - 2)\cdots(a_n - n + 1).$$

7.6.75 Suppose F is a Ferrers board with piles of heights $a_1, a_2, \ldots, a_n$. Use Theorem 7.5.5 to show that if $a_i = i$ for some i, $1 \le i \le n$ then

$$0 = \sum_{k=0}^{n} r_k(F)(n-k)!(-1)^k .$$

What is the combinatorial interpretation of this identity?

7.6.76 Verify that the following two boards have the same rook polynomials even though they are not equivalent.

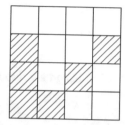 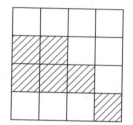

7.6.77 Two words $w = w_1 w_2 \cdots w_n$ and $v = v_1 v_2 \cdots v_n$ of length n over an alphabet are called *discordant* if $w_i \ne v_i$ for $i = 1, 2, \ldots, n$. For example, derangements of $[n]$ are precisely the permutations discordant with $1\,2\,3 \cdots n$.

(a) Calculate the number of permutations of [4] discordant with $1\,2\,3\,4$ and $2\,3\,4\,1$.
(b) Calculate the number of permutations of [5] discordant with $2\,1\,3\,4\,5$ and $1\,2\,4\,3\,5$.

7.6.78 Find the number of rearrangements of the word *AABBCC* discordant with *CCBBAA*.

7.6.79 Calculate the number of rearrangements of *AABBCC* discordant with *AAABBB* and *ACACBB*.

7.6.80 State the ménage problem in terms of discordant permutations.

7.6.81 Calculate the number of permutations of $[n]$ with an odd number of rises.

7.6.82 Let $\mathcal{F}_{m,n}$ denote the collection of functions mapping $[m]$ into $[n]$. We say that i is a *rise* of $g \in \mathcal{F}_{m,n}$ if $i \le g(i)$ and a *fall* of g if $i > g(i)$. Denote by $R(g)$ and $F(g)$ the number of rises and falls of g, respectively. Show that

(a)

$$\sum_{g \in \mathcal{F}_{n,n}} t^{R(g)} = nt((n-1)t+1)((n-2)t+2) \cdots (t+n-1) .$$

(b)

$$\sum_{g \in \mathcal{F}_{n,n}} x^{F(g)} = n(x+n-1)(2x+n-2) \cdots ((n-1)x+1) .$$

(c) More generally, show that

$$\sum_{g \in \mathcal{F}_{n,n}} t^{R(g)} x^{F(g)} = nt((n-1)t+x)((n-2)t+2x) \cdots (t+(n-1)x) .$$

(d) Construct the analogous generating functions for $\mathcal{F}_{m,n}$ when $m \neq n$.

7.6.83 Show that exactly half of the functions in $\mathcal{F}_{2n,2n}$ have an even number of rises.

7.6.84 Calculate the number of functions in $\mathcal{F}_{2n,2n}$ which have an odd number of falls.

7.6.85 Show that the permutation diagram of σ^{-1} can be obtained from the permutation diagram of σ by reflecting it across the secondary diagonal.

7.6.86 Prove the following:

(a) $sr(\sigma^{-1}) = f(\sigma)$
(b) $r(\sigma^{-1}) + sr(\sigma) = n$
(c) $r(\sigma^{-1}) - f(\sigma) = \#\{i \mid \sigma_i = i\}.$

7.6.87 Prove that

$$\sum_{\sigma \in P_n} t^{f(\sigma)+f(\sigma^{-1})} = \sum_{k=0}^{n} \binom{n}{k} D_k t^k,$$

where D_k is the number of derangements of $[k]$ with $D_0 = 1$.

7.6.88 Use Theorem 7.5.7 to construct the hit polynomials of the following boards:

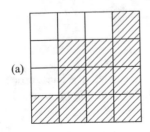

(a)

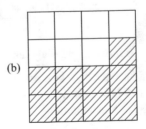

(b)

7.6.89 Construct the rook polynomials of the following boards:

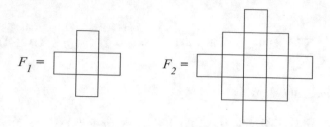

7.6.90 Let F_n denote the diamond shaped figure similar to the ones in problem 7.6.89 where the middle row consists of $2n + 1$ cells.

(a) Use Remark 7.5.1 and Theorem 7.5.5 to show that

$$\sum_{k=0}^{2n} r_{2n-k}(F_n)(x)_{k+1} = x^n(x+1)^{n+1} .$$

(b) Derive the formula

$$r_k(F_n) = \sum_{j=0}^{n+1} \binom{n+1}{j} S_{n+j, 2n+1-k} .$$

7.6.91 Find the number of ways of placing k nonattacking bishops on the black squares of the following board:

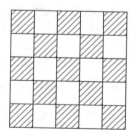

(The bishops move along diagonals).

7.6.92 Consider the $(2n+1) \times (2n+1)$ board shaded in a checkerboard fashion as in problem 7.6.91. In how many ways can k nonattacking bishops be placed on the black squares? White squares?

7.6.93 Let $R_{n,m}(x)$ denote the rook polynomial of an $n \times m$ board. Prove the following:

(a) $R_{n,m}(x) = R_{n-1,m}(x) + mx R_{n-1,m-1}(x)$

(b) $\frac{d}{dx} R_{n,m}(x) = nm R_{n-1,m-1}(x)$.

7.6.94 Let A_n denote the $n \times n$ square board.

(a) Use Theorem 7.5.5 to show that

$$(x + n + 1) \sum_{k=0}^{n} r_{n-k}(A_n)(x)_k = \sum_{k=0}^{n+1} r_{n+1-k}(A_{n+1})(x)_k .$$

(b) Derive the recursion

$$r_k(A_{n+1}) = r_k(A_n) + (2n + 2 - k)r_{k-1}(A_n)$$

(Hint: $(x + y)(x)_k = (x)_{k+1} + (y + k)(x)_k$).

(c) Show that part (b) implies

$$R(x, A_{n+1}) = (1 + (2n + 1)x)R(x, A_n) - x^2 \frac{d}{dx} R(x, A_n) .$$

(d) Finally, use problem 7.6.93 part (b) to show that

$$R(x, A_{n+1}) = (1 + (2n + 1)x)R(x, A_n) - n^2 x^2 R(x, A_{n-1}) .$$

7.6.95 Put

$$E_n(t) = \sum_{k=1}^{n} e_{n,k} t^k.$$

(a) Give a combinatorial proof of the recursion

$$e_{n,k} = k e_{n-1,k} + (n - k + 1)e_{n-1,k-1}.$$

(Hint: use (7.71)).

(b) Show that the Eulerian polynomials satisfy

$$E_n(t) = nt E_{n-1}(t) + (t - t^2) \frac{d}{dt} E_{n-1}(t) .$$

7.6.96 Verify that

$$E_4(t) = t + 11t^2 + 11t^3 + t^4$$

$$E_5(t) = t + 26t^2 + 66t^3 + 26t^4 + t^5 .$$

7.6.97 Calculate $E_6(t)$ using problem 7.6.95 part (b).

7.6.98 Verify that the Eulerian polynomials satisfy

$$E_n(-1) = \sum_{k=1}^{n} (-1)^{k+1} 2^{n-k} k! S_{n,k} .$$

7.6.99 Use Theorem 7.5.8 to show that

$$\sum_{n \geq} \frac{x^n}{n!} (-1)^n E_n(-1) = \tanh(x) .$$

7.6.100 Let $D_n(t)$ denote the hit polynomial of the shaded region in the following $n \times n$ board:

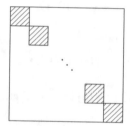

Show that

$$D_n(t) = \text{per} \begin{bmatrix} t & 1 & 1 & . & 1 \\ 1 & t & 1 & . & 1 \\ 1 & 1 & t & . & 1 \\ . & . & . & . & . \\ 1 & 1 & 1 & . & t \end{bmatrix} .$$

In particular the derangement numbers are given by

$$D_n = \text{per} \begin{bmatrix} 0 & 1 & 1 & . & 1 \\ 1 & 0 & 1 & . & 1 \\ 1 & 1 & 0 & . & 1 \\ . & . & . & . & . \\ 1 & 1 & 1 & . & 0 \end{bmatrix} .$$

The permanent of a square matrix is computed by summing the terms over all permutations just like the determinant, but without using the sign of the permutation as the coefficient of the resulting term.

7.6.101 Let A be a subset of the squares of an $n \times n$ board. Construct an $n \times n$ matrix $M_A = [a_{i,j}]$ by setting

$$a_{i,j} = \begin{cases} t & \text{if } (i, j) \in A, \\ 1 & \text{otherwise.} \end{cases}$$

Show that the hit polynomial of A is given by

$$H(t, A) = \operatorname{per} M_A .$$

7.7   Sample Quiz for Chapter 7

1. Calculate the number of nonnegative integers less than or equal to 300 and relatively prime to it.

2. Use inclusion-exclusion to calculate the number of permutations σ of [4] such that

$$\sigma_1 \neq 1$$
$$\sigma_2 \neq 1$$
$$\sigma_3 \neq 3$$
$$\sigma_4 \neq 4.$$

3. Compute the number of lattice paths (East and North unit steps) from A to B in the following mesh:

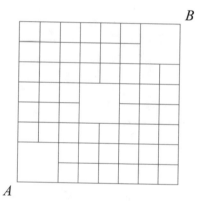

4. The wedges of the following dartboard

are to be colored with three colors R, B, Y using each color twice in such a way that no two adjacent wedges are given the same color.

(a) In how many ways can this be done?

(b) Construct the polynomial

$$e_0 + e_1 t + e_2 t^2 + e_3 t^3 \, .$$

(c) Calculate the number of colorings in which exactly two pairs of adjacent wedges are given the same color.

Chapter 8
Graphs, Chromatic Polynomials, and Acyclic Orientations

8.1 Graphs

A graph is usually represented as a set of points of the plane or space joined in pairs by a certain collection of arcs. The points are called *vertices,* and the arcs (although need not be drawn straight) are usually referred to as *edges.*

Thus for instance Figure 8.1 represents the graph with vertex set

$$V = \{1, 2, 3, 4, 5\} \tag{8.1}$$

and edge set

$$E = \{\, \{1, 2\}, \{1, 3\}, \{1, 4\}, \{2, 3\}, \{2, 4\}, \{4, 5\} \,\}\,. \tag{8.2}$$

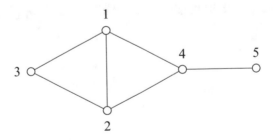

Fig. 8.1 A graph with 5 vertices and 6 edges.

Geometric representations of graphs can often be misleading. For instance, the graph in Figure 8.2 represents the same graph as that represented by Figure 8.1. Indeed, both Figures 8.1 and 8.2 yield the same vertex set and same edge set, and these are the only two things that matter in a graph.

© Springer Nature Switzerland AG 2021
Ö. Eğecioğlu, A. M. Garsia, *Lessons in Enumerative Combinatorics*, Graduate Texts in Mathematics 290, https://doi.org/10.1007/978-3-030-71250-1_8

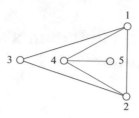

Fig. 8.2 Same graph as the one in Figure 8.1.

Thus, abstractly speaking a graph should be thought of as a pair $G = (V, E)$ with V a set and E a collection of pairs of elements of V. More precisely, we can represent a graph G on the vertex set $V = \{1, 2, \ldots, n\}$ by an $n \times n$ matrix $[a_{i,j}]$ where we set $a_{i,j} = 1$ or 0 according as the edge $\{i, j\}$ is in E or not. The matrix $[a_{i,j}]$ thus obtained is usually referred to as the *adjacency matrix* or *incidence matrix* of G. We shall denote it by $A(G)$.

For instance, the graph depicted in Figure 8.1 has adjacency matrix

$$A(G) = \begin{bmatrix} 0 & 1 & 1 & 1 & 0 \\ 1 & 0 & 1 & 1 & 0 \\ 1 & 1 & 0 & 0 & 0 \\ 1 & 1 & 0 & 0 & 1 \\ 0 & 0 & 0 & 1 & 0 \end{bmatrix}.$$

Such a matrix of course contains a great deal of redundant information. For purposes of storing a graph in computer memory, in some cases, it is more convenient and space saving to represent a graph by giving for each vertex only the list of vertices that are connected to it.

Thus we can represent the graph illustrated in Figure 8.1 by giving the sequence of lists

$$1 : 2, 3, 4$$

$$2 : 1, 3, 4$$

$$3 : 1, 2$$

$$4 : 1, 2, 5$$

$$5 : 4.$$

Sometimes multiple edges or loops are allowed. For instance, we have in Figure 8.3, respectively:

G_1:   *a graph with loops but no multiple edges,*
G_2:   *a graph with multiple edges but no loops,*
G_3:   *a graph with loops and multiple edges.*

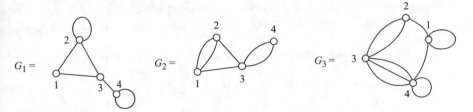

Fig. 8.3 Three types of graphs G_1, G_2, G_3.

Clearly such graphs can also be represented by adjacency matrices. In case G has a loop at the vertex i we simply set the diagonal element $a_{i,i}$ of $A(G)$ equal to 1. In case G has, say p distinct edges connecting i to j, we simply set $a_{i,j} = p$.

Thus for instance, the graphs of Figure 8.3 have adjacency matrices

$$
A(G_1) = \begin{bmatrix} 0 & 1 & 1 & 0 \\ 1 & 1 & 1 & 0 \\ 1 & 1 & 0 & 1 \\ 0 & 0 & 1 & 1 \end{bmatrix} \quad
A(G_2) = \begin{bmatrix} 0 & 2 & 1 & 0 \\ 2 & 0 & 1 & 0 \\ 1 & 1 & 0 & 2 \\ 0 & 0 & 2 & 0 \end{bmatrix} \quad
A(G_3) = \begin{bmatrix} 1 & 1 & 0 & 1 \\ 1 & 0 & 2 & 0 \\ 0 & 2 & 0 & 3 \\ 1 & 0 & 3 & 1 \end{bmatrix}.
$$

In our treatment here we shall restrict ourselves to graphs with no loops or multiple edges. These are usually referred to as *simple* graphs.

A further feature that may be added to a graph is a choice of orientation for each of its edges. This orientation is geometrically represented by *arrowheads* indicating the preferred direction on each of the edges. The resulting combinatorial object is usually referred to as an *oriented graph* or a *directed graph*. So in a directed graph, the edge set E can be thought of as a subset of $V \times V$ to account for the preferred direction.

An oriented graph may be visualized as a network of roads each of which has a one-way sign posted on it.

Choosing a preferred direction for each edge of a given graph G will be referred to as an *orientation* of G.

Thus for instance an orientation of the graph G of Figure 8.1 may be as shown in Figure 8.4.

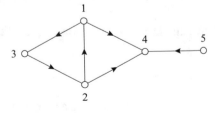

Fig. 8.4 An orientation of the graph G in Figure 8.1.

Again, we can use adjacency matrices to represent oriented graphs. Indeed, we simply let $a_{i,j} = 1$ in $A(G)$ if and only if i and j are connected by an edge that is oriented from i to j.

Thus for instance the oriented graph of Figure 8.4 has adjacency matrix

$$A(G) = \begin{bmatrix} 0 & 0 & 1 & 1 & 0 \\ 1 & 0 & 0 & 1 & 0 \\ 0 & 1 & 0 & 0 & 0 \\ 0 & 0 & 0 & 0 & 0 \\ 0 & 0 & 0 & 1 & 0 \end{bmatrix}.$$

Clearly a graph with m edges has a total of 2^m possible orientations.

By a *path* in a graph G we usually refer to a collection P of edges of G of the form

$$P = \{ \{i_1, i_2\}, \{i_2, i_3\}, \{i_3, i_4\}, \dots, \{i_{k-2}, i_{k-1}\}, \{i_{k-1}, i_k\} \}.$$

In this case we say that P *joins* or *goes from* vertex i_1 to vertex i_k. Thus for instance the path

$$P = \{ \{1, 3\}, \{3, 2\}, \{2, 4\}, \{4, 5\} \}$$

joins vertex 1 to 5 in the graph of Figure 8.1.

If the graph is oriented, then we may want to distinguish paths from *legal* paths, the latter being those that obey the direction of the arrows. Thus for instance

$$\{(1, 3), (3, 2), (2, 1)\}$$

is a legal path from vertex 1 back to itself in the graph of Figure 8.4.

In some applications, such as those coming from molecular chemistry, we may encounter combinatorial objects that are essentially graphs whose vertices are to be considered indistinguishable. These are usually referred to as *unlabeled* graphs. Geometrically they are usually depicted as graphs whose vertices have no identification numbers attached to them.

Mathematically speaking, these unlabeled graphs should be considered as *equivalence classes* of labeled graphs. Where here, two graphs G_1 and G_2 are said to be equivalent if there is a way of superimposing the vertices of G_1 in a one-to-one fashion upon the vertices of G_2 so that the edges of G_1 *fall* upon edges of G_2 and vice versa.

To be more precise, two graphs G_1 and G_2 with vertex set $V = \{1, 2, \dots, n\}$ and adjacency matrices

$$A(G_1) = [a_{i,j}], \quad A(G_2) = [b_{i,j}]$$

are said to be equivalent or isomorphic if there is a permutation

$$\sigma = \begin{pmatrix} 1 & 2 & \dots & n \\ \sigma_1 & \sigma_2 & \dots & \sigma_n \end{pmatrix}$$

such that for all $i, j = 1, 2, \ldots, n$ we have

$$b_{i,j} = a_{\sigma_i, \sigma_j} . \tag{8.3}$$

For instance, the reader may check that the two graphs in Figure 8.5 are equivalent, by constructing the permutation σ of $\{1, 2, 3, 4, 5\}$ for which the relation in (8.3) holds for all pairs $i, j = 1, 2, \ldots, 5$.

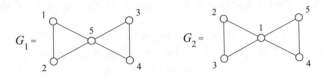

Fig. 8.5 Isomorphic (equivalent) graphs.

8.2 Chromatic Polynomials

In this section, we shall present a few basic facts concerning the coloring problem for graphs. To this end we need to make a few definitions.

First of all, by a *coloring* of a graph G we mean here a coloring of its vertices. Next, let us say that two vertices of G are *adjacent* if they are joined by an edge. This given, a coloring of G is said to be *admissible* (or *proper*) if any two adjacent vertices are of different colors.

For instance, we see that of the two colorings of the graph G of Figure 8.6 with B = blue, G = green, R = red, and Y = yellow, the first is admissible and the second is not.

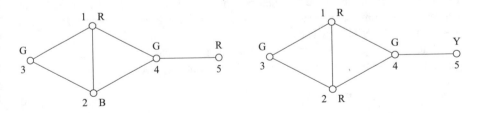

Fig. 8.6 Two colorings.

Given a graph $G = (V, E)$ and a coloring kit containing x distinct colors, the total number of admissible colorings of G that use only colors from the kit shall be denoted by $P(x, G)$.

Note for instance that for the graph G of Figure 8.1 we have

$$P(2, G) = 0 .$$

Indeed, there is no way we can produce an admissible coloring of G if we use only two colors on the vertices 1, 2, 3.

Our task here is to present an algorithm to facilitate the calculation of $P(x, G)$ for given x and G.

At this point the reader should experiment with some simple graphs before going any further. For instance, it might be good to try to calculate $P(5, G)$ for the graph G of Figure 8.1. This will lead to a better appreciation of the material we are about to present.

There are two extreme classes of graphs for which $P(x, G)$ is very easy to calculate, namely, those graphs that have no edges at all (free graphs) and those that have the maximum number of edges (complete graphs). For instance, in the case of 4 vertices, the corresponding two graphs may be depicted as in Figure 8.7.

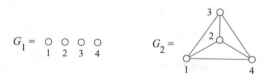

Fig. 8.7 The free graph on 4 vertices and the complete graph on 4 vertices.

We can easily see that

$$(a) \quad P(x, G_1) = x^4 \tag{8.4}$$

$$(b) \quad P(x, G_2) = x(x - 1)(x - 2)(x - 3) .$$

Indeed, for G_1 each vertex can be colored independently of the others and (8.4) a) must hold since there are x choices for each vertex. On the other hand for G_2 no two vertices can be given the same color. Thus, if we color the vertices 1, 2, 3, 4 in succession, we have x choices for 1, $x - 1$ choices for 2, $x - 2$ choices for 3, and $x - 3$ for 4. This yields (8.4) b).

More generally, if F_n and K_n, respectively, denote the free graph and the complete graph on n vertices, we can easily see that we have

$$a) \quad P(x, F_n) = x^n \tag{8.5}$$

$$b) \quad P(x, K_n) = x(x - 1)(x - 2) \cdots (x - n + 1) .$$

We note now that in both cases here $P(x, G)$ turns out to be a polynomial in the number of available colors x. The same holds true in the general case. To show this we need further terminology.

Let $G = (V, E)$ be a given graph and let us suppose for the sake of definiteness that

$$V = \{1, 2, \ldots, n\} \, .$$

We can also think of our *coloring kit* as the set

$$C = \{c_1, c_2, \ldots, c_x\} \, ,$$

where $c_1, c_2, \ldots, c_x$ is some chosen total order of the colors. This given, a coloring of G can be thought of as a map f of V into C. Indeed, we simply let $f(i) = c_i$ if the vertex i of G is given the color c_i. The admissible colorings of G thus can be identified as those maps of V into C that have the further property that

$$f(i) \neq f(j) \text{ whenever } \{i, j\} \in E \, .$$

To any coloring f of G, we shall now associate a partition

$$\Pi(f) = (F_1, F_2, \ldots, F_k) \tag{8.6}$$

of the set V. This partition is simply obtained by putting in the same part all vertices that are given the same color. We shall agree that the parts themselves are ordered not by the colors involved but by increasing smallest elements (as is usually done for partitions). For instance, if we denote by f_1 and f_2, respectively, the first and second colorings depicted in Figure 8.6, the associated partitions are

$$\Pi(f_1) = (\{1, 5\}, \{2\}, \{3, 4\}) \, ,$$

$$\Pi(f_2) = (\{1, 2\}, \{3, 4\}, \{5\}) \, .$$

Conversely, given a partition

$$\Pi = (F_1, F_2, \ldots, F_k) \tag{8.7}$$

of V, we can obtain a coloring f whose associated partition $\Pi(f)$ is equal to the given Π, by assigning all the elements of F_1 the same color, then all elements of F_2 some other color, all the elements of F_3 a third color, etc.

This shows that there is a total of

$$x(x - 1) \cdots (x - k + 1) = (x)_k$$

possible distinct colorings of G whose associated partition is the partition Π given in (8.6).

Clearly, none of the colorings thus obtained will be admissible if one of the parts F_i contains a single pair of adjacent vertices. Indeed, such a pair will be automatically given the same color. Conversely, if none of the parts F_i contains adjacent vertices, all of the maps thus constructed will be admissible.

We are thus led to a useful notion.

Let us say that a set of vertices F is *edge-free* if F does not contain a pair of adjacent vertices. Similarly, we shall say that a partition $\Pi = (F_1, F_2, \ldots, F_k)$ is *edge-free* if and only if each of its parts is edge-free.

This given, our observation leads to the following result.

Theorem 8.2.1 *Let $G = (V, E)$ be a given graph and let $F\Pi_k(G)$ denote the collection of all edge-free partitions of V with k parts. Then for all x we have*

$$P(x, G) = \sum_{k=1}^{n} (x)_k \#F\Pi_k(G) .   \tag{8.8}$$

Proof As we have noted, a coloring f is admissible if and only if its associated partition is edge-free. Moreover, given an edge-free partition $\Pi = (F_1, F_2, \ldots, F_k)$ of V, there are all together $(x)_k$ different admissible colorings f whose associated partition is the given Π. Thus the total number of admissible colorings whose associated partition has k parts is equal to

$$(x)_k \#F\Pi_k(G) .$$

Summing the contributions for $k = 1, 2, \ldots, n$ we obtain the total number of admissible colorings of G. Thus (8.8) must hold as asserted.   □

We have now derived that $P(x, G)$ is always a polynomial in x. It is usually referred to as the *chromatic polynomial* of G.

We should note that Theorem 8.2.1 implies our familiar identity

$$\sum_{k=1}^{n} S_{n,k}(x)_k = x^n .   \tag{8.9}$$

Indeed, when $G = F_n$ (the free graph on n vertices) all partitions of V are edge-free (there are no edges). Thus in this case $\#F\Pi_k(G)$ is equal to the Stirling number of the second kind $S_{n,k}$. Substituting this in (8.8) and comparing with (8.5) give (8.9) as asserted.

Our next goal is to present an algorithm for the construction of $P(x, G)$ that does not involve a direct enumeration of the collections $F\Pi_k(G)$. This algorithm is based on a very simple observation. To present it, we need further notation.

Let $G = (V, E)$ be a given graph and let e be an edge of G. We shall denote by $G - e$ and G/e two new graphs defined as follows. First of all, $G - e$ is simply the graph obtained by removing from G the edge e. The definition of G/e is somewhat more elaborate. Let us imagine that G is drawn in space or in the plane and that by a continuous deformation process we progressively shrink the edge e until it reduces to a point. As a result, the two vertices of G that originally were the endpoints of e will have collapsed into a single vertex, and a certain number of multiple edges will be created in the process. The graph G/e is then the graph obtained by collapsing multiple edges into single edge. This graph is referred to as the *contraction* of G along e.

This is best illustrated by an example. For instance, if G is the graph in Figure 8.8, and e is {3, 5} as indicated, then at the end of the deformation process we have the configuration in Figure 8.9.

$$G =$$
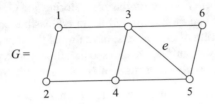

Fig. 8.8 Contracting the edge e of G.

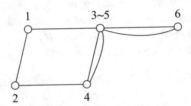

Fig. 8.9 The graph G of Figure 8.8 after the deformation.

It may be good to make the convention that the single vertex obtained by the collapse of the edge e should always be labeled by the smallest of the labels of the original vertices of e. Thus with this convention we have G/e as shown in Figure 8.10.

$$G/e =$$
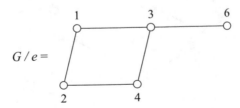

Fig. 8.10 G/e for the graph G of Figure 8.8.

It now turns out that the following identity holds.

Theorem 8.2.2 *If $G = (V, E)$ and e is an edge of G, then*

$$P(x, G - e) = P(x, G) + P(x, G/e) . \tag{8.10}$$

Proof Let i and j denote the endpoints of e. Since i and j are not adjacent in $G - e$, an admissible coloring f of $G - e$ may or may not assign the same color to i and j. If $f(i) \neq f(j)$, then f is an admissible coloring for G. On the other hand, if $f(i) = f(j)$, then f can be interpreted as an admissible coloring of G/e. Since every admissible coloring of $G - e$ must have one of these two properties and these properties are mutually exclusive, we can see that the right hand side of (8.10) gives the total number of admissible colorings of $G - e$. This proves our assertion. □

Formula (8.10) leads to a recursive algorithm for constructing the chromatic polynomial of a graph. This is perhaps better understood if we rewrite (8.10) in yet another way.

If G is a graph and e is not an edge of G, let $G + e$ denote the graph obtained by adjoining e to G. Furthermore, note that the construction of the graph G/e can be carried out as indicated above whether or not e is an edge of G (just bring together the two endpoints of e and collapse multiple edges). Thus we can adopt the convention that if e is not in G and G/e and $(G + e)/e$ should mean the same thing. This given, we can rewrite (8.10) in the form

$$P(x, G) = P(x, G + e) + P(x, G/e) . \tag{8.11}$$

This formula reduces the calculation of the chromatic polynomial of G to that of $G + e$ and G/e. It is easy to see that the operation of adding or contracting a given edge replaces the graph with a graph that is *closer* to a complete graph. The idea is then to carry out a sequence of such operations with the aim of producing complete graphs for which the calculation is immediate.

This process is best illustrated by an example. Let us suppose we are given to calculate $P(x, G)$ when G is the graph of Figure 8.11.

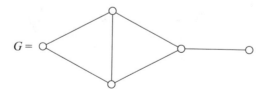

Fig. 8.11 The calculation of $P(x, G)$.

We have for this graph the identity given in Figure 8.12.

Repeated uses of (8.11) lead to a binary tree whose root is the given graph G and whose terminal nodes are all *complete graphs*. Each internal node branches off into two nodes by the addition and contraction of one of the edges. In Figure 8.13 the root of the tree is drawn at the top, and the edge in question at each step is

$$P(x, \text{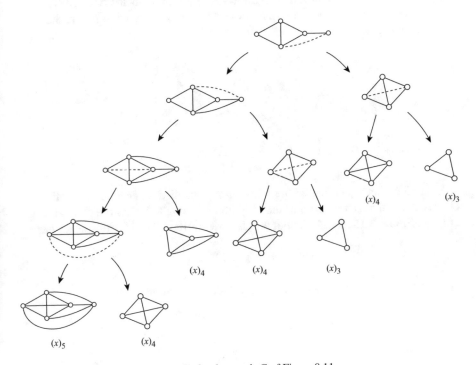}) = P(x, \quad) + P(x, \quad).$$

Fig. 8.12 $P(x, G)$ in terms of $P(x, G + e)$ and $P(x, G/e)$ for the graph G in Figure 8.11.

represented by a dotted line. The result of adding e is drawn as left child and that of contracting G along e is drawn as right child.

Carrying out this construction on the graph of Figure 8.11 yields the binary tree in Figure 8.13.

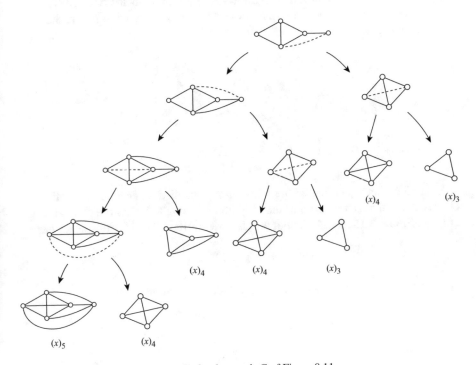

Fig. 8.13 The calculation of $P(x, G)$ for the graph G of Figure 8.11.

Adding up the chromatic polynomials that appear at the leaf nodes of this binary tree gives that

$$P(x, G) = (x)_5 + 4(x)_4 + 2(x)_3 = x(x - 1)^2(x - 2)^2. \tag{8.12}$$

In particular for $x = 5$, we get that there are all together

$$5 \cdot 4 \cdot 3 \cdot 2 \cdot 1 + 4 \cdot 5 \cdot 4 \cdot 3 \cdot 2 + 2 \cdot 5 \cdot 4 \cdot 3 = 720$$

distinct admissible colorings of the graph G of Figure 8.11 using a 5-color kit.

Note also that indirectly we have obtained at the same time also the numbers $\#F\Pi_k(G)$ appearing in formula (8.8). Thus we may conclude upon comparing (8.12) and (8.8) that the vertex set of this graph has, respectively, one 5-part, four 4-part, and two 3-part edge-free partitions.

8.3   Acyclic Orientations

A path through a graph is said to be *cyclic* if it ends where it starts. An orientation of a graph G is said to be *acyclic* if it allows no legal cyclic paths through G.

Thus for instance, of the two orientations of the graph shown in Figure 8.14, the first is acyclic and the second is not.

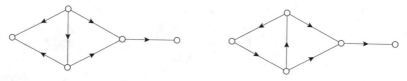

Fig. 8.14 Two orientations of the graph in Figure 8.11.

This given, let $a(G)$ denote the total number of distinct acyclic orientations of G. It is not difficult to see for instance (by going through all possible cases) that the graph in Figure 8.15 has 12 distinct acyclic orientations. Thus for this G we have

$$a(G) = 12 .$$

In point of fact we have a rather surprising fact due to Stanley [26]:

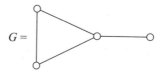

Fig. 8.15 Counting acyclic orientations.

Theorem 8.3.1 *Let* $G = (V, E)$ *be a graph on* n *vertices. Then the following identity holds*

$$a(G) = (-1)^n P(-1, G) . \tag{8.13}$$

The proof of this result will be given in the next section.

We should point out that formula (8.13) combined with (8.10) yields the recursion

$$a(G + e) = a(G) + a(G/e) ,\qquad(8.14)$$

where here e denotes an edge not in G. So for instance we have the identity shown in Figure 8.16.

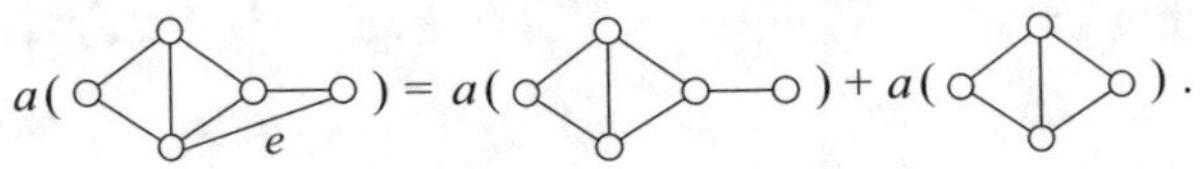

Fig. 8.16 Writing $a(G + e)$ in terms of $a(G)$ and $a(G/e)$.

To see this, note that if G has n vertices, then $G + e$ and G/e have n and $n - 1$ vertices, respectively. Thus from (8.13) and (8.10) we get

$$
\begin{aligned}
a(G) &= (-1)^n P(-1, G)\\
&= (-1)^n P(-1, G + e) + (-1)^n P(-1, G/e)\\
&= a(G + e) - a(G/e) ,
\end{aligned}
$$

which is (8.14).

Conversely it is not difficult to show, by reversing the process, that formulas (8.14) and (8.10) combined yield (8.13) back again. Thus a way to establish (8.13) might be to establish (8.14) first. However, we shall not do so here; since in the next section we shall obtain formula (8.13) as a simple consequence of a very beautiful and important result of general interest.

Nevertheless, the reader should find it challenging to obtain a direct combinatorial proof of (8.14).

8.4 The Cartier–Foata Languages

Let $X = \{x_1, x_2, \ldots, x_N\}$ be an alphabet and let X^n denote the collection of all words of length n over X. It is convenient to allow the possibility that a word may have no letters at all. As usual, we refer to this empty word as the *null word* and denote it by "1" or by ϵ.

It is customary to represent collections of words simply as formal sums of words. Furthermore, we define the *product* (concatenation) of two words w_1 and w_2 to be simply the word obtained by juxtaposing w_1 and w_2.

Thus for instance if $w_1 = x_1 x_2 x_1$ and $w_2 = x_2 x_3 x_1 x_2 x_1$, then

$$w_1 w_2 = x_1 x_2 x_1 x_2 x_3 x_1 x_2 x_1 .$$

This given, we may symbolically write

$$X^n = \sum_{w \in X^n} w = (x_1 + x_2 + \cdots + x_N)^n . \tag{8.15}$$

Furthermore, if X^* denotes the collection of all words over the alphabet X, then we may write as well

$$X^* = 1 + X^1 + X^2 + X^3 + \cdots = \sum_{n \geq 0} (x_1 + x_2 + \cdots + x_N)^n . \tag{8.16}$$

This formalism permits us to manipulate collections of words using the laws of elementary algebra. As we shall soon see, this fact may sometime lead us to some rather inspiring conclusions.

As an example in point, we note that it is very tempting to sum the geometric series in (8.16) and write

$$X^* = \frac{1}{1 - (x_1 + x_2 + \cdots + x_N)} . \tag{8.17}$$

However, this per se does not mean very much since we are not about to evaluate this expression for any particular values of $x_1 + x_2 + \cdots + x_N$. Thus we must interpret

$$\frac{1}{1 - (x_1 + x_2 + \cdots + x_N)}$$

as none other than a shorthand for

$$\sum_{n \geq 0} (x_1 + x_2 + \cdots + x_N)^n .$$

Nevertheless, we may also write (8.17) in the form

$$(1 - x_1 - x_2 - \cdots - x_N)X^* = 1 . \tag{8.18}$$

Now this identity may be given an interesting interpretation. Let C denote the collection of all pairs (u, v) where v is an arbitrary word in X^* and u is any one of the following $N + 1$ words:

$$1, x_1, x_2, x_3, \ldots, x_N .$$

Furthermore, let us define the *sign* of a pair $(u, v) \in C$ to be the number 1 if $u = \epsilon$ and -1 if $u = x_1, x_2, \ldots, x_{N-1}$, or x_N. Finally, for a given pair $(u, v) \in C$ we shall set

$$W(u, v) = uv$$

and refer to this word as the *weight* of the pair (u, v).

Thus for instance

$$W(x_2, x_1x_3x_1) = x_2x_1x_3x_1 .$$

Making use of these conventions, we see that we have

$$(1 - x_1 - x_2 - \cdots - x_N)X^* = \sum_{v \in X^*} (v - x_1v - x_2v - \cdots - x_Nv)$$

$$= \sum_{v \in X^*} (W(1, v) - W(x_1, v) - W(x_2, v) - \cdots - W(x_N, v))$$

$$= \sum_{(u,v) \in C} \text{sign}(u, v)W(u, v) .$$

Thus (8.18) becomes

$$\sum_{(u,v) \in C} \text{sign}(u, v)W(u, v) = 1 . \tag{8.19}$$

Now, it develops that this identity can be viewed as a consequence of the following very simple fact: Namely, we can assign to each element $(u, v) \in C$ (except when $(u, v) = (\epsilon, \epsilon)$) a *mate* (u', v') of equal weight and opposite sign. Of course, we have then

$$\text{sign}(u, v)W(u, v) + \text{sign}(u', v')W(u', v') = 0 . \tag{8.20}$$

Consequently, the terms in the sum will cancel in pairs, except for the term corresponding to (ϵ, ϵ) that contributes 1 to the sum, and (8.19) must necessarily hold true.

We can construct such an assignment $(u, v) \to (u', v')$ in a very simple manner. There are two cases depending on whether $u = \epsilon$ or $u \neq \epsilon$.

(a)   *In case $u = \epsilon$ and $v = x_iw$, we let the mate of (ϵ, x_iw) be (x_i, w)*

(b)   *In case $u = x_i$, we let the mate of (x_i, v) be (ϵ, x_iv) .*

In other words, the mate of (u, v) is simply obtained by moving the first letter of v to u if u is the null word, and moving u to v if u is not the null word.

Clearly for such an assignment we shall have (8.20) as desired.

This formalism may appear here a bit contrived, not to say extravagant, if it is to be put together just to prove such a simple identity as that in (8.18). Nevertheless, as we shall quickly see, there is a slightly more general context in which the same idea leads to a rather powerful and important result.

To see how this comes about, we need some further terminology. Note that in ordinary languages as well as computer languages we meet with pairs of words (synonyms we might say) or even pairs of sentences that although different as *strings* of letters, they do have the same semantic content.

These circumstances suggest that we should consider a *language* not as a collection of separate individual words but as a collection of *equivalence* classes of words, where two words w_1 and w_2 are put in the same equivalence class if and only if they have the same *meaning*. For instance, if the words w_1 and w_2 are C++ programs, we may say that they are equivalent, if given the same data they both produce the same output. Clearly, if we are asked to test all C++ programs of a given length, we need to list only one representative of each equivalence class of programs.

We can give a very striking illustration of this concept of language as equivalence classes of words over an alphabet by looking once more at our acyclic orientations.

Note that we may produce all possible acyclic orientations of the graph G in Figure 8.17 by the following very simple procedure. We mark points for the vertices of G on a straight line, for instance as indicated in Figure 8.18. Then redraw a flattened version of G using these points as vertices. This gives the representation in Figure 8.19.

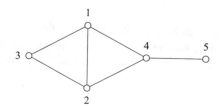

Fig. 8.17 A graph G.

Fig. 8.18 Placing the vertices of G on a straight line.

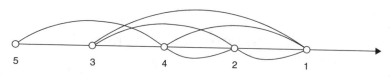

Fig. 8.19 Drawing G on a straight line.

Finally, we use the left to right orientation of the line as a guide for choosing an orientation for each of these edges. More precisely, we orient an edge $\{i, j\}$ from i to j if i occurs to the left of j and from j to i if j occurs to the left of i.

This gives the orientation in Figure 8.20.

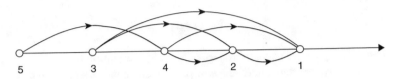

Fig. 8.20 Orienting the edges of G left to right.

Transferring this orientation back to Figure 8.17 we get then the configuration in Figure 8.21.

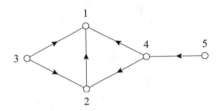

Fig. 8.21 The resulting acyclic orientation of G.

Now it is immediate that any orientation resulting from this construction will necessarily be acyclic. Indeed, following a legal path in Figure 8.20 can only bring us further and further to the right, and we cannot possibly ever end up where we started.

The resulting orientation clearly only depends on the relative order of the vertices $1, 2, \ldots, 5$ and not on the actual positions these vertices occupy on the line. Then we may symbolically represent such an arrangement of vertices by a word in the letters $x_1, x_2, \ldots, x_5$. In particular, the positions of the vertices of G in Figure 8.18 may be represented by the rearrangement

$$x_5 x_3 x_4 x_2 x_1 . \tag{8.21}$$

Given such a word w, we can then produce an orientation of the graph G by simply orienting an edge $\{i, j\}$ from i to j if x_i occurs before x_j in w and from j to i if x_j occurs before x_i.

Note however that there are $5! = 120$ such words and (as we shall see) not anywhere near so many different acyclic orientations of G. This is due to the fact that we may interchange certain pairs of adjacent letters without affecting the corresponding acyclic orientation of G.

Indeed, looking at Figure 8.20 we see that we may interchange the relative positions of 5 and 3 on the line without having to change any of the arrows. This of course would not have been the case if $\{3, 5\}$ had been an edge of G. For then moving 3 before 5 would affect the orientation of this edge.

Thus we see that the following word

$$x_3 x_5 x_4 x_2 x_1$$

and the word $x_5 x_3 x_4 x_2 x_1$ generate the same acyclic orientation of G.

More generally, we can interchange a pair of adjacent letters $x_i x_j$ in a word, without affecting the resulting orientation of G if and only if $\{i, j\}$ is *not* an edge of G.

For convenience, let us call two letters x_i, x_j a *commuting* pair if and only if $\{i, j\}$ is not an edge of G. This given, we see that if a word w_1 is obtained from w_2 by a sequence of interchanges involving commuting pairs of adjacent letters, then w_1 and w_2 represent the same orientation of G.

In the table below in (8.22) written as a 5×5 matrix, we have placed a 1 in position in row i and column j if $\{x_i, x_j\}$ is a commuting pair, and a 0 otherwise:

$$
\begin{bmatrix}
1 & 0 & 0 & 0 & 1 \\
0 & 1 & 0 & 0 & 1 \\
0 & 0 & 1 & 1 & 1 \\
0 & 0 & 1 & 1 & 0 \\
1 & 1 & 1 & 0 & 1
\end{bmatrix} . \tag{8.22}
$$

Clearly, this is none other than the logical complement of the adjacency matrix of G.

We are thus led to the following very general construction. Given an $N \times N$ symmetric $\{0, 1\}$-matrix $A = [a_{i,j}]$ with 1s down the diagonal as in (8.22), let us say that x_i, x_j are a commuting pair if and only if $a_{i,j} = 1$. Furthermore, let us call two words w_1, w_2 in X^* *equivalent* if and only if w_1 may be obtained from w_2 by a sequence of successive interchanges involving commuting pairs of adjacent letters.

This equivalence relation breaks up the collection X^* into equivalence classes. The collection of all these equivalence classes will be referred to as the *Cartier–Foata language* corresponding to the matrix A and will be denoted by the symbol $\mathcal{L}_A$. Let us assume next that in some manner we have selected from each equivalence class a representative word w. Let us also call $\mathcal{R}_A$ the collection of all such representative words. We may then symbolically write

$$\mathcal{L}_A = \sum_{w \in \mathcal{R}_A} w \tag{8.23}$$

and call the resulting expression the *representative* series of the language $\mathcal{L}_A$.

It so happens that there is a relatively simple formula for constructing the series $\mathcal{L}_A$ directly from the matrix A. To present it we need some notation.

Let us say that a set F of letters of the alphabet X is a *commuting* set if any pair of letters in F commute. The collection of all commuting sets will be denoted by C_A. It is convenient to include in C_A the empty set as well as all the sets consisting of a single letter.

Thus, for example, in the case A is the matrix appearing in (8.22), the collection C_A consists of 9 sets of letters: namely, we have

$$C_A = \{\emptyset, \{x_1\}, \{x_2\}, \{x_3\}, \{x_4\}, \{x_5\}, \{x_1, x_5\}, \{x_2, x_5\}, \{x_3, x_4\}, \{x_3, x_5\}\}.$$
$$(8.24)$$

Given a commuting set F, the word obtained by juxtaposing all the letters of F shall be denoted by $w(F)$. Clearly, the order in which these letters are juxtaposed is immaterial, for F being a commuting set, all arrangements of these letters of F lead to equivalent words of the language $\mathcal{L}_A$.

We now have the following remarkable identity.

Theorem 8.4.1

$$\left(\sum_{F \in C_A} (-1)^{|F|} w(F) \right) \left(\sum_{w \in \mathcal{R}_A} w \right) = 1, \tag{8.25}$$

where $|F|$ denotes the number of elements of F.

Before proceeding with the proof of this identity, it may be good to see what it says in some special cases. For instance, if we allow no pairs of letters to commute, then A reduces to the $N \times N$ identity matrix. In this case by our conventions we have

$$C_A = \{\emptyset, \{x_1\}, \{x_2\}, \dots, \{x_N\}\}$$

and (8.24) reduces to

$$(1 - x_1 - x_2 - \cdots - x_N) \sum_{w \in \mathcal{R}_A} w = 1$$

from which we deduce that

$$\sum_{w \in \mathcal{R}_A} w = \sum_{n \geq 0} (x_1 + x_2 + \cdots + x_N)^n.$$

Thus we see that our identity (8.16) is a special case of (8.25).

If we allow all pairs of letters to commute, then all subsets of letters are commuting sets. Thus in this case, for instance for $N = 3$ we get

$$\sum_{F \in C_A} (-1)^{|F|} w(F) = 1 - x_1 - x_2 - x_3 + x_1 x_2 + x_1 x_3 + x_2 x_3 - x_1 x_2 x_3$$

$$= (1 - x_1)(1 - x_2)(1 - x_3)$$

and (8.25) yields

$$\sum_{w \in \mathcal{R}_A} w = \frac{1}{(1 - x_1)(1 - x_2)(1 - x_3)} .$$

We thus recover the generating series of our *increasing words* of Chapter 1 from a completely different viewpoint.

In the case A is the matrix given in (8.22), using (8.25) we get

$$(1 - x_1 - x_2 - x_3 - x_4 - x_5 + x_1 x_5 + x_2 x_5 + x_3 x_4 + x_3 x_5) \sum_{w \in \mathcal{R}_A} w = 1 ,$$

or better

$$\sum_{w \in \mathcal{R}_A} w = \sum_{n \geq 0} (x_1 + x_2 + x_3 + x_4 + x_5 - x_1 x_5 - x_2 x_5 - x_3 x_4 - x_3 x_5)^n .$$

As a final example of the uses (8.25) may be put to, we shall show that we can derive from it our result concerning acyclic orientations. To this end note that if we isolate the term corresponding to the empty set, we may write (8.25) in the form

$$\sum_{w \in \mathcal{R}_A} w = \frac{1}{1 + \sum_{F \neq \emptyset} (-1)^{|F|} w(F)} ,$$

from which we derive that

$$\sum_{w \in \mathcal{R}_A} w = \sum_{k \geq 0} (-1)^k \left(\sum_{F \neq \emptyset} (-1)^{|F|} w(F) \right)^k .$$

This can be rewritten in the form

$$\sum_{w \in \mathcal{R}_A} w = \sum_{k \geq 0} (-1)^k \sum_{F_1 \neq \emptyset} \sum_{F_2 \neq \emptyset} \cdots \sum_{F_k \neq \emptyset} (-1)^{|F_1| + |F_2| + \cdots + |F_k|} w(F_1) w(F_2) \cdots w(F_k) .$$

$$(8.26)$$

Now let G be a given graph with N vertices and let A be the logical complement of the adjacency matrix of G .

From the consideration at the beginning of this section it is not difficult to infer that the words w_1 and w_2 containing the letters $x_1, x_2, \ldots, x_N$ correspond to the same acyclic orientation if and only if they are equivalent as words of the language $\mathcal{L}_A$. Thus, to enumerate acyclic orientations of G we need only to take account of the terms on the right hand side of (8.26), which produce words that are rearrangements of $x_1, x_2, \ldots, x_N$.

Note now that A being the complement of the adjacency matrix of G, a set F is in C_A if and only if F is edge-free in the sense of Section 8.2. We also see that $w(F_1)w(F_2)\cdots w(F_k)$ will be an arrangement of $x_1, x_2, \ldots, x_N$ if and only if $F_1, F_2, \ldots, F_k$ are disjoint sets whose union is the set $[N] = \{1, 2, \ldots, N\}$, a fact that we can express by writing

$$F_1 + F_2 + \cdots + F_k = [N] \, .$$

Thus, from (8.26) we deduce that the number $a(G)$ of acyclic orientations of G is given by the expression

$$a(G) = \sum_{k=1}^{N}(-1)^k \sum_{F_1+F_2+\cdots+F_k=[N]} (-1)^{|F_1|+|F_2|+\cdots+|F_k|} \, , \tag{8.27}$$

where the second sum here is to be carried out over all k-tuples of disjoint edge-free sets whose union is $[N]$. However, the number of such k-tuples is simply $k!$ times the number of edge-free partitions of G.

Combining all these observations, we can easily see that (8.27) reduces to the formula

$$a(G) = \sum_{k=1}^{N}(-1)^k k!(-1)^N \#F\Pi_k(G) \, ,$$

and this is precisely what (8.8) becomes if we replace n by N and x by -1, and multiply through by $(-1)^N$. Thus formula (8.13) must hold true as asserted.

We have thus seen that Theorem 8.4.1 implies Theorem 8.3.1.

We shall now proceed with the proof of the identity in (8.25). Our approach will be entirely analogous to that we followed in deriving formula (8.18). We work with the collection C of pairs (u, v) of the form

$$(u, v) = (w(F), w),$$

where F can be any one of the commuting sets in C_A and w can be any element of the language $\mathcal{L}_A$. Of course, two pairs (u, v) and (u, v') will be considered identical if v and v' are equivalent words of the language $\mathcal{L}_A$. Just as we did in the proof of (8.18), we set

$$\operatorname{sign}(w(F), w) = (-1)^{|F|} \tag{8.28}$$

$$W(w(F), w) = w(F)w \, .$$

Thus for instance, if A is the matrix in (8.22) and

$$(u, v) = (x_5, x_5x_1x_2x_4x_3) \tag{8.29}$$

then

$$\text{sign}(u, v) = (-1)^1 = -1 \qquad (8.30)$$

$$W(u, v) = x_5 x_5 x_1 x_2 x_4 x_3 .$$

This given, we see that the left hand side of (8.25) can be rewritten in the form

$$\sum_{F \in C_A} \sum_{w \in R_A} (-1)^{|F|} w(F) w = \sum_{(u,v) \in C} \text{sign}(u, v) W(u, v) = 1 .$$

Thus we are again reduced to show the identity

$$\sum_{(u,v) \in C} \text{sign}(u, v) W(u, v) = 1 . \qquad (8.31)$$

Just as before we shall obtain our result by constructing for each element $(u, v) \in C$ (except when $(u, v) = (\epsilon, \epsilon)$) a mate (u', v') in such a manner that

$$\text{sign}(u, v) W(u, v) + \text{sign}(u', v') W(u', v') = 0 . \qquad (8.32)$$

To this end we need an auxiliary construction. Given a word $w \in X^*$ there is a set of letters that may be brought to the beginning of w by a succession of interchanges involving pairs of commuting adjacent letters. This set will be referred to as the *initial part* of w and will be denoted by $I(w)$. More precisely, a letter x_i belongs to $I(w)$ if and only if there is a word equivalent to w (in the language $\mathcal{L}_A$), which is of the form

$$x_i w' .$$

For instance, when A is the matrix of (8.22) and w is the word appearing in (8.30), we have

$$I(x_5 x_5 x_1 x_2 x_4 x_3) = \{x_1, x_5\} . \qquad (8.33)$$

Note in particular that x_2 is not a member of $I(w)$ in this case because x_1 and x_2 do not commute. It is thus clear that the initial part of a word is always a commuting set of letters.

This given, our assignment $(u, v) \rightarrow (u', v')$ is constructed as follows. We first determine the set $F_0 \in C_A$ that is the initial part of uv. Thus in symbols

$$F_0 = I(W(u, v)) .$$

Let us assume that our alphabet is totally ordered in a manner consistent with our labeling. That is we set

$$x_1 < x_2 < \cdots < x_N .$$

With this in mind, let x_i be the smallest letter in F_0. There are two cases:

(a) *If x_i is a letter of u, then we move it from u to v. We do this regardless of whether or not x_i is also in v.*

(b) *If x_i is not a letter of u, then of course x_i must occur in v at least once. In this case we simply transfer the leftmost x_i from v to u.*

Whatever may be the case, the resulting pair (u', v') will be taken to be the *mate* of (u, v).

It may be good to illustrate our construction with some examples. For instance, let (u, v) be as in (8.29). In view of (8.33), the smallest letter of F_0 is x_1 and we are in case b). Thus the mate of

$$(u, v) = (x_5, x_5x_1x_2x_4x_3) \tag{8.34}$$

is

$$(u', v') = (x_1x_5, x_5x_2x_4x_3) . \tag{8.35}$$

Note also that if we start with the pair in (8.35), our construction leads as back to the pair in (8.34).

We should also note that in case all the letters commute and we take

$$(u, v) = (x_2x_1, x_2x_3x_1x_2)$$

then

$$I(uv) = \{x_1, x_2, x_3\} .$$

So we are in case b) and the mate of (u, v) is the pair

$$(u', v') = (x_2, x_1x_2x_3x_1x_2) .$$

Clearly the correspondence $(u, v) \rightarrow (u', v')$ we have just defined satisfies the requirement in (8.32). Our argument can now proceed as in the case we previously considered. The terms in the sum (8.30) will cancel in pairs, leaving only the term (ϵ, ϵ) that has no mate and contributes a 1 to the sum.

This completes our proof of formula (8.25).

Theorem 8.4.1 has further interesting consequences that are worth mentioning here. Let C_A and $\mathcal{R}_A$ have the same meaning as before and let $\mathcal{R}_A(n)$ denote the collection of words of $\mathcal{R}_A$ that have exactly n letters. Thus $\#\mathcal{R}_A(n)$ can be interpreted as the number of distinct (nonequivalent) words of length n in the language $\mathcal{L}_A$. Theorem 8.4.1 yields us immediately a formula for the generating function of these numbers.

This can be stated as follows.

Theorem 8.4.2

$$\sum_{n \geq 0} t^n \#R_A(n) = \frac{1}{\displaystyle\sum_{F \in C_A} (-1)^{|F|} t^{|F|}} \,. \tag{8.36}$$

Proof If we replace in formula (8.24), each $x_i \in X$ by a t, a word $w(F)$ is replaced by $t^{|F|}$ and therefore each of the words in $R_A(n)$ produces a t^n. Thus after this replacement formula (8.25) becomes

$$\left(\sum_{F \in C_A} (-1)^{|F|} t^{|F|} \right) \left(\sum_{n \geq 0} t^n \#R_A(n) \right) = 1 \,,$$

and this gives (8.36) upon division.   □

Formula (8.36) may lead to some surprising conclusions. For example let A be as in (8.22). Then

$$\sum_{F \in C_A} (-1)^{|F|} t^{|F|} = 1 - 5t + 4t^2 \,.$$

Thus in this case

$$\sum_{n \geq 0} t^n \#R_A(n) = \frac{1}{1 - 5t + 4t^2} = \sum_{k \geq 0} (5t - 4t^2)^k \,.$$

In particular we derive that the coefficient of t^n in this series is always a nonnegative integer! This fact would be difficult to guess without the present theory. The reader may find it amusing to derive further applications of formula (8.36).

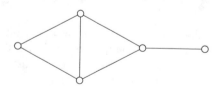

Fig. 8.22 A graph G.

It might be good to point out, before closing, that if we combine Theorem 8.3.1 with (8.12) and carry out the calculations in Figure 8.23, we find that the graph in Figure 8.22 has a total of 36 acyclic orientations.

We see then that this is substantially less than the 120 words we may use to represent acyclic orientations.

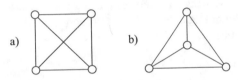

$$a( ) = (-1)^5 P(-1,) = 2^2 \cdot 3^2 = 36$$

Fig. 8.23 The number of acyclic orientations of the graph G of Figure 8.22.

8.5   Planar Graphs

8.5.1   Kuratowski's Theorem

Recall that the abstract definition of a graph G involves a set of vertices V and a set E of pairs of elements of V that are the edges of G. This definition involves no points or lines that we use to draw pictures of graphs. In fact, we have seen in Section 8.1 that the geometric representation of a graph as a collection of points (vertices) and arcs joining these points (edges) in the plane is quite misleading. For instance the two drawings in Figure 8.24 are both representations of the complete graph K_4 as geometric figures drawn in the plane.

a)

b)

Fig. 8.24 Two geometric representations of the complete graph K_4.

Notice however that the first one of these figures has a crossing of the edges represented. If we require a faithful representation of a graph in the plane where the points corresponding to the vertices of G are actually points of the plane rather than small circles as we have drawn them in Figure 8.24, then we would be forced to consider the crossing point of the two diagonals in figure $a)$ as a vertex. Since this is not actually a part of K_4 as an abstract graph, we conclude that in some sense the representation in Figure 8.24b is the more natural of the two.

We would like to make these notions a little more precise.

Let us call the figures in Figure 8.24 *representations* of K_4 in the plane. We shall call the representation in Figure 8.24b an *embedding* of K_4 in the plane. More generally, by an embedding of a graph G in some Euclidean space, we mean an identification of the vertex set V of G with certain distinguished points, and an identification of the edges of G with arcs connecting the relevant vertices in such a way that

the only crossings between these arcs occur at the points corresponding to the vertices of G.

Relying on our intuition, it is not difficult to see that any graph has an embedding in three-dimensional space. The corresponding statement for the plane, however, does not hold. The reader can convince herself that the complete graph K_5 and the so-called utility graph $K_{3,3}$ in Figure 8.25 do not have embeddings in the plane. Thus, K_5 and $K_{3,3}$ are examples of *nonplanar* graphs. In contrast to this, a graph is called *planar* if it has an embedding in the plane.

$$K_{3,3} = $$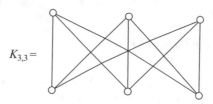

Fig. 8.25 The utility graph $K_{3,3}$.

We will state a central result in graph theory that characterizes planar graphs. First we need the following terminology.

Two graphs are called *homeomorphic* if one is obtained from the other by inserting a number of vertices into some of the edges, or by discounting a number of vertices of degree 2. For example K_4 is homeomorphic to the graph in Figure 8.26 where the added vertices are marked.

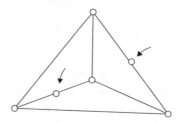

Fig. 8.26 A graph homeomorphic to K_4.

Clearly, if a graph is planar, then so are all the graphs homeomorphic to it. Thus planarity is a property of graphs that is invariant under homeomorphism.

We have remarked that the complete graph K_5 and the utility graph $K_{3,3}$ are nonplanar graphs. Hence any graph that contains either one of these as a subgraph cannot be planar.

It turns out that this property alone actually characterizes all nonplanar (therefore all planar) graphs [19]. More precisely we have Kuratowski's theorem:

Theorem 8.5.1 *G is nonplanar if and only if it contains a subgraph that is homeomorphic to K_5 or to $K_{3,3}$.*

The nonplanarity of K_5 and $K_{3,3}$ can be shown fairly easily by making use of Euler's formula that will be derived in the next section. The reader is referred to problem 8.7.41 for the idea of the proof of Theorem 8.5.1.

8.5.2   Euler's Formula

Suppose G is a connected planar graph. Any embedding of G in the plane breaks up the plane into *regions* whose boundaries are determined by the arcs representing the edges of G.

For instance the embedding of K_4 pictured in Figure 8.24(b) gives rise to four such regions R_1, R_2, R_3, R_4 as shown in Figure 8.27. Of these regions, R_1, R_2, and R_3 are bounded, whereas R_4 is unbounded.

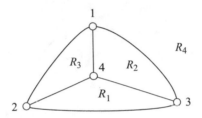

Fig. 8.27   An embedding of K_4 with the regions labeled.

For convenience, let us first fix an embedding of G in the plane and identify it with this particular embedding (among many other possible embeddings). In this manner, we shall think of G as a figure in the plane rather than an abstract graph. It will also be convenient to assume for now that G has no vertices of degree 1.

Traditionally, Euler's formula is stated in terms of three parameters v, e, and f, where $v = |V|$ and $e = |E|$ denote the number of vertices and the number of edges of the graph and where its embedding splits the plane into f regions (counting the unbounded region).

The following remarkable theorem of Euler [8] relates the quantities v, e, and f:

Theorem 8.5.2   *For any connected planar graph G the following relation holds*

$$f - e + v = 2 .$$ (8.37)

Proof   Before we proceed with the proof it will be instructive to consider the embedding of K_4 given in Figure 8.27 in detail:

Note that for this example, we have

$$v = 4, \ e = 6, \ f = 4$$

so that (8.37) indeed holds.

To prove the general case of Euler's theorem we shall establish an intuitive characterization of the regions determined by G in terms of certain cycles. In the case of K_4, we see that if we treat the unbounded region as a special case, the bounded regions R_1, R_2, and R_3 are determined by cycles of K_4 that contain no points of K_4 in their interior. We may refer to these cycles as the *bounding cycles*.

For instance for Figure 8.27 the bounding cycles are as shown in Figure 8.28.

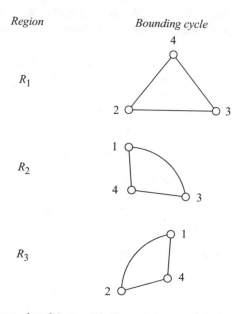

Fig. 8.28 The bounding cycles of the graph in Figure 8.27.

In general, put $F = f - 1$ and suppose that the F bounded regions $R_1, R_2, \ldots, R_F$ are determined by the corresponding bounding cycles $C_1, C_2, \ldots,$ C_F of G. Denote the unbounded region of G by $R (= R_f)$.

Now pick a region R_i of G that shares an edge with the unbounded region R. Renaming if necessary, we may assume that $R_i = R_F$. This is shown in Figure 8.29. Let us now remove from G that part of C_F that is incident only to R_F and R as depicted in Figure 8.30.

If the removed part of C_F contains x vertices, then, as we can easily see, it must contain $x + 1$ edges. Thus the new graph G' we obtain from G by this process contains x fewer vertices and $x + 1$ fewer edges than G. In other words

$$v' = v - x , \quad e' = e - (x + 1) . \tag{8.38}$$

Moreover, R_F and R coalesce to become the unbounded region R' of G'. Since this decreases the number of regions by 1, we have

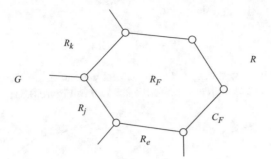

Fig. 8.29 R_F shares an edge with the unbounded region R.

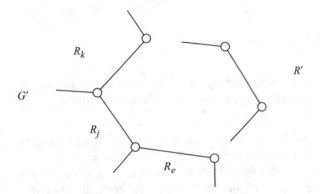

Fig. 8.30   Removing the common edges and the new unbounded region R'.

$$f' = f - 1 .$$ (8.39)

Combining (8.39) with (8.38), we see that

$$f' - e' + v' = f - 1 - (e - (x + 1) + v - x = f - e + v .$$ (8.40)

This means that whatever the value of $f - e + v$ may be for the original graph, it is unchanged by this process of reduction.

Now the number of regions determined by G is finite, and at each stage of the removal process it decreases by 1. Thus if we continue removing the parts of the bounding circuits as we did above, sooner or later we will end up with a graph that has only 2 regions, i.e., a circuit C.

Our remarks show that

$$f(C) - e(C) + v(C) = f - e + v .$$ (8.41)

Now clearly the number of vertices of a circuit is the same as the number of edges so that $-e(C) + v(C) = 0$. Since C splits the plane into 2 regions, we have $f(C) = 2$.

In view of (8.41) we conclude that

$$f - e + v = 2 \,.$$

If G happens to have vertices of degree 1, first we can delete these nodes from G together with the edges incident to them as shown in Figure 8.31.

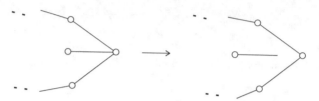

Fig. 8.31 Eliminating the vertices of degree 1.

This leaves the number of regions f unaltered but decreases both e and v by 1. Since this has no net effect on $f - e + v$ our previous argument still goes through. This establishes Theorem 8.5.2.   □

Euler's formula can be interpreted as relating the number of vertices, edges, and faces of a simple polyhedron in a 3-dimensional space.

A *polyhedron* P can be defined intuitively as the surface of a solid whose faces consist of polygons. If all the faces of P are congruent regular polygons, then P is called a *regular polyhedron*. Thus in a regular polyhedron the angles formed by adjacent edges meeting at a vertex are all equal.

P is called *simple* if it has no "holes" in it. Thus intuitively, a simple polyhedron can be deformed continuously to become the surface of a sphere.

As examples, we have in Figure 8.32 three polyhedra P_1, P_2, and P_3.

Fig. 8.32 Examples of polyhedra.

Here P_1 and P_3 are simple, whereas P_2 is not. P_3 is also a regular polyhedron since its faces are congruent equilateral triangles.

If we imagine the polyhedra P_1 and P_3 to be made out of thin rubber, then we could inflate them until they become spherical. For instance for P_3 this process would be as shown in Figure 8.33.

Thus a simple polyhedron can be represented as a graph embedded in the surface of the sphere.

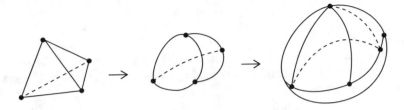

Fig. 8.33 Inflating P_3 of Figure 8.32.

Now it turns out that any graph that can be embedded in the sphere can be embedded in the plane and vice versa. This is done by a construction known as *stereographic projection* as shown in Figure 8.34.

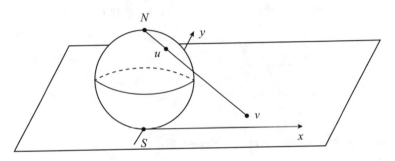

Fig. 8.34 Stereographic projection.

Imagine that the unit sphere is placed on the plane with its south pole S incident to the origin. Suppose u is a point on the surface of the sphere. We draw a line segment joining the north pole N and u and then extend it until it meets the xy-plane. The point of intersection v of this line segment and the plane is the image of the point u under stereographic projection.

This gives a one-to-one correspondence between the plane and the points on the sphere excluding the north pole N. Thus we see that without loss of generality, we could have defined a planar graph as a graph that admits an embedding on the sphere.

Now if we adjust the north pole so that it meets no edges or vertices of a simple polyhedron P that we imagine to be deformed into the surface of the sphere, then stereographic projection yields a planar embedding G of P whose unbounded region corresponds to the face of P that contains the north pole. Moreover the other faces of P correspond exactly to the bounded regions of the plane determined by G. We conclude that the vertices, edges, and faces of P correspond to the vertices, edges, and the regions of G, respectively. Thus if P has n vertices, m edges, and f faces, then we have

$$f - e + v = 2 \qquad\qquad (8.42)$$

by Theorem 8.5.2.

At this point we may wonder whether a formula similar to (8.42) holds for polyhedra that are not simple. Experimenting with the polyhedron P_2 depicted in Figure 8.32, we have

$$v = 16, \ e = 32, \ f = 16$$

so that

$$f - e + v = 0 . \qquad\qquad (8.43)$$

Now note that P_2 cannot be continuously deformed into the sphere, but it can be deformed into a *torus*, i.e., the surface of a doughnut as in Figure 8.35.

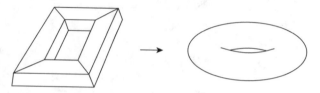

Fig. 8.35 Continuous deformation of P_2 of Figure 8.32.

A torus in turn can be deformed continuously into a sphere with a handle as indicated in Figure 8.36.

Fig. 8.36 P_2 deformed into a sphere with a handle.

Since this much information does not give as a clue as to what the generalization of (8.42) to arbitrary polyhedra might look like, we consider another example. Take for instance the polyhedron P_4 in Figure 8.37.

$P_4 =$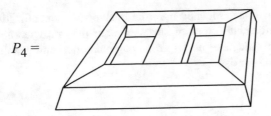

Fig. 8.37 A polyhedron P_4.

For P_4 we have the counts

$$v = 24, \ e = 48, \ f = 22 \, .$$

This gives

$$f - e + v = -2 \, . \tag{8.44}$$

In this case P_4 can be deformed continuously into a sphere with two handles as shown in Figure 8.38.

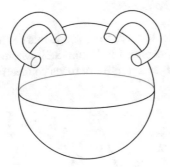

Fig. 8.38 $P4$ of Figure 8.37 deformed into a sphere with two handles.

This experimentation suggests that if a polyhedron P can be deformed into a sphere with g handles, then

$$f - e + v = 2 - 2g \, . \tag{8.45}$$

For if P is simple, then $g = 0$ and (8.45) reduces to (8.42). When $g = 1$, we have $2 - 2g = 0$, which agrees with (8.43), and when $g = 2, 2 - 2g = -2$, in agreement with (8.44).

Using techniques similar to the one we have employed in the proof of Euler's theorem, it can be shown that (8.45) indeed holds for all polyhedra.

In this formula the number of handles g is called the *genus* of P. The quantity $2 - 2g$ is called the *Euler characteristic* of P.

The notions we have touched upon here have vast generalizations and applications in connection with types of surfaces other than polyhedra. This part of the subject though would take us far from combinatorics and graph theory into the domain of topology proper.

8.6   Miscellaneous Applications and Examples

8.6.1   Spanning Trees and the Matrix-Tree Theorem

Graph theory has many applications ranging from electrical networks to routing telephone calls and distribution of mail. As we shall later see, these considerations often require graph algorithms to detect and count particular subgraphs of a graph: circuits, shortest paths between two nodes, complete subgraphs, subtrees, etc.

Central to a number of existing graph algorithms is the notion of a *spanning tree*. Formally, T is a spanning tree of a connected graph G if

$\qquad$ (*i*)   *T is a subgraph of G with* $V(T) = V(G)$,

$\qquad$ (*ii*)   *T is a tree.* $\hspace{5cm}$ (8.46)

The following example should serve to motivate this definition:

Consider the graph G_0 depicted in Figure 8.39 where we imagine the nodes to represent towns and the edges to represent roads connecting these towns.

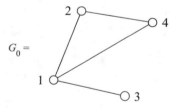

Fig. 8.39 Towns and connecting roads.

Suppose we wish to build a highway system over the existing network of roads so that any two towns are connected by a highway route. It is clear that the road $\{1, 3\}$ connecting the towns 1 and 3 has to be part of any such system if town 3 is to be connected to any town at all. On the other hand, replacing all of the roads $\{1, 2\}, \{2, 4\}, \{4, 1\}$ by highways would be too costly. Indeed, rebuilding only two of these roads would still give us a network of highways connecting all four towns and any such network would cost less than replacing all of the roads by highways. Thus we conclude that the most economical highway system must look like one of the three possible configurations in Figure 8.40.

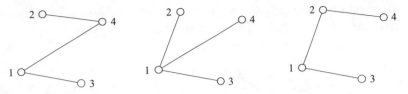

Fig. 8.40 Spanning trees of the graph in Figure 8.39.

By inspection, we see that these are exactly the spanning trees of G_0. Of course the actual cost corresponding to each of the configurations in Figure 8.40 would be determined by the distances between the towns, but we are not interested in this aspect of the problem here.

Note that in the example above, the spanning trees of G_0 were obtained by breaking the circuit formed by the edges $\{1, 2\}, \{2, 4\}, \{4, 1\}$ by removing an edge from it all possible ways. This suggests the following procedure to construct a spanning tree of an arbitrary connected graph G.

If G happens to be a tree, then do nothing—it is already a spanning tree. Otherwise G must contain a circuit. If we remove an edge from this circuit, then the remaining graph is still connected and it has one less circuit than the original graph. In the case that the new graph has no circuits then it is a spanning tree of G. Otherwise we pick a circuit and remove an edge from it. Continuing this way, sooner or later there will be no circuits left and the final graph will be a spanning tree of G.

Our aim in this section is to give a general formula of theoretical interest that will produce all spanning trees of a connected graph G at once. As a byproduct, we shall also obtain a classical determinantal formula known as the *matrix-tree* theorem, giving the number of spanning trees of such a graph.

The machinery necessary to derive our results here was developed in Chapter 4, Section 4.6. The reader is advised to review various definitions and conventions used there. Specifically, we will need the concept of the antisap orientation and the antisap word *(asw)* of a Cayley tree together with Theorem 8.6.1 of that section that characterizes the generating series of the antisap words in the following form:

$$\sum_{T \in C_{n+1}(n+1)} asw(T) = \det \begin{bmatrix} R_1 & -a_{12} & -a_{13} & \cdots & -a_{1n} \\ -a_{21} & R_2 & -a_{23} & \cdots & -a_{2n} \\ -a_{31} & -a_{32} & R_3 & \cdots & -a_{3n} \\ \vdots & \vdots & \vdots & & \vdots \\ -a_{n1} & -a_{n2} & -a_{n3} & \cdots & R_n \end{bmatrix} \qquad (8.47)$$

where $R_i = a_{i1} + a_{i2} + \cdots + a_{i,n+1}$ and $a_{ii} = 0$ for $i = 1, 2, \ldots, n + 1$.

At this point the notion of a spanning tree of a connected graph might seem quite unrelated to Cayley trees and antisap words. But a closer look at the 16 trees depicted in Figure 4.21 of Chapter 4 should reveal that these are precisely the spanning trees of the complete graph K_4. Two of these trees together with their antisap orientations are reproduced in Figure 8.41.

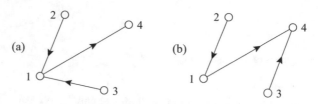

Fig. 8.41 Two spanning trees of K_4 with their antisap orientation.

In general the spanning trees of the complete graph K_{n+1} are the Cayley trees on $n + 1$ nodes. Thus we have the following count:

Theorem 8.6.1 *The number of spanning trees of K_{n+1} is $(n + 1)^{n-1}$.*

Now suppose G is a connected graph. It will be convenient to assume that G has $n + 1$ vertices. This given, we can consider G as a subgraph of the complete graph K_{n+1}. For example we shall think of the graph G_0 pictured in Figure 8.42 as a subgraph of K_4 obtained by removing the edges $\{2, 3\}$ and $\{3, 4\}$.

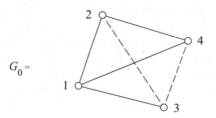

Fig. 8.42 Graph obtained from K_4 by removing the edges $\{2, 3\}$ and $\{3, 4\}$.

In this setup, we easily see that spanning trees of G are also spanning trees of K_{n+1}. The converse of course does not always hold. Indeed, the spanning tree in Figure 8.41 (b) of K_4 is not a spanning tree of G_0 since $\{3, 4\}$ is not an edge of G_0. More generally a spanning tree T of K_{n+1} will be a spanning tree of G provided all the edges of T are also edges of G.

Note that since each spanning tree of G is a Cayley tree on $n + 1$ nodes, it makes sense to talk about antisap words of spanning trees of G. Let us denote by $S_G(n+1)$ those Cayley trees in $C_{n+1}(n + 1)$ that are actually spanning trees of G. Then we have:

Theorem 8.6.2 *Suppose G is a connected graph on $n + 1$ nodes. For $1 \leq i, j \leq n + 1$ put*

$$b_{i,j} = \begin{cases} a_{i,j} & \text{if } \{i, j\} \in E(G), \\ 0 & \text{otherwise.} \end{cases}$$

Then

$$\sum_{T \in S_G(n+1)} asw(T) = \det \begin{bmatrix} R_1 & -b_{12} & -b_{13} & \cdots & -b_{1n} \\ -b_{21} & R_2 & -b_{23} & \cdots & -b_{2n} \\ -b_{31} & -b_{32} & R_3 & \cdots & -b_{3n} \\ \vdots & \vdots & \vdots & & \vdots \\ -b_{n1} & -b_{n2} & -b_{n3} & \cdots & R_n, \end{bmatrix} \tag{8.48}$$

where $R_i = b_{i1} + b_{i2} + \cdots + b_{i,n+1}$.

Proof As we have noted before, each word $asw(T)$ appearing on the left hand side of (8.48) also appears in the sum

$$\sum_{T \in C_{n+1}(n+1)} asw(T). \tag{8.49}$$

On the other hand, an antisap word $w \in C_{n+1}(n+1)$ corresponds to a spanning tree of G if and only if all the directed edges (i, j) coded by the variables $a_{i,j}$ appearing in w are also edges of G. Thus setting $a_{i,j} = a_{j,i} = 0$ for all $\{i, j\}$ not in $E(G)$ kills off all the unwanted words in (8.49), leaving only those words that are in $S_G(n+1)$. In view of (8.47) our proof is complete. $\qquad\square$

The reader is urged to verify that the determinant of the following matrix

$$\begin{bmatrix} a_{12} + a_{13} + a_{14} & -a_{12}, & -a_{13} \\ -a_{21} & a_{21} + a_{24} & 0 \\ -a_{31} & 0 & a_{31} \end{bmatrix}$$

is precisely the sum

$$a_{14}a_{21}a_{31} + a_{12}a_{24}a_{31} + a_{14}a_{24}a_{31}$$

giving the antisap words of the three spanning trees of G_0 that are depicted in Figure 8.40. In particular setting

$$b_{i,j} = \begin{cases} 1 & \text{if } \{i, j\} \in E(G_0), \\ 0 & \text{otherwise} \end{cases}$$

we obtain

$$\#S_{G_0}(4) = \det \begin{bmatrix} 3 & -1 & -1 \\ -1 & 2 & 0 \\ -1 & 0 & 1 \end{bmatrix} = 3,$$

giving the number of spanning trees of G_0. This is an instance of the matrix-tree formula that we shall prove next.

Theorem 8.6.3 *Let G be a connected graph with vertices*

$$v_1, v_2, \ldots, v_{n+1}$$

having degrees (valences, i.e., the number of neighbors)

$$d_1, d_2, \ldots, d_{n+1},$$

respectively. Let $A(G)$ denote the adjacency matrix of G and let D be the $(n+1) \times (n+1)$ diagonal matrix

$$D = [d_i \delta_{i,j}] \,.$$

Then the number of spanning trees of G is given by any cofactor of the matrix

$$D - A(G) \,.$$

Proof As a consequence of Theorem 4.6.2 of Chapter 4, we see that the cofactor of any element of the kth row of the $(n+1) \times (n+1)$ matrix

$$
\begin{bmatrix}
R_1 & -b_{12} & -b_{13} & \ldots & -b_{1n} \\
-b_{21} & R_2 & -b_{23} & \ldots & -b_{2n} \\
-b_{31} & -b_{32} & R_3 & \ldots & -b_{3n} \\
\vdots & \vdots & \vdots & & \vdots \\
-b_{n1} & -b_{n2} & -b_{n3} & \ldots & R_n,
\end{bmatrix}
\tag{8.50}
$$

where $b_{i,j}$ and R_i are defined as in Theorem 4.6.2, is equal to the sum of all antisap words corresponding to spanning trees of G rooted at v_k. If we set

$$
b_{i,j} = \begin{cases} 1 & \text{if } \{i, j\} \in E(G), \\ 0 & \text{otherwise} \end{cases}
$$

this sum reduces to the number of spanning trees of G no matter what k is. Also with this specialization of the $b_{i,j}$'s, R_i becomes the degree of the vertex v_i.

Thus the diagonal entries of (8.50) become $d_1, d_2, \ldots, d_{n+1}$, and an off-diagonal entry in row i and column j becomes -1 or 0 depending on whether $\{i, j\}$ is an edge of G or not. Hence the matrix in (8.50) reduces to

$$D - A(G) \,.$$

This completes the proof of Theorem 8.6.3.   □

We should also remark that our definition of a spanning tree given in (8.46) makes perfectly good sense even for a connected graph with multiple edges. Theorem 8.6.3 holds in this more general situation as well.

8.6.2 Computation of Chromatic Polynomials

As we have observed in Section 8.2, the chromatic polynomial $P(x, G)$ of G can be obtained by repeated application of the identity

$$P(x, G) = P(x, G + e) + P(x, G/e) \tag{8.51}$$

resulting in a binary tree whose root is the given graph G and whose terminal nodes are complete graphs. This construction allows us to express $P(x, G)$ as a linear combination of the lower factorial polynomials $(x)_k$ corresponding to the complete graphs at the leaf nodes. Instead of (8.11), if we make use of the recursion

$$P(x, G) = P(x, G - e) - P(x, G/e) \tag{8.52}$$

given in Theorem 8.2.2, we can obtain the expansion of $P(x, G)$ in terms of the power basis $\{x^k\}$ directly.

To see how this comes about, first note that the effect of repeated use of the identity (8.52) is to steadily decrease the number of edges at each stage. Thus the terminal nodes of the binary tree constructed by using (8.52) will be free graphs rather than complete graphs. Also note that the right son of each internal node in this tree must be tagged with a sign that is *opposite* to that of its parent, while the left sons carry the sign of their parents. This rule comes from the recursion (8.52) itself. The root node G is attached a "+" sign.

For instance, carrying out this construction on the graph G in Figure 8.43, we obtain the binary tree in Figure 8.44.

$G =$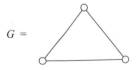

Fig. 8.43 The graph $G = K_3$.

Here the edge to be removed (and contracted) at each internal node is indicated by e. Adding up the signed chromatic polynomials of the free graphs that appear as terminal nodes, we have

$$P(x, G) = x^3 - 3x^2 + 2x .$$

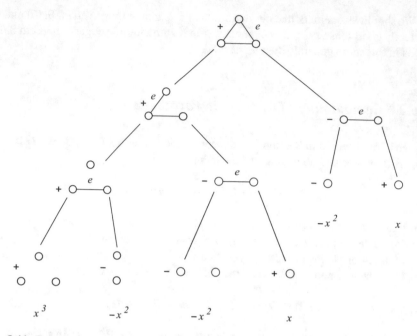

Fig. 8.44 Computation of $P(x, G)$ in terms of the power basis.

Of course here $G = K_3$ so that $P(x, G)$ is just the expansion of $P(x, K_3) = (x)_3$ in terms of the power basis, as we can easily verify.

It turns out that for various classes of graphs, the chromatic polynomial can be expressed in a closed form. In some cases we can find reduction rules that facilitate the computation of $P(x, G)$. In particular, in the construction of the binary tree using either (8.51) or (8.52), we can stop generating further subtrees as soon as we come across a graph whose chromatic polynomial is known to us.

The families of graphs we shall consider in this context are *trees*, the *circuit graphs* C_n, and the *wheel graphs* W_n.

8.6.3   Trees

Recall that a tree T is a connected graph without any circuits. Since in this definition we neither require a root nor a total order on the neighbors of a given node of T, these objects are quite different from the planar trees we have considered in Chapter 3. Nevertheless, for our present purposes it will be convenient to picture a tree as a rooted object with an up-down orientation. For instance in Figure 8.45, we have a tree T with 6 nodes. where we have selected the identification labels arbitrarily. Note that if we color the vertices 1, 2, 3, 4, 5, 6 in succession, we have x choices for vertex 1 , $x - 1$ choices for each one of the vertices 2, 3, 4 (since the

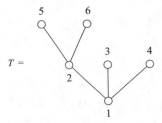

Fig. 8.45 A tree T with 6 nodes.

color used for vertex 1 should be avoided), and $x - 1$ choices for each one of the vertices 5 and 6 (the color used for vertex 2 should be avoided). Thus we conclude that

$$P(x, T) = x(x - 1)^5.$$

Reasoning along the same lines, in the general case we first color the root one of x ways and then proceed to color all the children of the root. Each one of them can be colored $x - 1$ ways, avoiding the color assigned to the root. After this, we color the children of all the children of the root. Every node considered has to avoid the color assigned to its parent. Continuing this way level by level, we see that the chromatic polynomial of a tree T with n nodes is given by

$$P(x, T) = x(x - 1)^{n-1} . \tag{8.53}$$

8.6.4 The Circuit Graphs

For each $n \geq 3$, the circuit graph C_n is simply the graph of a necklace with n beads. C_n is also called the n-cycle. Thus $C_3 = K_3$ and C_4 and C_6 are as given in Figure 8.46.

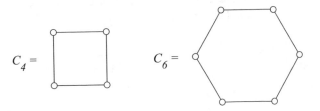

Fig. 8.46 The circuit graphs C_4 and C_6.

We claim that

$$P(x, C_n) = (x - 1)^n + (-1)^n(x - 1) . \tag{8.54}$$

Indeed for $n = 3$, (8.54) gives

$$P(x, C_3) = (x - 1)^3 - (x - 1) = (x - 1)(x^2 - 2x) = x(x - 1)(x - 2) = (x)_3 .$$

Using (8.52) we can write

$$P(x, C_n) = P(x, C_n - e) - P(x, C_n/e) \tag{8.55}$$

for any edge of C_n.

Clearly $C_n - e$ is a tree with n nodes and $C_n/e = C_{n-1}$ whenever $n > 3$. By induction on n, we can assume

$$P(x, C_{n-1}) = (x - 1)^{n-1} + (-1)^{n-1}(x - 1)$$

and by (8.53) we have

$$P(x, C_n - e) = x(x - 1)^{n-1} .$$

Substituting these two expressions in (8.55) we obtain

$$P(x, C_n) = x(x - 1)^{n-1} - \left((x - 1)^{n-1} + (-1)^{n-1}(x - 1) \right)$$

$$= (x - 1)(x - 1)^{n-1} + (-1)^n(x - 1) .$$

Thus (8.54) holds as asserted.

8.6.5   The Wheel Graphs

For $n \geq 4$ the wheel graph W_n is constructed from the circuit graph C_{n-1} by adding an extra vertex v that is connected to all the nodes of C_{n-1}. For example the graphs W_4 and W_5 are as shown in Figure 8.47.

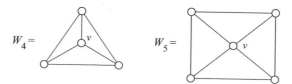

Fig. 8.47 The wheel graphs W_4 and W_5.

Knowing the chromatic polynomial of C_n enables us to compute $P(x, W_n)$ fairly easily.

First of all the center vertex v of W_n can be colored with any one of the x available colors. Having done this, we can use the remaining $x - 1$ colors to color the rim of W_n. Since this can be done in $P(x - 1, C_{n-1})$ ways, we must have

$$P(x, W_n) = xP(x - 1, C_{n-1}) \, . \tag{8.56}$$

Using the expression for $P(x, C_n)$ given in (8.55) with x replaced by $x - 1$, (8.56) becomes

$$P(x, W_n) = x \left((x - 1 - 1)^{n-1} + (-1)^{n-1}(x - 1 - 1) \right) \, .$$

Therefore the chromatic polynomial of W_n is given by

$$P(x, W_n) = x \left((x - 2)^{n-1} + (-1)^{n-1}(x - 2) \right) \, . \tag{8.57}$$

8.6.6 A Reduction Rule

In some cases the removal of a vertex v of a connected graph G together with all the edges incident to it disconnects G into two or more subgraphs. The vertices with this property are called *articulation vertices* of G. For instance the node labeled 3 is an articulation vertex of the graph in Figure 8.48, since its removal disconnects G into the two graphs in Figure 8.49.

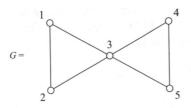

Fig. 8.48 The node labeled 3 is an articulation vertex of G.

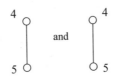

Fig. 8.49 Removal of vertex 3 from the graph in Figure 8.48.

Note that in this case if we define the graphs G_1 and G_2 as in Figure 8.50, then we can recover G by gluing G_1 and G_2 at the marked vertices.

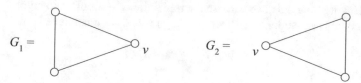

Fig. 8.50 G_1 and G_2 meeting at v.

This is true in general: If v is an articulation vertex of G, then G is constructed from two graphs G_1 and G_2 by identifying a vertex of G_1 and a vertex of G_2. This common vertex then becomes the articulation vertex v of G. This is shown in pictures in Figure 8.51.

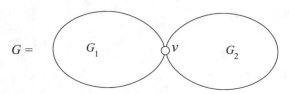

Fig. 8.51 Identifying the vertex v in G_1 and G_2.

We would like to relate the chromatic polynomial of G to those of G_1 and G_2.

To this end, let us first fix a color (say c_i) from our coloring kit, and denote by $n_v(G)$ the number of admissible colorings of G where the vertex v is colored with this fixed color. Note that these colorings of G give rise to colorings of G_1 and G_2 where the copy of the vertex v in each is colored with c_i. On the other hand, two such colorings of G_1 and G_2 yield an admissible coloring of G where v has color c_i. This correspondence yields the formula

$$n_v(G) = n_v(G_1)n_v(G_2) .$$
$$(8.58)$$

By problem 8.7.13 we have

$$n_v(G) = \frac{P(x, G)}{x}, \quad n_v(G_1) = \frac{P(x, G_1)}{x}, \quad n_v(G_2) = \frac{P(x, G_2)}{x} .$$

Substituting these expressions in (8.58) we obtain

$$P(x, G) = \frac{P(x, G_1)P(x, G_2)}{x} .$$
$$(8.59)$$

In particular for the graph G depicted in Figure 8.48 this gives the expression in Figure 8.52 for the chromatic polynomial without having to go through the binary tree construction described in Section 8.2.

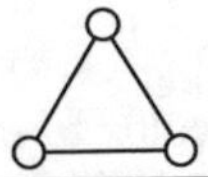

$$P(x, G) = \frac{P(x, \triangle\ )^2}{x} = \frac{(x)_3^2}{x} = x(x-1)^2(x-2)^2 .$$

Fig. 8.52 Calculation of $P(x, G)$ for the graph in Figure 8.48.

We leave the various generalizations of the formula (8.59) to the exercises.

8.6.7 Colorings and Inclusion-Exclusion

Let $G = (V, E)$ be a given graph with $V = [n]$ and assume that our coloring kit consists of x colors

$$C = \{c_1, c_2, \ldots, c_x\} ,$$

as before. A coloring of the vertices of G can be thought of as a map f of V into C where $f(i) = c_j$ if the vertex i of G is given the color c_j. For the computation of $P(x, G)$ we are interested in admissible colorings in which the endpoints i and j of every edge $e = \{i, j\} \in E$ are assigned different colors by f. We can recognize this as another setting for inclusion-exclusion.

Suppose $E = \{e_1, e_2, \ldots, e_m\}$, and say a coloring f has property i if the endpoints of the edge e_i are assigned the same color by f. As before we indicate by $A_1, A_2, \ldots, A_m$, the collections of colorings f having properties $1, 2, \ldots, m$, respectively.

Then $P(x, G)$ is the number d of colorings that satisfy none of the properties $1, 2, \ldots, m$. For any subset $T \subseteq \{1, 2, \ldots, m\}$ we let a_T refer to the number of configurations having property i, at least for each i in T. Thus

$$a_T = \left| \bigcap_{i \in T} A_i \right| .$$

Then the inclusion-exclusion principle gives that

$$P(x, G) = d = \sum_{T \subseteq [m]} (-1)^{|T|} a_T .$$

It remains to understand the nature of the sets

$$\bigcap_{i\in T} A_i \ .$$

Suppose $T = \{i_1, i_2, \ldots, i_k\}$ so that $f \in \bigcap_{i\in T} A_i$ if and only if f assigns the same color to the endpoints of e_{i_1}, the same color to the endpoints of e_{i_2}, and so on. However the colors f assigns to the endpoints of two edges in T cannot always be picked arbitrarily: if e_{i_1} and e_{i_2} have a common vertex, for example, then all 3 vertices involved must be assigned the same color by f. Not only that, any other edge e_{i_3} that shares a vertex with one of the endpoints of e_{i_1} or one of the endpoints of e_{i_2} must have its other endpoint assigned the same color that was already used for the endpoints of e_{i_1} (and of e_{i_2}) as well.

Corresponding to each $T = \{i_1, i_2, \ldots, i_k\}$, let us consider the graph G_T that has the same vertex set as G, but its edge set consists only of the edges $\{e_{i_1}, e_{i_2}, \ldots, e_{i_k}\}$. It follows that a coloring f is in $\bigcap_{i\in T} A_i$ if and only if f assigns the same color to each *connected component* of G_T. The value of f can be picked independently on each connected component one of x possible ways. If we use $c(G_T)$ to denote the number of connected components of the graph G_T, we then have

$$\left|\bigcap_{i\in T} A_i\right| = x^{c(G_T)}.$$

Therefore

$$P(x, G) = \sum_{T\subseteq[m]} (-1)^{|T|} x^{c(G_T)} \ . \tag{8.60}$$

We already know that the degree of $P(x, G)$ is n, and in fact it has leading coefficient 1 from our previous treatment.

The expression (8.60) also implies immediately that the degree of $P(x, G)$ is n and the coefficient of x^{n-1} in $P(x, G)$ is $-m$.

8.6.8   The Platonic Solids

As an application of Euler's formula (8.37), we can show that there are exactly *five* regular polyhedra. They are pictured in Figure 8.53.

The existence of these five figures has been known since ancient times. They are also referred to as the *Platonic solids*.

Suppose now that P is a regular polyhedron whose vertices are regular s–gons. Denote the number of vertices, edges, and faces of P by v, e, and f, respectively. Also suppose that each vertex of P has degree n. This means that each vertex of P has n edges incident to it.

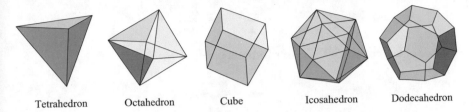

Tetrahedron Octahedron Cube Icosahedron Dodecahedron

Fig. 8.53 The five Platonic solids.

In this context it is convenient to think of P as drawn on the sphere. We know that the sum of the degrees of the vertices of P is equal to twice the number of edges (Exercise 8.7.5). Thus

$$nv = 2e. \tag{8.61}$$

Next, we imagine that we replace each edge in P by 2 edges as shown in Figure 8.54.

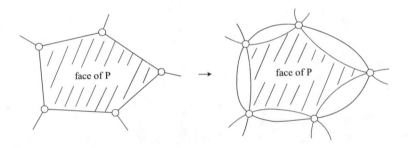

Fig. 8.54 Doubling the edges.

In this situation with the duplicated set of edges, each face contributes s edges and the edge sets of the new faces are all disjoint. Thus

$$sf = 2e . \tag{8.62}$$

Combining (8.61) and (8.62) we have that

$$2e = nv = sf . \tag{8.63}$$

By Euler's formula

$$f - e + v = 2 \tag{8.64}$$

so when we make the substitutions

$$f = \frac{2e}{s}, \quad v = \frac{2e}{n}$$

that we obtain from (8.63), Euler's formula yields

$$\frac{2e}{s} - e + \frac{2e}{n} = 2,$$

which can be rewritten as

$$\frac{1}{s} + \frac{1}{n} = \frac{1}{2} + \frac{1}{e}. \tag{8.65}$$

Since the faces of P are regular s-gons, we have $s \geq 3$. Also note that at least three faces must meet at a vertex. This gives $n \geq 3$.

The possibility $s > 3$ and $n > 3$ is ruled out by (8.65) since such values of s and n would make the left hand side of (8.65) at most $1/2$, whereas the right hand side is clearly larger than $1/2$. Thus we conclude that either $n = 3$ or $s = 3$.

For $n = 3$, (8.65) becomes

$$\frac{1}{s} = \frac{1}{6} + \frac{1}{e},$$

or equivalently

$$s = 6 - \frac{36}{e+6}. \tag{8.66}$$

Since s is an integer we see that $e + 6$ must divide 36 . This gives the values $s = 3, 4, 5$ corresponding to the three possibilities $e = 6, 12, 30$.

For $s = 3$, the symmetry of (8.65) in s and n yields $n = 3, 4, 5$ corresponding to the possible values $e = 6, 12, 30$ again.

Making use of the relation (8.63) we end up with the following distinct cases:

	n	s	v	e	f	
I	3	3	4	6	4	TETRAHEDRON
II	3	4	8	12	6	HEXAHEDRON (Cube)
III	3	5	20	30	12	DODECAHEDRON
IV	4	3	6	12	8	OCTAHEDRON
V	5	3	12	30	20	ICOSAHEDRON.

Thus we see that the regular polyhedra depicted in Figure 8.53 exhaust the possibilities so that what we have in Figure 8.53 is a complete list of the Platonic solids.

Remark 8.6.1 Marking the centers of the faces of the octahedron and connecting these centers by edges we obtain the cube (see Figure 8.55).

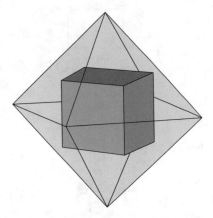

Fig. 8.55 Marking the centers of the faces of the octahedron.

The same construction applied to the cube gives the octahedron back as shown in Figure 8.56.

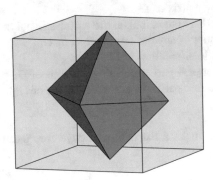

Fig. 8.56 Marking the centers of the faces of the cube.

The octahedron and the cube are *dual* Platonic solids. Similarly the icosahedron and the dodecahedron form a dual pair and the tetrahedron is self-dual.

8.7 Exercises for Chapter 8

8.7.1 Show that the number of edges of the complete graph K_n is $\binom{n}{2}$.

8.7.2 Construct the adjacency matrices of the following graphs

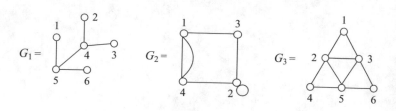

8.7.3 Let G_0 be the following graph

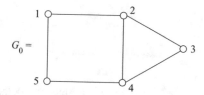

(a) Give the adjacency matrix of G_0.
(b) Use the recursion (8.11) to calculate the chromatic polynomial of G_0.
(c) Calculate the number of acyclic orientations of G_0.
(d) Construct an acyclic orientation of G_0.
(e) What is the number of 4-part edge-free partitions of G_0?
(f) Give a 3-part edge-free partition of the vertex set of G_0.

8.7.4

(a) For $k = 1, 2, 3, 4$, make a list of k-part edge-free partitions of the following
graph

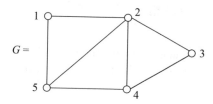

(b) What is $P(x, G)$?

8.7.5

(a) If G is a simple graph, show that the sum of the degrees of the vertices of G is equal to twice the number of edges of G.

(b) Conclude from part (a) that the number of vertices of odd degree in G is even.

8.7.6 If G has two connected components G_1 and G_2, show that G is equivalent to a graph whose adjacency matrix is of the form

$$\begin{bmatrix} A(G_1) & 0 \\ 0 & A(G_2) \end{bmatrix}$$

where the zero matrices are of the appropriate dimensions.

8.7.7 Show that the degree of $P(x, G)$ is equal to the number of vertices of G.

8.7.8

(a) If G has two connected components G_1 and G_2, prove that

$$P(x, G) = P(x, G_1)P(x, G_2).$$

(b) Argue that in general the chromatic polynomial of a graph is the product of the chromatic polynomials of its connected components.

8.7.9 Suppose G is the following graph:

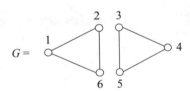

(a) Construct the adjacency matrix of G.

(b) Construct the adjacency list of G.

(c) Give an example of an admissible coloring of the vertices of G using colors from the coloring kit $\{R, B, Y\}$.

(d) Calculate the chromatic polynomial $P(x, G)$.

8.7.10 Calculate the number of ways the following flag

can be colored in red, blue, yellow, and green so that no two adjacent stripes are given the same color.

8.7.11 Show that the (i, j)th entry in $A(G)^k$ is the number of paths of length k connecting the nodes i and j.

8.7.12

(a) Show that the number of acyclic orientations of the complete graph K_n is $n!$.
(b) Suppose $\sigma = \sigma_1\sigma_2\cdots\sigma_n$ is a permutation of $[n]$. Orient the edge (σ_i, σ_j) in K_n away from σ_i whenever $i < j$. Show that this defines a bijection between the permutations of $[n]$ and acyclic orientations of K_n.
(c) Using (b) construct the acyclic orientation of K_5 corresponding to the permutation $\sigma = 25143$.
(d) Construct the permutation that corresponds to the following acyclic orientation of K_4:

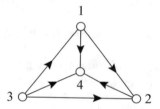

8.7.13 Let $n_v(G)$ denote the number of admissible colorings of G where the vertex v gets a fixed color. Show that

$$n_v(G) = \frac{P(x, G)}{x}.$$

8.7.14 Find the values of the constants $a, b,$ and c if the chromatic polynomial of G is of the form

$$P(x, G) = x^5 + ax^3 + bx^2 + cx.$$

Justify your answer.

8.7.15 Give a direct combinatorial proof of the relation

$$a(G + e) = a(G) + a(G/e)$$

where e is an edge not in $E(G)$.

8.7.16

(a) The following dart board

is to be painted with four colors so that each color is used at least once and adjacent wedges do not get the same color. In how many ways can this be done?

(b) How many such colorings are there if we remove the condition that each color must be used at least once?

8.7.17 Find the number of ways the following graphs can be colored with the given number of colors:

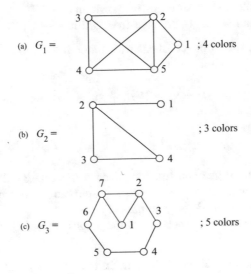

(a) $G_1 =$; 4 colors

(b) $G_2 =$; 3 colors

(c) $G_3 =$; 5 colors

8.7.18 Use the recursion (8.52) and the construction described in Section 8.6.2 to calculate the chromatic polynomials of the graphs in problem 8.7.17 in terms of the power basis.

8.7.19 Calculate the number of acyclic orientations of the graphs G_1, G_2, G_3 in problem 8.7.17.

8.7.20 Let G be the following graph:

$$G =$$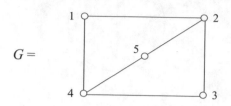

(a) Construct $P(x, G)$.
(b) Calculate the number of admissible colorings of G using a coloring kit with 5 colors.
(c) Give an example of an acyclic orientation of G.
(d) Calculate the number of acyclic orientations of G.
(e) Give an example of a 4-part edge-free partition of $V(G)$.
(f) Calculate the number of 4-part edge-free partitions of $V(G)$.

8.7.21 Find the values of the constants $a, b,$ and c if the chromatic polynomial of G is of the form

$$P(x, G) = x^5 - 10x^4 + ax^3 + bx^2 + cx .$$

Justify your answer.

8.7.22

(a) Modify the binary tree construction given in Section 8.6.2 to show that the chromatic polynomial of a connected graph can be expressed as a linear combination of chromatic polynomials of trees.
(b) Express the chromatic polynomials of the graphs G_1 and G_2 in problem 8.7.17 in this way.

8.7.23 Use recursion (8.52) and induction on the number of edges of G to show that

$$P(x, G) \bigg|_{x^{n-1}} = -\#E(G)$$

where $n = deg P(x, G)$.

8.7.24 Use the recursion (8.52) and induction on the number of edges of G to show that the terms of $P(x, G)$ alternate in sign.

8.7.25

(a) Show that a graph G is a tree if and only if any two vertices of G can be connected by a unique path.

(b) Prove that for any tree T and an edge e not in $E(T)$, $T + e$ necessarily contains a circuit.

8.7.26 Given $G_1 = (V_1, E_1)$, $G_2 = (V_1, E_1)$, their *sum* is the graph $G_1 + G_2 = (V_1 \cup V_2, E_1 \cup E_2 \cup \{\{u, v\} \mid u \in V_1, v \in V_2\})$. For instance $K_2 + K_2$ is constructed as follows:

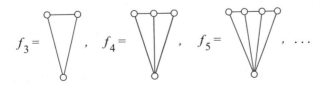

(a) Show that $K_n + K_m = K_{n+m}$ for $n, m \geq 1$.
(b) Describe an algorithm to determine the adjacency matrix $A(G_1 + G_2)$ directly from the adjacency matrices $A(G_1)$ and $A(G_2)$.
(c) Show that

$$P(x, G + K_n) = (x)_n P(x - n, G) .$$

(d) Suppose

$$P(x, G_1) = \sum_{k=1}^{n} F_k(x)_k.$$

Show that

$$P(x, G_1 + G_2) = \sum_{k=1}^{n} F_k(x)_k P(x - k, G_2).$$

(e) The wheel graph W_n can be written as $W_n = K_1 + C_{n-1}$, where C_{n-1} is the circuit graph on $n - 1$ nodes (see Section 8.6.4). Use part (c) to calculate $P(x, W_n)$.
(f) The graph $F_n + F_m$ is referred to as the *complete bipartite graph* of type n, m and is denoted by $K_{n,m}$. Use the result obtained in part (c) to calculate $P(x, K_{2,3})$.

8.7.27 Consider the sequence of graphs f_n where

$$f_3 = \quad , \quad f_4 = \quad , \quad f_5 = \quad , \quad \ldots$$

We may call these the *fan-out* graphs.

(a) Use the recursion (8.52) to show that for $n \geq 4$

$$P(x, f_n) = (x - 1)P(x, f_{n-1}) - P(x, f_{n-1}).$$

(b) Since $f_3 = K_3$, $P(x, f_3) = (x)_3$. Conclude that

$$P(x, f_n) = (x - 2)^{n-2}x(x - 1).$$

(c) Alternately obtain this result by noting that

$$f_n = K_1 + \text{O—O— -- —O}$$

and using the formula in problem 8.7.26 (c) for the chromatic polynomial of the sum of two graphs.

(d) Verify the relation that for $n \geq 5$,

$$P(x, W_n) = P(x, f_n) - P(x, W_{n-1}).$$

Use this to obtain the expression (8.57) for $P(x, W_n)$.

8.7.28 Calculate the number of acyclic orientations of the graphs f_n and W_n.

8.7.29 Calculate the chromatic polynomials of the following graphs

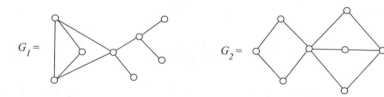

using (8.59).

8.7.30

(a) Prove the following generalization of (8.59): If G_1 and G_2 both have K_n as a subgraph for some n, then the chromatic polynomial of the graph G constructed by overlapping these copies of K_n is given by

$$P(x, G) = \frac{P(x, G_1)P(x, G_2)}{(x)_n}.$$

(b) Calculate the chromatic polynomials and the number of acyclic orientations of the following graphs:

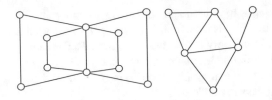

8.7.31

(a) G is called *k-colorable* if $P(k, G) > 0$. Verify that the following graph

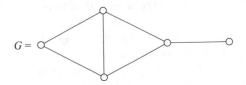

is 3-colorable.

(b) If G is k-colorable but not $(k - 1)$-colorable, then k is called the *chromatic number* of G and denoted by $\chi(G)$. Find the chromatic numbers of the graphs in problem 8.7.17.

8.7.32 Calculate $\chi(G)$ where G is:

(a) a tree on n nodes,
(b) the complete graph K_n,
(c) the circuit graph C_n,
(d) the wheel graph W_n,
(e) the complete bipartite graph $K_{n,m}$,
(f) the fan-out graph f_n.

8.7.33 Suppose G is a graph whose largest vertex degree is d. Use induction on $\#V(G)$ to show that

$$\chi(G) \leq d + 1.$$

8.7.34 Show that $P(2, G) > 0$ if and only if G has no circuits of odd length.

8.7.35 Use problem 8.7.34 to give a characterization of all graphs G with $\chi(G) = 2$.

8.7.36 Are the following statements true or false? Justify your answers.

(a) $P(x, G) = x(x - 1)^{n-1}$ implies G is a tree on n nodes.
(b) $P(x, G) = (x)_n$ implies $G = K_n$.
(c) Every nonnegative integral combination of the lower factorial polynomials is the chromatic polynomial of some graph.

434

(d) Equivalent graphs have identical chromatic polynomials.
(e) Homeomorphic graphs have identical chromatic polynomials.
(f) $P(1, G) > 0$ if and only if G is the free graph F_n for some n.
(g) $x + 1$ never divides a chromatic polynomial.
(h) If G_1 and G_2 are planar graphs, then so is $G_1 + G_2$.
(i) $P(x, G)$ is always divisible by x.
(j) $P(m, G) > 0$ implies $P(m + 1, G) > 0$ for any positive integer m.
(k) $P'(0, G) \neq 0$ implies G is connected.
(l) $\chi(F_n + F_m) = \chi(F_n) + \chi(F_m)$.
(m) $\chi(K_n + K_m) = \chi(K_n) + \chi(K_m)$.
(n) $\chi(G_1 + G_2) = \chi(G_1) + \chi(G_2)$.
(o) $P(x, G_1) = P(x, G_2)$ implies G_1 is equivalent to G_2.
(p) $P(x, G_1)P(x, G_2)$ is always a chromatic polynomial.
(q) $P(x, G_1) + P(x, G_2)$ is always a chromatic polynomial.
(r) $P(P(x, G_1), G_2)$ is always a chromatic polynomial.
(s) $x^4 - 3x^3 + 3x^2$ is not the chromatic polynomial of any graph.
(t) If $G \neq F_n$, then $P(-1, G)$ is even.

8.7.37 Prove that if G is a connected graph on n vertices, then

$$P(x, G) \leq x(x - 1)^{n-1}$$

for all positive integral values of x.

8.7.38 Suppose G is a simple graph with n nodes. For $x \leq n$ define $D(x, G)$ as the number of admissible colorings of G from a coloring kit with x colors where each color appears at least once.

(a) Use the inclusion-exclusion principle to prove that

$$D(x, G) = \sum_{k=0}^{n} \binom{x}{k}(-1)^k P(x - k, G).$$

(Hint: Take Ω as all admissible colorings of G. Let p_i be the property that the ith color does not appear)
(b) Take $G = F_n$ to prove the formula

$$m!S_{n,m} = \sum_{k=0}^{m} \binom{m}{k}(-1)^k (m - k)^n.$$

(c) Suppose

$$P(x, G) = \sum_{k=0}^{n} F_k(x)_k.$$

Show that

$$D(x, G)\bigg|_{(x)_m} = \sum_{k=0}^{m} \frac{(-1)^k}{k!} F_{m-k},$$

for $m = 1, 2, \ldots, n$.

(d) Suppose $\{a_m\}$ and $\{b_m\}$ are two sequences such that

$$a_m = \sum_{k=0}^{m} \binom{m}{k} (-1)^k b_{m-k}.$$

Show that

$$b_m = \sum_{k=0}^{m} \binom{m}{k} a_{m-k}.$$

(e) Using your results from parts (a) and (d), conclude that

$$P(x, G) = \sum_{k=0}^{n} \binom{x}{k} D(x - k, G)$$

whenever $x \le n$.

(f) Suppose T is a tree on 4 nodes. Use the expansion

$$P(x, T) = (x)_2 + 3(x)_3 + (x)_4$$

together with part (c) to show that

$$D(x, T) = (x)_2 + 2(x)_3 - \frac{3}{2}(x)_4 .$$

Verify that the formula in part (e) for $P(x, T)$ holds for $x = 0, 1, 2, 3, 4$.

8.7.39 Consider the problem of tiling the plane with congruent regular n-gons.

(a) Show that the number of polygons meeting at a vertex must be

$$\frac{2n}{n - 2}.$$

Conclude that $n = 3, 4,$ or 6.

(b) Construct the tilings of the plane corresponding to these values of n.

8.7.40 Suppose G is planar with s connected components. Prove the following generalization of Euler's formula:

$$f - e + v = s + 1$$

where $v = |V|$, $e = |E|$, and f is the number of regions of the plane determined by G as before.

8.7.41

(a) Suppose G is a connected planar graph with $v = |V| > 2$. Show that

$$3f \leq 2e.$$

Apply Euler's formula to conclude that

$$e \leq 3v - 6.$$

(b) Use this to prove that K_5 is nonplanar.
(c) Show that for $K_{3,3}$, the sharper inequality

$$4f \leq 2e$$

must hold. Use Euler's formula to conclude that $K_{3,3}$ is nonplanar.

8.7.42 Show that K_5 and $K_{3,3}$ can be drawn on the torus. Verify the relation (8.45) (with $g = 1$) for the embeddings you have constructed.

8.7.43 G is called *regular* of degree k if each vertex of G has degree k.

(a) Show that a connected graph G is regular of degree 2 if and only if it is the circuit graph C_n for some n.
(b) If G is a connected graph with $|V| = n$ and regular of degree $n - 1$, show that G must be the complete graph K_n.

8.7.44

(a) Use problem 8.7.41 to show that if G is planar and regular of degree k, then

$$12 \leq (6 - k)v.$$

(b) Conclude that if a graph is regular of degree > 5, then it is nonplanar.
(c) Prove that a planar graph G must have a vertex of degree ≤ 5.
(d) Use induction on $|V|$ to prove that any planar graph is 6-colorable. It is in fact known that any planar graph is 4-colorable.

8.7.45 Show that homeomorphism is an equivalence relation on graphs.

8.7.46 Verify that any circuit graph is homeomorphic to K_3.

8.7.47 Is the chromatic polynomial invariant under homeomorphism? Explain.

8.7.48

(a) Suppose T is a tree with $n + 1$ nodes. Use the identity (8.9) together with (8.53) to show that

$$P(x, T) = \sum_{k=1}^{n} S_{n,k}(x)_{k+1},$$

where $S_{n,k}$ are the Stirling numbers of the second kind.

(b) Compare this with the expression (8.8) for $P(x, T)$ obtained by taking $G = T$ in Theorem 8.2.1. Conclude that the number of edge-free partitions of $V(T)$ into $k + 1$ parts is $S_{n,k}$.

(c) Note that the result you have obtained in part (b) holds for an arbitrary tree T on $n + 1$ nodes. Take

$$T = \overset{1}{\circ}\!\!-\!\!\overset{2}{\circ}\!\!-\!\!\overset{3}{\circ}\!\!- \cdots -\!\!\overset{n}{\circ}\!\!-\!\!\overset{n+1}{\circ}$$

Show that the number of partitions of $[n + 1]$ into $k + 1$ parts where no part contains consecutive integers is equal to $S_{n,k}$.

(d) Let $\sigma = \sigma_1 \sigma_2 \cdots \sigma_{n+1}$ be a permutation of $[n + 1]$. Prove the following generalization of (c): The number of partitions of $[n + 1]$ into $k + 1$ parts where σ_i and σ_{i+1} belong to different parts $(i = 1, 2, \ldots, n)$ is equal to the numbers of partitions $[n]$ into k parts.

8.7.49 Calculate the number of partitions of $[n + 1]$ into $k + 1$ parts where i and $\lfloor \frac{i}{2} \rfloor$ belong to different parts for all $i > 1$.

8.7.50 Consider the following algorithm to construct the edge-free partitions of G: In the construction of the binary tree to calculate $P(x, G)$ using (8.11), assign a partition Π of $V(G)$ to each node created as follows:

(a) the root node is assigned the partition $\{1\}, \{2\}, \ldots, \{n\}$, where we put $n = \#V(G)$.

(b) The left son of each node is assigned the partition Π of its parent.

(c) If the right son is obtained by contracting the edge (i, j), then its partition Π' is obtained from Π by collapsing the parts containing i and j.

(a) Show that the partitions obtained at the leaf nodes are exactly the edge-free partitions of G.

(b) Take

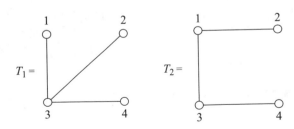

Contract the edges in lexicographic order. Read the partitions at the terminal nodes from left to right to obtain the following correspondence between the edge-free partitions of T_1 and T_2:

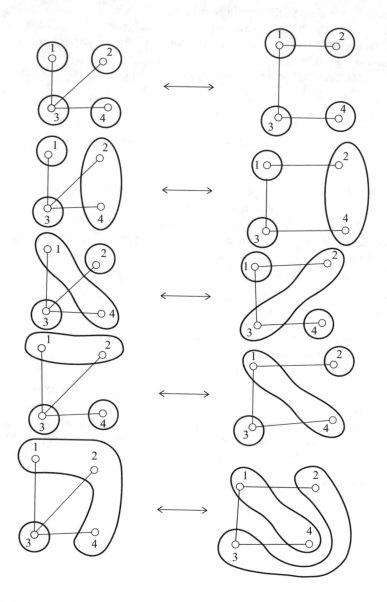

(c) Indicate how to obtain a bijection between admissible colorings of T_1 and T_2.

(d) Suppose G_1 and G_2 have the same chromatic polynomial. Describe an algorithm that gives a bijection between the admissible colorings of G_1 and those of G_2.

8.7.51 Define a coloring of the faces $f_1, f_2, \ldots, f_n$ of polyhedron P admissible if no two faces sharing an edge are given the same color. Put $Q(x, P) =$ the number of admissible colorings of the faces of P with x available colors. Verify the following:

(a) $Q(x, octahedron) = P(x, G)$ where

$$G = $$

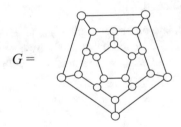

(b) $Q(x, icosahedron) = P(x, G)$ where

$$G = $$

(c) $Q(x, tetrahedron) = (x)_4$.

8.7.52 Find the genus of the following polyhedron by using (8.45).

(Image of cube)

8.7.53 For each of the graphs G in problem 8.7.17 do the following:

(a) Construct the matrix A giving the Cartier–Foata language $\mathcal{L}_A$ which can then be used to represent the acyclic orientations of G.

(b) Determine all the commuting sets of letters in the language $\mathcal{L}_A$ of part (a).

(c) Construct the generating series

$$\sum_{w \in R_A} w$$

using Theorem 8.4.1.

8.7.54 Show that the coefficient of t^n in each of the following series is a nonnegative integer.

(a) $\displaystyle\sum_{k \geq 0} (5t - 3t^2)^k$.

(b) $\displaystyle\sum_{k \geq 0} (5t - 6t^2 + 2t^3)^k$.

(c) $\displaystyle\sum_{k \geq 0} (6t - 8t^2 + t^3)^k$.

8.7.55 Use the matrix-tree theorem to calculate the number of spanning trees of the graphs G_1 and G_3 of problem 8.7.17.

8.7.56 Think of the complete bipartite graph $K_{n,m}$ as $F_n + F_m$ where the vertices in F_n are labeled with $1, 2, \ldots, n$ and those in F_m by $n + 1, n + 2, \ldots, n + m$. Let S denote the spanning trees of $K_{n,m}$ rooted at $n + m$ with weight $w(T)$ defined as in problem 4.8.23, Chapter 4.

(a) Show that

$$\sum_{T \in S} W(T) = m^{n-1} q^{\frac{1}{2}(m-1)(2n+m)} (t + t^2 + \cdots + t^n)^{m-1}.$$

(b) Conclude in particular that the number of spanning trees of $K_{n,m}$ is given by

$$m^{n-1} n^{m-1}.$$

8.7.57 Label the vertices of the circuit graph C_n with $1, 2, \ldots, n$ consecutively moving clockwise. Let $S_n(i)$ denote the spanning trees of C_n rooted at i. For each $T \in S_n(i)$, define $F(T)$ as in problem 4.8.19, Chapter 4.

(a) Show that

$$\sum_{T \in C_n(n)} q^{F(T)} = 2 + q + q^2 + \cdots + q^{n-2}$$

$$\sum_{T \in C_n(1)} q^{F(T)} = q + q^2 + \cdots + q^{n-2} + 2q^{n-1}.$$

(b) For each $n \geq 1$, indicate how to construct a graph G rooted at its largest labeled vertex such that

$$\sum_{T \in S_G(m)} q^{F(T)} = 2(2+q)(2+q+q^2) \cdots (2+q+q^2+\cdots+q^n).$$

(c) Similarly, indicate how to construct a graph G such that

$$\sum_{T \in S_G(1)} q^{F(T)} = 2q^{n+1}(1+2q)(1+q+2q^2) \cdots (1+q+\cdots+q^{n-1}+2q^n).$$

8.7.58 Calculate the number of spanning trees of the following graph

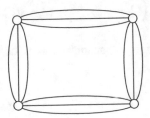

(a) directly by inspection,
(b) by using the matrix-tree theorem.

8.7.59 Suppose we replace each edge of the complete graph K_n by two edges. Call the resulting multiple-edged graph G_n. Argue that

$$\#S(G_n) = 2(2n)^{n-2}.$$

8.7.60 Let $G_{n,m}$ denote the graph obtained from the complete bipartite graph $K_{n,m}$ by replacing each edge with k edges. Prove that

$$\#S(G_{n,m}) = k(km)^{n-1}(kn)^{m-1}.$$

8.7.61 Prove that an edge $e \in E(G)$ is a part of every spanning tree of G if and only if $G - e$ is disconnected.

8.7.62 Denote by $G//e$ the graph obtained by identifying the vertices incident to e (as in G/e) but keeping the multiple edges created in this process.

(a) Show that

$$\#S(G) = \#S(G - e) + \#S(G//e).$$

(b) Calculate the number of spanning trees of the graphs G_1 and G_3 of problem 8.7.17 using this recursion.

8.7.63 Show that the number of spanning trees of the $1 \times n$ rectangular grid

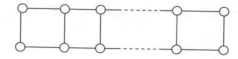

is given by

$$\frac{1}{2\sqrt{3}} \left((2 + \sqrt{3})^{n+1} - (2 - \sqrt{3})^{n+1} \right).$$

8.7.64 Show that the number of spanning trees of the fan-out graph f_n (see problem 8.7.27) is

$$\frac{1}{2^{n-1}\sqrt{5}} \left((3 + \sqrt{5})^{n-1} - (3 - \sqrt{5})^{n-1} \right).$$

8.7.65 Suppose $P_1, P_2, \ldots, P_m$ are $n \times n$ permutation matrices of derangements. Show that the cofactors of the matrix

$$2mI - \sum_{i=1}^{m} (P_i + P_i^T)$$

are all equal.

8.7.66 Show that for each $n \geq 3$, the adjacency matrix of the circuit graph C_n can be expressed as a sum of two permutation matrices.

8.7.67 Prove that for a simple connected graph G

$$\det(D - A(G)) = 0,$$

where D is the diagonal matrix defined in Theorem 8.6.3. What happens if G is disconnected?

8.7.68 Show that G is disconnected if and only if the cofactors of the matrix

$$D - A(G)$$

are all equal to zero.

8.7.69

(a) Prove that the number of spanning trees of the *complete tripartite graph*

$$F_{n_1} + F_{n_2} + F_{n_3}$$

is given by

$$(n_1 + n_2 + n_3)(n_1 + n_2)^{n_3 - 1}(n_1 + n_3)^{n_2 - 1}(n_2 + n_3)^{n_1 - 1}.$$

(b) Show that in general the number of spanning trees of

$$F_{n_1} + F_{n_2} + \cdots + F_{n_k}$$

is given by

$$S^{k-2} \prod_{i=1}^{k} (S - n_i)^{n_i - 1},$$

where $S = n_1 + n_2 + \cdots + n_k$.

8.8   Sample Quiz for Chapter 8

1. Find the number of ways of colorings the countries A_1, A_2, A_3, A_4, A_5 using x colors, such that no two countries sharing a common border are given the same color.

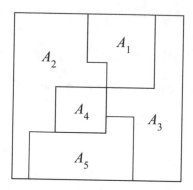

2. Let G be the following graph

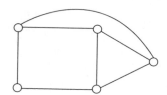

   (a) Give an example of an acyclic orientation of G.
   (b) Calculate the number of acyclic orientations of G.
   (c) Find the number of 4-part edge-free partitions of the vertices of G.

3. Construct the chromatic polynomial and find the number of acyclic orientations of the following graphs

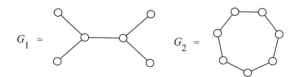

4. Let $\mathcal{L}$ be the language consisting of all words in the letters a, b, c, d where the only pairs allowed to commute are ab and ac. Let $\mathcal{L}_n$ be a set of distinct

representatives of the equivalence classes of words of $\mathcal{L}$. Find the generating function of the sequence

$$\{\#\mathcal{L}_n\}.$$

Chapter 9
Matching and Distinct Representatives

9.1 Binary Matrices, Marked Checkerboards, Rook Domains, and Planks

The material presented here really concerns $n \times m$ *binary matrices* $[a_{i,j}]$, i.e., matrices whose entries are all zeros or ones. However, several different interpretations of such matrices are useful for the applications as well as for a better understanding of the proofs.

Each such matrix can be visualized as an $n \times m$ checkerboard with a certain number of marked squares (those corresponding to the ones in the matrix).

Alternatively, each binary matrix defines a map from the integers $[m] = \{1, 2, \ldots, m\}$ to the subsets S of $[n] = \{1, 2, \ldots, n\}$ by simply setting for every $j \in [m]$,

$$S_j = \{i \mid a_{i,j} = 1\} .$$

Conversely, if we have m subsets

$$S_1, S_2, \ldots, S_m,$$

each of $[n]$, we can construct a matrix $[a_{i,j}]$ that completely describes these subsets by setting

$$a_{i,j} = \begin{cases} 1 \text{ if } i \in S_j, \\ 0 \text{ otherwise.} \end{cases}$$

In other words, the jth column of $[a_{i,j}]$ tells which points belong to S_j, or looking at it another way, the ith row of $[a_{i,j}]$ tells which subsets S_j contain the element i.

© Springer Nature Switzerland AG 2021
Ö. Eğecioğlu, A. M. Garsia, *Lessons in Enumerative Combinatorics*, Graduate
Texts in Mathematics 290, https://doi.org/10.1007/978-3-030-71250-1_9

For instance, if $m = 4$, $n = 5$, and we are given the subsets

$$S_1 = \{1, 3, 5\}$$
$$S_2 = \{2, 3\}$$
$$S_3 = \{4\}$$
$$S_4 = \{1, 2, 4\}$$

of the set $\{1, 2, 3, 4, 5\}$, then the corresponding binary matrix is

$$\begin{bmatrix} 1 & 0 & 0 & 1 \\ 0 & 1 & 0 & 1 \\ 1 & 1 & 0 & 0 \\ 0 & 0 & 1 & 1 \\ 1 & 0 & 0 & 0 \end{bmatrix}. \tag{9.1}$$

But we can visualize such matrices in yet another way. Indeed, let R be the rectangle of lattice points

$$R = \{(i, j) \mid i = 1, 2, \ldots, n, \ j = 1, 2, \ldots, m\},$$

and set

$$A = \{(i, j) \in R \mid a_{i,j} = 1\}.$$

In this manner, each binary matrix corresponds to a definite subset A of R. Conversely, if $A \subseteq R$, then it simply corresponds to the matrix $[a_{i,j}(A)]$ given by

$$a_{i,j}(A) = \begin{cases} 1 \text{ if } (i, j) \in A \\ 0 \text{ otherwise.} \end{cases}$$

Going back to the checkerboard visualization, by a "plank," we mean a *row* or a *column* of squares.

By a plank through the point (i_0, j_0) of R, we mean either of the two sets

$$R(i_0, j_0) = \{(i, j) \mid i = i_0, \ j = 1, 2, \ldots, m\} \quad \text{(row through } (i_0, j_0)\text{)},$$
$$C(i_0, j_0) = \{(i, j) \mid i = 1, 2, \ldots, n, \ j = j_0\} \quad \text{(column through } (i_0, j_0)\text{)}.$$

By a *rook domain*, we mean the union of a *row* and a *column* of R. By the rook domain of (i_0, j_0), we simply mean the set of squares attacked by a rook placed at (i_0, j_0). In other words,

$$RD(i_0, j_0) = R(i_0, j_0) \cup C(i_0, j_0).$$

This given, to each binary matrix $[a_{i,j}]$ or equivalently to each subset $A \subseteq R$, we associate two numbers $p(A)$ and $r(A)$, called, respectively, the *plank number of A* and the *rook number of A*. These are defined as follows:

(a) $p(A)$ *is equal to the least number of planks that are needed to cover the elements of A (i.e., to cover all the ones of the matrix $[a_{i,j}]$).*

(b) $r(A)$ *is equal to the maximum number of mutually nonattacking rooks that can be placed on the elements of A (i.e., on the marked squares of the corresponding checkerboard),*

where, of course, two rooks are in "attacking" position if and only if one is in the rook domain of the other, i.e., if and only if they lie in the same plank.

Some further terminology will simplify our exposition.

A cover ρ of A by planks will be called *pure* if it consists only of *vertical* or only of *horizontal* planks. All other covers will be called *mixed*.

Finally, a *minimal* plank cover ρ of A is one that consists of $p(A)$ planks.

We are now in a position to prove the following remarkable fact.

Theorem 9.1.1 (König–Frobenius) *For any subset $A \subseteq R$, we have*

$$p(A) = r(A) .$$

Proof Clearly, we must have

$$p(A) \geq r(A) . \tag{9.2}$$

Indeed, any plank that covers one of the nonattacking rooks can cover no other, i.e., each rook needs its very own plank, and so we shall need at least $r(A)$ planks to cover all of A.

To prove the reverse inequality, we proceed by induction on $|A|$, the number of elements of A.

The theorem is trivially true for $|A| = 1$. So let us assume that it is true for $|A| \leq N - 1$.

This given, suppose first that A with $|A| = N$ has a mixed minimal cover ρ. Clearly, each plank $P \in \rho$ has points of A that are in no other plank of ρ, for otherwise it would not be needed. Let A_1 consist of the points of A that are in the horizontal planks and not in the vertical ones. Similarly, let A_2 consist of the points of A that are in the vertical but not in the horizontal planks.

A look at Figure 9.1 may help visualize the situation. Let ρ have p_1 horizontal planks. Note that we must have $p(A_1) = p_1$, for otherwise any minimal cover of A_1 together with the p_2 vertical planks of ρ would give a cover of A with less than $p(A)$ planks. Similarly, we must have $p(A_2) = p_2$. Applying the induction hypothesis, we obtain

$$p(A) = p_1 + p_2 = p(A_1) + p(A_2) = r(A_1) + r(A_2) \leq r(A) .$$

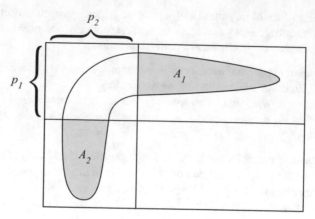

Fig. 9.1 The point sets A_1 and A_2.

The last inequality holding true since no rook placed on A_1 can take any rook placed on A_2.

Thus, the theorem is established in this case.

Suppose now that A has no mixed minimal covers, and let p be any point of A (see Figure 9.2).

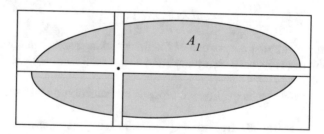

Fig. 9.2 p and the points that are not in the rook domain of p.

Let A_1 consist of the points of A that are not in the rook domain of p. Only two cases are possible.

$$p(A_1) = p(A) \quad \text{or} \quad p(A_1) = p(A) - 1 .$$

Indeed, $p(A_1) < p(A) - 1$ is excluded, for otherwise the rook domain of p (two planks) together with the minimal cover for A_1 would yield a mixed minimal cover for A.

However, in the first case, the induction hypothesis gives

$$p(A) = p(A_1) = r(A_1) \leq r(A) .$$

In the second case, again by induction, we can find exactly $r(A_1) = p(A_1) = p(A) - 1$ nonattacking rooks in A_1. Clearly, a further rook can be safely placed on p, and we have again

$$r(A) \geq p(A_1) + 1 = p(A).$$

This completes our proof.   □

Remark 9.1.1 Given $A \subseteq R$, the König–Frobenius theorem assures that we can place $p(A)$ nonattacking rooks on A. However, it says nothing about the number $M(A)$ of different ways these rooks can be so placed. It turns out that by making some further assumptions about A, the proof of Theorem 9.1.1 can be used to obtain a very good lower bound for $M(A)$. This further result can be stated as follows.

Theorem 9.1.2 *Suppose every vertical plank covers k points of A. Then there are at least*

$$\frac{k!}{(k-m)!} \tag{9.3}$$

distinct ways of placing $p(A)$ nonattacking rooks on A. (Here we adopt the convention that $a! = 1$ whenever $a \leq 0$.)

Proof We follow the lines of the previous proof and proceed by induction on $|A|$ (i.e., assume that (9.3) is true for all k, m, and $|A| \leq N - 1$). Only now, we are going to keep track of the choices available.

In the "mixed minimal cover" case, we must have

$$k \leq p_1 \leq m - p_2 < m,$$

and so by induction, we can already put the $p(A_1)$ nonattacking rooks on A_1 in at least

$$k! = \frac{k!}{(k-m)!}$$

different ways.

In the "pure minimal cover" case, we have to proceed in a more organized way. First of all we choose p in a definite column, say the first. Note that because of our hypotheses we still have k distinct choices for p. This given, for each choice of p, we fall either in the first case ($p(A_1) = p(A)$) or in the second case ($p(A_1) = p(A) - 1$). In either case, A_1 lies in $m - 1$ columns, and each column intersects A_1 in at least $k - 1$ points. Thus the induction hypothesis guarantees at least

$$\frac{(k-1)!}{(k-1-m+1)!} = \frac{(k-1)!}{(k-m)!}$$

ways of placing the $p(A)$ (first case) or $p(A) - 1$ (second case) nonattacking rooks on A_1. So we see again that we have at least

$$k \frac{(k-1)!}{(k-m)!} = \frac{k!}{(k-m)!}$$

ways of placing the rooks on A.   □

Remark 9.1.2 Note that the bound (9.3) cannot be improved. Indeed, if A consists of the first k rows of the rectangle R, then when $k \geq m$ we see that the m nonattacking rooks can be placed in exactly

$$k(k-1) \cdots (k-m+1) = \frac{k!}{(k-m)!}$$

ways.

Therefore the best we can say in general is that this number $M(A)$ of placing the rooks satisfies the inequalities

$$\frac{k!}{(k-m)!} \leq M(A) \leq k^m .$$

9.2 The Distinct Representative Theorem

Let $S = [n]$ and $S_1, S_2, \ldots, S_m$ be subsets of S, not necessarily distinct. We shall say that $x_1, x_2, \ldots, x_m \in S$ are a *system of distinct representatives* (briefly an SDR) for $S_1, S_2, \ldots, S_m$ if and only if

$$(a) \quad x_i \in S_i, \ i = 1, 2, \ldots, m, \tag{9.4}$$

$$(b) \quad x_i \neq x_j \text{ when } i \neq j.$$

In many applications, an SDR is required for a given collection of sets $S_1, S_2, \ldots, S_m$, where each x_i in addition belongs to some special subset M of S. However, the existence of such a *special* SDR follows from the theory we shall present here upon replacing the sets $S; S_1, S_2, \ldots, S_m$, respectively, by

$$M \cap S; \ M \cap S_1, \ M \cap S_2, \ldots, \ M \cap S_m .$$

Suppose now that the sets $S_1, S_2, \ldots, S_m$ admit an SDR $x_1, x_2, \ldots, x_m$. Then we see that for each choice of $1 \leq i_1 < i_2 < \cdots < i_v \leq m$, we have

$$S_{i_1} \cup S_{i_2} \cup \cdots \cup S_{i_v} \supseteq \{x_{i_1}, x_{i_2}, \ldots, x_{i_v}\} ,$$

and thus

$$|S_{i_1} \cup S_{i_2} \cup \cdots \cup S_{i_\nu}| \geq \nu \quad \text{for each } 1 \leq i_1 < i_2 < \cdots < i_\nu \leq m. \tag{9.5}$$

For convenience, let us say that $S_1, S_2, \ldots, S_m$ satisfy the *Hall condition* if (9.5) holds for all $\nu = 1, 2, \ldots, m$. The remarkable fact is that this rather trivial necessary condition for the existence of an SDR for $S_1, S_2, \ldots, S_m$ is also sufficient [16, 24]. More precisely, we have the following theorem.

Theorem 9.2.1 (P. Hall) *The sets $S_1, S_2, \ldots, S_m$ admit an SDR if and only if they satisfy the Hall condition. Furthermore, if each S_i has at least k points, then there are at least*

$$\frac{k!}{(k-m)!} \tag{9.6}$$

different systems of distinct representatives for $S_1, S_2, \ldots, S_m$.

Proof As may be guessed from the bound (9.6), we shall derive this result from Theorem 9.1.2. We simply set

$$a_{i,j} = \begin{cases} 1 \text{ if } i \in S_j, \\ 0 \text{ otherwise,} \end{cases}$$

or equivalently

$$A = \{(i, j) \mid i \in S_j\},$$

and then proceed to show that the Hall condition implies

$$p(A) = m. \tag{9.7}$$

The theorem will then follow from the trivial fact that any collection of m nonattacking rooks on A gives precisely a recipe for constructing an SDR for $S_1, S_2, \ldots, S_m$.

Let then ρ be a minimal cover for A. Let $i_1, i_2, \ldots, i_\nu$ be the indices of the vertical planks that are not in ρ, and let $j_1, j_2, \ldots, j_\mu$ be the indices of the horizontal planks of ρ. Then the total number of planks of ρ is

$$p(A) = m - \nu + \mu.$$

By the very construction of A and the definition of ρ, we get

$$S_{i_1} \cup S_{i_2} \cup \cdots \cup S_{i_\nu} = \{j_1, j_2, \ldots, j_\mu\}.$$

Thus, if $p(A) < m$, we get $\mu < \nu$ or

$$|S_{i_1} \cup S_{i_2} \cup \cdots \cup S_{i_\nu}| = \mu < \nu,$$

but this contradicts (9.5) and the theorem is established.   □

Philip Hall's theorem is colorfully referred to as *Hall's marriage theorem* in the literature as it has an interpretation in terms of marrying off compatible pairs.

Remark 9.2.1 Theorem 9.2.1 seems weaker than Theorem 9.1.2. The remarkable fact is that they are equivalent. Indeed, suppose that for a certain $A \subseteq R$ we have a minimal cover ρ with m_1 vertical planks and n_1 horizontal planks. Let again A_1 consist of the points of A that are in the horizontal but not in the vertical planks of ρ and A_2 consist of the points of A that are in the vertical but not in the horizontal planks of ρ. To visualize the situation, see Figure 9.3.

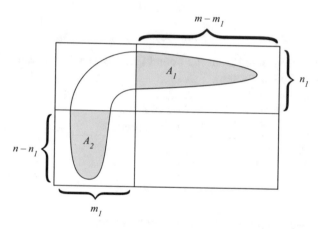

Fig. 9.3 The sets A_1 and A_2.

Introduce the subsets $S_1, S_2, \ldots, S_{m_1}$ by setting

$$S_j = \{i \mid (i, j) \in A_2\}.$$

Note that here we are tacitly assuming that ρ consists of the first m_1 vertical and first n_1 horizontal planks. It should be clear that such a situation can always be reached upon permutations of rows and columns.

Now we see that $S_1, S_2, \ldots, S_{m_1}$ satisfy the Hall condition. Indeed, if not, say

$$S_{i_1} \cup S_{i_2} \cup \cdots \cup S_{i_\nu} = \{j_1, j_2, \ldots, j_\mu\}$$

and $\mu < \nu$, then replacing the ν vertical planks of ρ with indices $i_1, i_2, \ldots, i_\nu$ by the μ horizontal planks with indices $j_1, j_2, \ldots, j_\mu$, we can still cover A and ρ would not be minimal. Thus by Theorem 9.2.1, we can find an SDR for $S_1, S_2, \ldots, S_{m_1}$, in other words, a collection of m_1 nonattacking rooks in A_2.

Using the same procedure on A_1, by working on the transposed matrix, we can find n_1 nonattacking rooks in A_1 and we are done.

9.3 Applications

Philip Hall's theorem and the rook-plank theorem have innumerable applications. We shall state and prove here some of the most classical ones.

To this end, let us recall that an $n \times n$ binary matrix $P = [p_{i,j}]$ is said to be a *permutation matrix* if and only if in each row and each column of P, there is one and only one entry $p_{i,j}$ equal to one.

More colorfully, such a matrix represents a way to place n nonattacking rooks on an $n \times n$ checkerboard.

Of course, we obtain one such matrix for each permutation

$$\sigma = \begin{pmatrix} 1 & 2 & \ldots & n \\ \sigma_1 & \sigma_2 & \ldots & \sigma_n \end{pmatrix}$$

of $[n]$ by setting

$$p_{i,j} = \begin{cases} 1 \text{ if } j = \sigma_i, \\ 0 \text{ otherwise.} \end{cases}$$

An $n \times n$ matrix $Q = [q_{i,j}]$ with nonnegative entries is said to be *doubly stochastic* if and only if

$$\sum_{j=1}^{n} q_{i,j} = 1 \quad \text{for } i = 1, 2, \ldots, n, \tag{9.8}$$

$$\sum_{i=1}^{n} q_{i,j} = 1 \quad \text{for } j = 1, 2, \ldots, n .$$

The doubly stochastic $n \times n$ matrices form a *convex* set in the n^2-dimensional Cartesian space. It turns out that the extreme points of this convex set are precisely the permutation matrices.

More precisely, the following result holds.

Theorem 9.3.1 (G. Birkhoff) *Given any doubly stochastic $n \times n$ matrix $Q = [q_{i,j}]$, there are $N (\leq n^2)$ permutation matrices $P_1, P_2, \ldots, P_N$ and N positive real numbers $c_1, c_2, \ldots, c_N$ such that*

$$Q = c_1 P_1 + c_2 P_2 + \cdots + c_N P_N, \tag{9.9}$$

with $c_1 + c_2 + \cdots + c_N = 1$.

We usually express (9.9) by saying that a doubly stochastic matrix Q can be written as a *convex combination* of permutation matrices.

We shall obtain this theorem as a consequence of the following (only apparently) more general result.

Theorem 9.3.2 *Let $Q = [q_{i,j}]$ be an $n \times n$ matrix with nonnegative entries. Suppose that its row sums and column sums are all equal to a (>0). Then there are N ($\leq n^2$) permutation matrices $P_1, P_2, \ldots, P_N$ and positive constants $c_1, c_2, \ldots, c_N$ such that*

$$Q = c_1 P_1 + c_2 P_2 + \cdots + c_N P_N. \tag{9.10}$$

Furthermore, if the entries of Q are all integers, then $c_1, c_2, \ldots, c_N$ can also be taken to be integers.

Proof Define the subset A of the $n \times n$ square of lattice points by setting

$$A = \{(i, j) \mid q_{i,j} > 0\}.$$

Then we must have

$$p(A) = n. \tag{9.11}$$

Indeed, suppose we can cover A with p_1 vertical and p_2 horizontal planks. Then the sum of all entries in Q is majorized by

$$(p_1 + p_2)a,$$

since each plank can only contribute a to the sum.

On the other hand, we clearly have

$$\sum_{i=1}^{n} \sum_{j=1}^{n} a_{i,j} = \sum_{i=1}^{n} a = na,$$

so $n \leq p_1 + p_2$ and (9.11) necessarily follows.

We can thus place n nonattacking rooks on A. Let P_1 denote the corresponding permutation matrix, and let c_1 be equal to the least nonzero $q_{i,j}$. The matrix $Q_1 = P - c_1 P_1$ has only nonnegative entries, and its row sums and column sums are all equal to $a - c_1$.

Thus we can work on Q_1 as we did on Q. Proceeding in this manner, we produce a sequence of permutation matrices $P_1, P_2, \ldots, P_\nu$ and positive constants $c_1, c_2, \ldots, c_\nu$ such that

$$Q_\nu = Q - c_1 P_1 - c_2 P_2 - \cdots - c_\nu P_\nu$$

has all nonnegative entries and all row sums and column sums are equal to $a - c_1 - c_2 - \cdots - c_v$. If at some stage $a = c_1 + c_2 + \cdots + c_v$, then $Q_v = 0$ and we are through.

But this must happen after no more than $|A| \leq N^2$ steps since each time we decrease at least by one the number of nonzero entries. Thus our argument is complete.   □

Theorem 9.3.2 has the following interesting application.

Suppose that there are n jobs available and there are n applicants who are looking for jobs. Suppose further that for each available job, there are exactly k applicants who are suitable to perform that job; and assume that each applicant is suitable to perform exactly k of the n available jobs. Then it is possible to assign to each applicant a job among those for which the applicant is suitable.

To see this set

$$q_{i,j} = \begin{cases} 1 & \text{if applicant } i \text{ is suitable for job } j, \\ 0 & \text{otherwise.} \end{cases}$$

From our hypothesis, it follows that the row sums and column sums of the matrix $Q = [q_{i,j}]$ are all equal to k. Now Theorem 9.3.2 is applicable, and indeed already the permutation matrix P_1 gives the desired assignment.

Remark 9.3.1 Suppose we only know that each job has *no more than k applicants* suitable for it and each applicant is suitable for *no more than k jobs*. Then, it is still possible to make such an assignment for the applicants who are suitable with the maximum (i.e., k) number of jobs.

The result that is need here can be stated as follows.

Theorem 9.3.3 *Let $Q = [q_{i,j}]$ be a binary matrix, and suppose that the row sums and column sums are all greater than or equal to 1 and less than or equal to k. Then if $A = \{(i, j) \mid q_{i,j} > 0\}$, we can place nonattacking rooks on A at least on each of the rows and columns with sums equal to k.*

Proof We put together a $2n \times 2n$ binary matrix $[a_{i,j}]$ as indicated in the figure below:

$$[a_{i,j}] = \begin{bmatrix} Q & B \\ C & Q^T \end{bmatrix},$$

where $B = [b_{i,j}]$ and $C = [c_{i,j}]$ are the $n \times n$ diagonal matrices defined by setting

$$b_{i,i} = k - \sum_{j=1}^{n} q_{i,j} ,$$

$$c_{i,i} = k - \sum_{j=1}^{n} q_{j,i} \,.$$

Then clearly, $[a_{i,j}]$ has only nonnegative entries, and its row sums and column sums are all equal to k. So by Theorem 9.3.2, we can place $2n$ nonattacking rooks on $\{(i, j) \mid a_{i,j} > 0\}$.

Some of the rooks of course may fall on the nonzero elements of B and C, but these may only fall in those rows or columns of Q whose sum is less than k. □

Remark 9.3.2 We end this section with one more interesting application. Let us suppose that both $\pi = (S_1, S_2, \ldots, S_m)$ and $\pi' = (S_1', S_2', \ldots, S_m')$ are partitions of $[n]$. Suppose that we want to find m points $x_1, x_2, \ldots, x_m \in S$ and a permutation $\sigma_1 \sigma_2 \cdots \sigma_m$ of $[m]$ such that

$$x_i \in S_i \cap S_{\sigma_i}', \qquad i = 1, 2, \ldots, m \,.$$

Usually, $x_1, x_2, \ldots, x_m$ are called a *common set of representatives* (CSR) for π and π'.

Clearly, if such a set exists, for any $1 \le i_1 < \cdots < i_v \le n$, we have

$$S_{i_1} \cup S_{i_2} \cup \cdots \cup S_{i_v} \supseteq \{x_{i_1}, x_{i_2}, \ldots, x_{i_v}\},$$

and then such a union can never be contained in a union of the form

$$S_{j_1}' \cup S_{j_2}' \cup \cdots \cup S_{j_\mu}'$$

with $\mu < v$. In other words, the following condition is satisfied:

> *No union of elements of π is contained in a union of a* (9.12)
> *fewer number of elements of π' and vice versa.*

It turns out that this condition is also sufficient.

Theorem 9.3.4 *If π and π' are two partitions of $S = [n]$ with m elements each, and if π and π' satisfy (9.12), then a CSR can be found.*

Proof We shall only give a sketch of the proof.

The idea is to introduce for each i the subset T_i of $[n]$ consisting of all j such that $S_i \cap S_j' \ne \emptyset$. The claim is that $T_1, T_2, \ldots, T_m$ satisfy Hall's condition. Indeed, it is easy to see that if this is not the case, condition (9.12) is violated. Thus we must be able to find $\sigma_i \in T_i$ such that $i \ne j$ implies $\sigma_i \ne \sigma_j$. This means that for each i,

$$S_i \cap S_{\sigma_i}' \ne \emptyset,$$

and $x_1, x_2, \ldots, x_m$ are then easily chosen. $\square$

Theorem 9.3.4 has the following remarkable consequence.

Corollary 9.3.1 *If G is a finite group and H is a subgroup of G, then it is possible to find elements $x_1, x_2, \ldots, x_m$ $(m = |G|/|H|)$ in G such that*

$$G = x_1 H + x_2 H + \cdots + x_m H = H x_1 + H x_2 + \cdots + H x_m .$$

Proof Let π be the partition of G defined by the equivalence $\gamma_1 \equiv \gamma_2$ if and only if $\gamma_1 \gamma_2^{-1} \in H$. Let π' be the partition of G defined by the equivalence $\gamma_1 \equiv \gamma_2$ if and only if $\gamma_2^{-1} \gamma_1 \in H$. Then both π and π' have exactly $m = |G|/|H|$ parts, and in addition each part has exactly $|H|$ elements. This latter fact guarantees (9.12), so Theorem 9.3.4 is applicable.

This yields the desired coset representatives $x_1, x_2, \ldots, x_m$. $\square$

9.4   The First Available Matching Algorithm

In our proof of the König–Frobenius theorem, we were not concerned with the problem of actually constructing minimal plank covers or maximal rook placements. Of course these configurations could, in principle, be produced by an exhaustive search through all possibilities. However, we can do much better than that. More precisely, in this section, we shall present an algorithm that produces, given an $n \times n$ binary matrix $[a_{i,j}]$ with N ones, a maximal rook placement in a number of operations which grows only linearly as a function of the product nN.

To facilitate our exposition, we shall carry out our presentation in the setting of the example we have used before. Namely, let A and J denote, respectively, the collections of "applicants" and "jobs" that we are given. Let us also say that an applicant a and a job j form a *compatible* pair if and only if the applicant a is suitable to perform the job j. The whole situation can then be represented by a bipartite graph P with vertex set $A + J$ and edge set E consisting of all compatible pairs $\{a, j\}$.

We shall change our point of view slightly here and our problem will be to find a compatible job for each and every one of the applicants. Given an applicant $a \in A$, it will be convenient to denote by $J(a)$ the collection of jobs that are compatible with a. More generally, given a subset $A_0 \subseteq A$, we denote by $J(A_0)$ the collection of jobs that are compatible with at least one of the applicants in A_0. Thus

$$J(A_0) = \bigcup_{a \in A_0} J(a) .$$

So our problem is then to find for each $a \in A$ a job $j(a) \in J(a)$. Since an applicant can only be assigned one job, we see that we are in fact looking for a set of distinct representatives for the collection of subsets $\{J(a)\}_{a \in A}$. We shall refer to any solution of this problem as a *perfect matching* (PM). From our previous considerations, we know that a PM is possible if and only if for every set $A_0 \subseteq A$, we have

$$|A_0| \leq |J(A_0)| . \tag{9.13}$$

This is the Hall condition in our setting.

The algorithm we present here can be used without previous knowledge of whether or not a PM is possible. Indeed, when carried out in full, it will either construct the desired PM or produce a set of applicants A_0 for which (9.13) is violated.

Our first step is to choose a linear order for the applicants and the jobs. We are not to be concerned here with the manner in which this order is to be chosen, even though a judicious choice may sometimes lead to a further reduction in computation.

This given, the algorithm proceeds in the most naive and optimistic way. We simply process the applicants one by one in the given order, each time matching an applicant with the first "available" compatible job. Proceeding in such a blind manner will in great probability lead to a state where all compatible jobs of the next applicant a to be processed have already been assigned and are not any more available. At this point, however, we follow a correcting procedure that will seek to produce a job for a by performing a few rematches. The crucial feature of the algorithm is that when a match is possible, the procedure will always succeed.

To give a more precise description of the algorithm, let us denote by

$$a_1, a_2, \ldots, a_n$$

the 1st, 2nd, $\ldots$, nth applicant in the given order. Let us suppose also that after the first $i - 1$ steps of the algorithm, we have matched the applicants

$$a_1, a_2, \ldots, a_{i-1}$$

with the jobs

$$j(a_1), j(a_2), \ldots, j(a_{i-1})$$

satisfying the requirements

$$j(a_k) \in J(a_k) \text{ for } k = 1, 2, \ldots, i - 1 .$$

This given, at the ith step of the algorithm, we find a match for a_i according to the following procedure.

We construct first the set $R(a_i)$ obtained by removing from $J(a_i)$ all of the jobs that have already been assigned. In symbols,

$$R(a_i) = J(a_i) - \{j(a_1), j(a_2), \ldots, j(a_{i-1})\}.$$

If $R(a_i)$ is not empty, we simply let $j(a_i)$ be the first job in $R(a_i)$, assign this job to applicant a_i, and go on to process a_{i+1}.

If $R(a_i)$ is empty, we look for rematches that will produce a compatible job for a_i in the following manner.

We construct a tree-like structure T with a_i as the root. The descendants of a_i in T are simply taken to be the jobs in $J(a_i)$ (see Figure 9.4). This set of jobs forms what we call the first "layer" of nodes of T. We shall denote it by J_1.

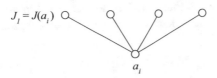

Fig. 9.4 The descendants of a_i.

By our assumptions, all of the jobs in J_1 have been matched. We then let each job in J_1 have the applicant it has been matched with as its only successor in T. This adds a second layer of nodes to T, namely the set A_1 of the applicants of the jobs in J_1 (see Figure 9.5 where the matchings are depicted by heavy lines).

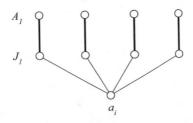

Fig. 9.5 The second layer and the matchings.

We go next through the applicants in A_1 one by one in the given order, and for each $a' \in A_1$, we place in T, as descendants of a', all the jobs in $J(a')$ which are not already placed in T. After we process all the jobs in A_1 in this manner, we obtain a set of jobs J_2 which will be the third layer of nodes in T (see Figure 9.6).

Now, something very interesting can happen. It may be that as we construct the descendants of a job $a' \in A_1$, we may find that one of the jobs j' we are to place in J_2 is actually not matched. Such a circumstance (depicted in Figure 9.7) can save the situation. Indeed, by breaking the match made of a' to $j(a')$, we can make this

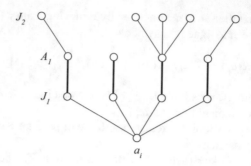

Fig. 9.6 Third layer of nodes in T.

job available to a_i. We can thus match a_i to $j(a')$, and in order not to lose ground, we now match a' to j'. This done we can go on to process a_{i+1}.

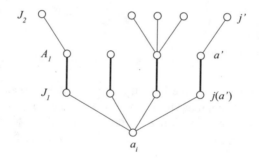

Fig. 9.7 Rematching.

However, we must take account of the possibility that all of the jobs in J_2 have already been matched. In this case we build up the fourth layer of T by adding to it the set of nodes A_2 consisting of the matched applicants of all the jobs in J_2. Of course, again we let the successor of a job in J_2 be the applicant matched to that job. We thus reach the configuration illustrated in Figure 9.8.

We repeat this process and construct successive sets of jobs and applicants

$$J_1, A_1; \ J_2, A_2; \ldots; \ J_s, A_s; \ J_{s+1}, \ldots,$$

where A_s is the set of applicants matched with the jobs in J_s and J_{s+1} consists of the jobs that are compatible with some applicant in A_s and who do not already appear in the tree. Using our notation, this latter set of jobs can be described by the formula

$$J_{s+1} = J(A_s) - J_1 - J_2 - \cdots - J_s . \tag{9.14}$$

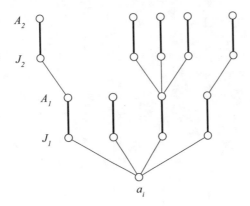

Fig. 9.8 The next layer in T.

This process cannot go on forever, eventually we should run out of people, and the whole thing must come to a stop. However, this can happen in one and only one of the two ways.

(A) *In the process of constructing A_{s+1}, we discover that one of the jobs in J_{s+1} has not been matched.*

    We have depicted such a situation in Figure 9.9 when $s = 3$, the unmatched job being j''. We have also labeled all the nodes we encounter as we go down T along the path that leads from j'' to the root a_i. These are

    a'' *the predecessor of j'' in T,*
    $j(a'')$ *the job matched with a'',*
    a' *the predecessor of $j(a'')$ in T, and finally*
    $j(a')$ *the job matched with a'.*

    Under these circumstances, we break up the two matches $a'' \to j(a'')$ and $a' \to j(a')$. This frees $j(a')$ for a_i. We then match a_i to $j(a')$ and rematch a' to $j(a'')$ and a'' to j''. We are thus done processing a_i and we can stop the construction of T. However, in the alternative that all the jobs in J_{s+1} are matched, we can increase T by adding to it all the applicants matched with the jobs in J_{s+1}. There is only one circumstance that can force us to stop the construction of T this way. That happens when after having constructed A_s we discover that

(B) *The set J_{s+1} given by (9.14) turns out to be empty.*

    If this is the case, let A_0 and J_0 denote, respectively, the sets of applicants and jobs appearing in T. In symbols,

$$A_0 = \{a_i\} + A_1 + A_2 + \cdots + A_s,$$

$$J_0 = J_1 + J_2 + \cdots + J_s .$$

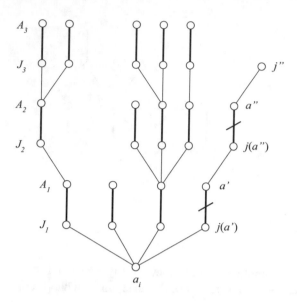

Fig. 9.9 The job j'' is unmatched.

Now, it is easy to see that we have

$$J_0 = J(A_0) . \tag{9.15}$$

Indeed, let j be any job that is compatible with an applicant in A_0. If j is compatible with a_i, then it must be in J_1 since $J_1 = J(a_i)$. On the other hand, if j is compatible with some applicant in A_k for some $k \leq s$, then j must appear in one of the sets

$$J_1, J_2, \ldots, J_k, J_{k+1} .$$

This is because the way T is constructed, if j was not placed in J_{k+1}, it is simply because this job had already been placed in one of the earlier sets

$$J_1, J_2, \ldots, J_k .$$

However, note now that for each $k = 1, 2, \ldots, s$, the sets A_k and J_k have the same number of elements, and this gives that

$$|A_0| = 1 + |A_1| + |A_2| + \cdots + |A_s|$$
$$= 1 + |J_1| + |J_2| + \cdots + |J_s| = 1 + |J_0| .$$

Combining this with (9.15), we must conclude that

$$|A_0| > |J(A_0)|.$$

This of course contradicts (9.13).

Thus if a PM is possible, case B) can never occur during the construction of the tree T, and we will be able to process each and every one of the applicants.

To summarize, we see that the algorithm we have just presented will, in any case, either construct a PM or produce a set of applicants who simply do not have a sufficiently numerous set of compatible jobs among them.

The reader should find it challenging to estimate the number of operations that will be needed to carry out this algorithm in the worst possible case.

9.5   Exercises for Chapter 9

9.5.1 Use the first available matching algorithm to try and find a perfect matching between the following sets of applicants and jobs.

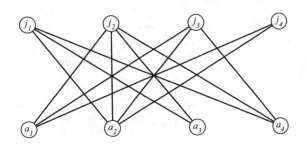

9.5.2 Consider the sets $S_1 = \{a, b, c\}$, $S_2 = \{a, b, d\}$, $S_3 = \{a, b, e\}$, $S_4 = \{c, d, e, f\}$, and $S_5 = \{c, d, e, f\}$. Find an SDR for this collection: that is, find s_1, s_2, s_3, s_4, and s_5 such that $s_i \in S_i$ for $i = 1, 2, \ldots, 5$ and the s_i's are all distinct.

9.5.3 In the following bipartite graph, the bottom vertices represent applicants, the top vertices represent jobs, and the edges represent compatible pairs. Use the first available matching algorithm to find a compatible job for each one of the applicants.

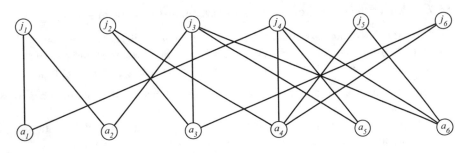

9.5.4 Find the maximum number of nonattacking rooks that can be placed on the darkened squares of the following checkerboard.

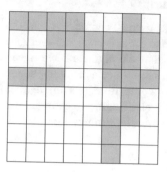

9.5.5 A *vertex cover* in a graph $G = (V, E)$ is a subset $S \subseteq V$ such that every edge has at least one endpoint in S. Show that if G is bipartite, the cardinality of a minimum vertex cover is equal to the cardinality of a maximum matching. This result is known as König's theorem.

9.5.6 An array $a_{i,j}$ of positive integers ($i = 1, 2, \ldots, r$; $j = 1, 2, \ldots, n$) is an $r \times n$ *Latin rectangle* if each row consists of the numbers $[n]$ in some order and no column contains an integer more than once. So for example,

$$1\ 2\ 3\ 4\ 5$$
$$3\ 1\ 5\ 2\ 4$$
$$4\ 3\ 1\ 5\ 2$$

is a 3×5 Latin rectangle.

(a) For the example above, construct a bipartite graph G with vertex sets $V_1 = \{1, 2, 3, 4, 5\}$ and $V_2 = \{c_1, c_2, c_3, c_4, c_5\}$ and edges $\{i, c_j\}$ iff i does not appear in column c_j. Show that G must have a perfect matching.

(b) Show how the perfect matching you found in part (b) can be used to add another row to the array to obtain a 4×5 Latin rectangle.

(c) Show that in general any $r \times n$ Latin square with $r < n$ can be extended to an $n \times n$ Latin square.

9.5.7 Calculate the number of $2 \times n$ Latin rectangles (see problem 9.5.6).

9.5.8 A *diagonal* of an $n \times n$ matrix A is a collection of entries taken one from each row and one from each column. Every diagonal corresponds to a permutation of $[n]$, and the diagonal associated with a permutation σ of $[n]$ is $a_{1,\sigma_1}, a_{2,\sigma_2}, \ldots, a_{n,\sigma_n}$.

(a) Show that an $n \times n$ binary matrix A has a diagonal that consists of only ones iff $p(A) = n$.

(b) Suppose A is an $n \times n$ binary matrix. Show that every diagonal of A contains a zero if and only if A has an $s \times t$ zero submatrix with $s + t = n + 1$.

9.5.9 The permanent of an $n \times n$ matrix $A = [a_{i,j}]$ is defined as

$$\text{per}(A) = \sum_{\sigma} a_{1,\sigma_1} a_{2,\sigma_2} \cdots a_{n,\sigma_n},$$

where the summation is over all permutations of $[n]$ (see Chapter 7, problem 7.6.100). If we have n subsets $S_1, S_2, \ldots, S_n$ of $[n]$ with the corresponding binary matrix A as constructed in (9.1), show that the total number of SDRs for the S_i is given by $\text{per}(A)$.

9.5.10 The entries of a 2×2 doubly stochastic matrix can always be written using a single parameter $0 \le \lambda \le 1$ in the form

$$\begin{bmatrix} \lambda & 1 - \lambda \\ 1 - \lambda & \lambda \end{bmatrix}.$$

Write down the entries of the most general 3×3 doubly stochastic matrix using 5 variables $\lambda_1, \ldots, \lambda_5$ with $1 \le \lambda_i \le 1$ ($i = 1, 2, \ldots, 5$).

9.5.11 Write each of the following matrices:

$$A = \begin{bmatrix} 4 & 2 & 3 \\ 2 & 4 & 3 \\ 3 & 3 & 3 \end{bmatrix} \qquad B = \begin{bmatrix} 5 & 4 & 3 \\ 4 & 6 & 2 \\ 3 & 2 & 7 \end{bmatrix}$$

as a positive integral combination of permutation matrices.

9.5.12 Write the following doubly stochastic matrix:

$$A = \begin{bmatrix} \frac{1}{2} & \frac{1}{5} & \frac{3}{10} \\[4pt] \frac{1}{10} & \frac{3}{5} & \frac{3}{10} \\[4pt] \frac{2}{5} & \frac{1}{5} & \frac{2}{5} \end{bmatrix}$$

as a convex combination of permutation matrices.

9.5.13 [10, pp.69-70]

(a) Suppose $S_j \subseteq [n]$ for $j = 1, 2, \ldots, m$. Let $s_j = |S_j|$ and define r_i to be the number of S_j that contain the element $i \in [n]$. Let $S = \sum_{j=1}^{m} s_j \ (= \sum_{i=1}^{n} r_i)$ and $M = \max\{r_1, \ldots, r_n, s_1, \ldots, s_m\}$. Show that if $(m - 1)M < S$, then $S_1, S_2, \ldots, S_m$ have an SDR.

(b) Suppose $S_j \subseteq [n]$ for $j = 1, 2, \ldots, m$ such that $|S_j| = k > 0$ for every j and each $i \in [n]$ occurs in k sets. Prove that $S_1, S_2, \ldots, S_m$ have an SDR.

(c) Show that if A is an $n \times n$ binary matrix with row and column sums equal to $k > 0$, then A can be written as a sum of k permutation matrices.

9.5.14 Suppose A is an $n \times n$ doubly stochastic matrix. Show that A can be written as a convex combination of at most $(n - 1)^2 - 1$ permutation matrices. This result is known as Carathéodory's theorem.

9.6 Sample Quiz for Chapter 9

1. For the following marked checkerboard:

   (a) find a minimal pure plank cover and
   (b) find a minimal plank cover.

2. (a) Use the first available matching algorithm to try and find a perfect matching
       between the following sets of applicants and jobs.

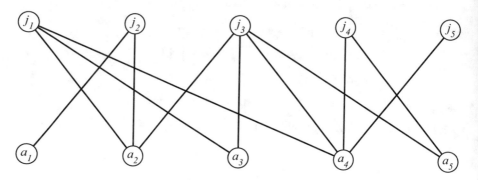

   (b) Use your result from part (a) to find a way of placing the maximum number
       of nonattacking rooks on the shaded squares of the following checkerboard.

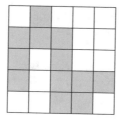

3. A *diagonal* of an $n \times n$ matrix A is a collection of entries taken one from each
   row and one from each column. Given n and positive integers s, t with $s + t = n$
   describe how to construct an $n \times n$ binary matrix A such that A contains an $s \times t$
   zero submatrix and a diagonal consisting of ones only.
4. Show directly (i.e., without using any of the theorems of Chapter 9) that if an
   $n \times n$ binary matrix A contains a $3 \times (n-2)$ submatrix of zeros, then $\operatorname{per}(A) = 0$.

Reference Books

As mentioned in the Preface to this book, for further study we point the reader to Richard P. Stanley's encyclopedic two-volume work *Enumerative Combinatorics* and the wealth of references therein:

Stanley, R. P., *Enumerative Combinatorics* Volume 1 (2nd ed.). Cambridge Studies in Advanced Mathematics, 2011.

Stanley, R. P., *Enumerative Combinatorics* Volume 2. Cambridge Studies in Advanced Mathematics, 2001.

There are numerous other treatments of the subject at various levels of sophistication and focus, employing varying pedagogical approaches. An alphabetically ordered and necessarily incomplete list is as follows:

Bogart, K. P., *Introductory Combinatorics* (3rd ed.) Harcourt/Academic Press, San Diego 2000.

Bóna, M., *Introduction to Enumerative Combinatorics*. McGraw-Hill Higher Education, Boston, 2007.

Brualdi, R. A., *Introductory Combinatorics* (5th ed.). Pearson, NY, 2010.

Charalambides, C. A., *Enumerative Combinatorics*. CRC Press, Florida, 2002.

Comtet, L., *Advanced Combinatorics: The Art of Finite and Infinite Expansions* (rev. enl. ed.). Reidel, Dordrecht, Netherlands, 1974.

Cohen, D. I. A., *Basic Techniques of Combinatorial Theory*. Wiley, NY, 1979.

Erickson, M. J., *Introduction to Combinatorics*. (2nd ed.) Wiley, NJ, 2013.

Hall, M. Jr., *Combinatorial Theory* (2nd ed.). Wiley, NY, 1998.

van Lint, J. H., Wilson, R. M., *A Course in Combinatorics*. Cambridge University Press, 1992.

Liu, C. L., *Introduction to Combinatorial Mathematics*. McGraw-Hill, NY, 1968.

Merris, R. *Combinatorics* (2nd ed.). Wiley, NY, 2003.

Riordan, J., *An Introduction to Combinatorial Analysis*. Princeton University Press, NJ, 1978.

© Springer Nature Switzerland AG 2021
Ö. Eğecioğlu, A. M. Garsia, *Lessons in Enumerative Combinatorics*, Graduate Texts in Mathematics 290, https://doi.org/10.1007/978-3-030-71250-1

Ryser, H. J., *Combinatorial Mathematics*. Wiley, NY, 1963.
Stanton D., White, D., *Constructive Combinatorics*. Springer, NY, 1986.
Tucker, A., *Applied Combinatorics* (6th ed.). Wiley, NY, 2012.

Bibliography

1. Andrews, G.E.: The Theory of Partitions Chapter 7. Encyclopedia of Mathematics and Its Applications. Addison-Wesley, London (1976)
2. Catalan, E.: Note sur une Équation aux difffences finies. J. Math. Pures Appl. **3**, 508–516 (1838)
3. Cayley, A.: A theorem in trees. Q. J. Math. **23**, 376–378 (1889); Collected Papers, Cambridge **13**, 26–28 (1897)
4. Cayley, A.: A memoir on the theory of matrices. Philos. Trans. R. Soc. of Lon. **148**, 17–37 (1858)
5. Eğecioğlu Ö., Remmel, J.: Bijections for Cayley trees, spanning trees, and their q-analogues. J. Combin. Theory **42**, 15–30 (1986)
6. Euler, L.: Evolutio producti infiniti $(1 - x)(1 - xx)(1 - x^3)(1 - x^4)(1 - x^5)(1 - x^6)$ etc. Acta Acad. Sci. Imperialis Petropolitanae **4**, 47–55 (1783). Opera Omnia: Series 1 **3**, 472–479
7. Euler, L.: Remarques sur un beau rapport entre les series des puissances tant directes que reciproques. Memoires de l'academie des sciences de Berlin **17**, 83–106 (1768). Opera Omnia: Series 1 **15**, 70–90
8. Euler, L.: Elementa doctrinae solidorum. Novi Commentarii academiae scientiarum Petropolitanae **4**, 109–140 (1758). Opera Omnia: Series 1 **26**, 71–93
9. Foata, D.: Distribution Euleriennes et Mahoniennes sur le Groupe des Permutations. In: Aigner, M. (ed.), Higher Combinatorics, pp. 27–49. Dordrecht, Reidel (1977)
10. Ford, L.R. Jr., Fulkerson, D.R.: Flows in Networks. Princeton University Press, Princeton (1962)
11. Franklin, F.: Sur le développement du produit infini $(1 - x)(1 - x^2)(1 - x^3)\cdots$. C. R. **82**, 448–450 (1881)
12. Frobenius, G.: Über lineare Substutionen und bilineare Formen. J. Reine Angew. Math. **84**, 1–63 (1878)
13. Garsia, A.M., Milne, S. C.: A Rogers-Ramanujan bijection. J. Combin. Theory Ser. A **31**, 289–339 (1981)
14. Garsia, A.M.: An exposé of the Mullin–Rota theory of polynomials of binomial type. Linear Multilinear Algebra **1**, 47–65 (1973)
15. Goldman, J.R., Joichi, J.T., White, D.E.: Rook theory: I. Rook equivalence of Ferrers boards. Proc. Am. Math. Soc. **52**, 485–492 (1975)
16. Hall, P.: On representatives of subsets. J. Lond. Math. Soc. **10**, 26–30 (1935)
17. Hamilton, W.R.: Lectures on Quaternions. Hodges and Smith, Dublin (1853)
18. Joyal, A.: Une théorie combinatoire des séries formelles. Adv. Math. **42**, 115–122 (1981)

© Springer Nature Switzerland AG 2021
Ö. Eğecioğlu, A. M. Garsia, *Lessons in Enumerative Combinatorics*, Graduate Texts in Mathematics 290, https://doi.org/10.1007/978-3-030-71250-1

19. Kuratowski, K.: Sur le problème des courbes gauches en topologie. Fund. Math. **15**, 271–283 (1930)
20. Lagrange, J.-L.: Nouvelle méthode pour résoudre des équations littérales par le moyen des séries. Mém. Acad. Roy. Sci. Belles-Lettres de Berlin 24 (1770)
21. MacMahon, P.A.: Combinatory Analysis. Cambridge Univ. Press, London (1916); reprinted by Chelsea, NY, 1960
22. Mullin, R., Rota, G. C.: On the theory of binomial enumeration. In: Graph Theory and its Applications. Academic Press, New York (1970)
23. Prüfer, H.: Never Beweis eines Satzes über Permutationen. Arch. Math. Phys. Sci. **27**, 742–744 (1918)
24. Ryser, H. J.: Combinatorial mathematics. M. A. A. Carus math. monographs, No. 14. Wiley, New York (1963)
25. Stanley, R. P.: Catalan Numbers. Cambridge University Press, Cambridge (2015)
26. Stanley, R. P.: Acyclic orientations of graphs. Discrete Math. **5**, 171–178 (1973)
27. Straubing, H.: A combinatorial proof of the Cayley–Hamilton theorem. Discrete Math. **43**, 273–279 (1983)
28. Touchard, J.: Sur un problème de permutations. C. R. Acad. Sci. Paris **198**, 631–633 (1934)
29. Young, A.: On Quantitative substitutional analysis. Proc. Lond. Math. Soc. **33**, 97–146 (1900)

Index

© Springer Nature Switzerland AG 2021
Ö. Eğecioğlu, A. M. Garsia, *Lessons in Enumerative Combinatorics*, Graduate
Texts in Mathematics 290, https://doi.org/10.1007/978-3-030-71250-1

Printed in the United States
by Baker & Taylor Publisher Services